T0238607

Lecture Notes of the Institute for Computer Sciences, Social-Informatics and Telecommunications Engineering 40

Petros Daras   Oscar Mayora Ibarra (Eds.)

User Centric Media

First International Conference, UCMedia 2009
Venice, Italy, December 9-11, 2009
Revised Selected Papers

 Springer

Volume Editors

Petros Daras
Informatics and Telematics Institute
Centre for Research and Technology Hellas
1st Km Thermi-Panorama Road
57001 Thermi-Thessaloniki, Greece
E-mail: daras@iti.gr

Oscar Mayora Ibarra
CREATE-NET
Via alla Cascata 56D
38100 Trento, Italia
E-mail: ibarra@ieee.org

Library of Congress Control Number: 2010924631

CR Subject Classification (1998): H.3, H.4, H.2, H.2.8, I.4, H.5.1

ISSN   1867-8211
ISBN-10   3-642-12629-4 Springer Berlin Heidelberg New York
ISBN-13   978-3-642-12629-1 Springer Berlin Heidelberg New York

springer.com

© ICST Institute for Computer Sciences, Social-Informatics and Telecommunications Engineering 2010
Printed in Germany

Typesetting: Camera-ready by author, data conversion by Scientific Publishing Services, Chennai, India
Printed on acid-free paper   SPIN: 06/3180   5 4 3 2 1 0

Preface

The new International Conference on User-Centric Media, UCMedia is dedicated to addressing emerging topics and challenges in various user-centric media-related areas. The scope of the conference is to improve our understanding of the changing landscape of media. It has a particular focus on forms of media that are user-centered and on the way they will be delivered over a Future Media Internet. It aims to provide a high-profile, leading-edge forum for researchers, engineers, and standards developers to present their latest research, techniques, and experiences in the field of user-centric media.

The First UCMedia conference was held during December 9–11, 2009 at Hotel Novotel Venezia Mestre Castellana in Venice, Italy. The conference`s focus was on forms and production, delivery, access, discovery, and consumption of user-centric media. Delegates from industry and academia joined the conference for three exciting days of new idea dissemination and technical discussions.

After a thorough review process of the papers received, 23 were accepted from an open call for the main conference and 24 papers for the workshops. The technical program of UCMedia 2009 reflected well the current priorities in overwhelming volumes of content, user-generated content, changing distribution mechanisms, and changing representations of media, and was organized in seven sessions. In addition, a considerable part of the technical program was devoted to consolidated and emerging application areas for mobile services, multimedia and user experiences, multimedia search, and retrieval and interactive TV. Lower-layer issues such as identity management in peer-to-peer networks and legal issues in virtual worlds also attracted the attention of conference authors.

Keynote speeches were given by three very prominent members of the user-centric community: The keynote speech titled "Towards Truly User-Centric Convergence of Multimedia" was delivered by Nadia M. Thalmann from the University of Geneva, Switzerland. The keynote speech entitled "Challenges and Opportunities in Multimedia Content Delivery" was delivered by Aggelos Katsaggelos from the Northwestern University of Chicago, USA. The keynote speech entitled "Networks and Media: Trends and Prospects on EU Research" was delivered by Luis Rodriguez-Rosello from the European Commission. Moreover, an invited speech titled "NEM: Networked and Electronic Media European Technology Platform" was delivered by Pierre-Yves Danet from France Telecom/Orange Labs, so as to highlight the industry perspective on user-centric media aspects.

Apart from the main track all delegates were invited to participate in in-depth discussions in the five pre-conference workshops, covering a wide range of topics including: virtual worlds, digitally mediated home-on-demand for neonomads, personalization of media delivery platforms, mining user-generated content for security and experience design, and evaluation of social user-centric media applications.

On behalf of the Organizing Committee of UCMedia 2009, we would like thank the volunteer efforts of all the Workshop Chairs, TPC members and reviewers, and the strong support of the Steering Committee. This conference was organized under the sponsorship of ICST and CREATE-NET. We are grateful to Springer for their financial sponsorship, to EC-funded project nextMEDIA for technical cooperation, and the Networked Media Unit of the European Commission and especially the Head of Unit Luis Rodriguez-Rosello and the Project Officer Isidro Laso-Ballesteros for their great support!

We hope that the delegates found the symposium both enjoyable and valuable, and also enjoyed the architectural, cultural, and natural beauty of Venice.

UCMedia 2010 is scheduled to be held in September 2010 in Palma de Mallorca, Spain. In addition to the technical sessions, UCMedia 2010 is also soliciting tutorials and workshop proposals.

Organization

Steering Committee Chair

Imrich Chlamtac   CREATE-NET, Italy

Steering Committee

Petros Daras   Centre for Research and Technology
Hellas/Informatics and Telematics Institute,
Greece

General Chair

Petros Daras   Centre for Research and Technology
Hellas/Informatics and Telematics Institute,
Greece

General Co-chair

Aggelos Katsaggelos   Northwestern University, USA

Program Chairs

Oscar Mayora   CREATE-NET, Italy
Federico Alvarez   University Politecnica de Madrid, Spain
Antonio Camurri   University of Genoa, Italy
Theodore Zahariadis   TEI Chalkidas, Greece

Workshops Chairs

Cristina Costa   CREATE-NET, Italy
Aljoscha Smolic   Fraunhofer HHI, Germany
Peter Stollenmayer   Eurescom, Germany
Doug Williams   British Telecom, UK

Advisory Committee

Isidro Laso   European Commission
Nadia Magnenat-Thalmann   University of Geneva, Switzerland
Ebroul Izquierdo   Queen Mary University of London, UK

Publicity Chair

Apostolos Axenopoulos   Centre for Research and Technology
                                 Hellas/Informatics and Telematics Institute,
                                 Greece

Keynote Speakers

Luis Rodriguez-Rosello   European Commission
Aggelos Katsaggelos   Northwestern University, USA
Nadia Magnenat-Thalmann   University of Geneva, Switzerland

Technical Program Committee

Amar Aggoun   University of Brunnel, UK
Jesus Vegas Hernandez   University of Valladolid, Spain
A. Murat Tekalp   Koc University, Turkey
Andrea Sanna   Politecnico di Torino, Italy
Ben Knapp   Queens University of Belfast, UK
Carmen Guerrero   University Carlos III Madrid, Spain
Carmen Mc Williams   Grassroot
Cristina Costa   CREATE-NET, Italy
David Griffin   University College London, UK
Dimitrios Tzovaras   CERTH/ITI, Greece
Emanuele Ruffaldi   Scuola Superiore S.Anna
Emilio Corchado   University of Burgos, Spain
Fernando Kuipers   Delft University of Technology, The Netherlands
Frank Pollick   University of Glasgow, UK
Giorgio Ventre   University of Napoli Federico II, Italy
Giovanni Pau   UCLA, USA
Giulia Boato   Universita di Trento, Italy
Gonzalo Camarillo   Ericsson
Gualtiero Volpe   University of Genoa, Italy
Jan Bouwen   Alcatel-Lucent
Kate Wac   University of Geneva, Switzerland
Kovacs Peter Tamas   Holografika
Luca Celetto   STMicroelectronics
Margarita Anastassova   CEA, France
Michela Spagnuolo   CNR-IMATI
Mikolaj Leszczuk   AGH University of Science and Technology
Nelly Leligou   TEI Chalkidas, Greece
Nikos Nikolaou   Synelixis
Pablo Cesar   CWI Netherlands, The Netherlands
Paul Marrow   BT, UK
Paul Moore   Atos Origin
Safak Dogan   University of Surrey, UK

PerMeD Steering Committee

Paul Moore	Atos Origin
David Salama	Atos Origin
Theodore Zahariadis	Synelixis
Charalampos Z. Patrikakis	National Technical University Athens, Greece
Nikolaos Chr. Papaoulakis	National Technical University Athens, Greece
Stefan Poslad	Queen Mary University, UK
Mario Nunes	INOV

PerMeD Conference General Chair

Paul Moore	Atos Origin

PerMeD First Session Chair

Stefan Poslad	Queen Mary University, UK

PerMeD Second Session Chair

Theodore Zahariadis	Synelixis

TrustVWs Program Committee

Vytautas Čyras	Vilnius University, Lithuania (chair)
Peeter Laud	Cybernetica, Estonia
Gerald Spindler	University of Göttingen, Germany
Francesco Zuliani	Nergal, Italy

TrustVWs Workshop Organizers

Vytautas Čyras	Vilnius University, Lithuania
Marianna Panebarco	Panebarco & C., Italy

NSA Steering and Organizing Committee

Nashid Nabian	GSD and MIT Senseable City Laboratory
Giusy Di Lorenzo	MIT Senseable City Laboratory
Francesco Calabrese	MIT Senseable City Laboratory

MinUCS Organizing Committee

Ulf Brefeld   Technische Universität Berlin, Germany
Jakub Piskorski   FRONTEX Research&Development
Roman Yangarber   University of Helsinki, Finland

MinUCS Program Committee

Fabio Crestani   University of Lugano (USI) - Faculty of Informatics,
                              Switzerland
Gregory Grefenstette   Exalead, France
Marko Grobelnik   Jožef Stefan Institute, Slovenia
Ben Hachey   Macquarie University, Australia
David L. Hicks   Aalborg University, Denmark
Mijail Kabadjov   Joint Research Centre of the European Commission,
                              Italy
Sadao Kurohashi   Kyoto University, Japan
Udo Kruschwitz   University of Essex, UK
Nasrullah Memon   The Maersk Mc-Kinney Moller Institute, Denmark
Maria Milosavljevic   Capital Markets CRC, Australia
Marie-Francine Moens   Katholieke Universiteit Leuven, Belgium
Horacio Saggion   University of Sheffield, UK
Satoshi Sekine   New York University, USA
Ralf Steinberger   Joint Research Centre of the European Commission,
                              Italy
Mark Stevenson   University of Sheffield, UK

ExpDes Workshop Organizers

Joke Kort
Rob Willems
Peter Ljungstrand

ExpDes Program Committee

Arnold Vermeeren   Technical University Delft, The Netherlands
David Geerts   University of Leuven, Belgium
Henri ter Hofte   Novay, The Netherlands
Huib de Ridder   Technical University Delft, The Netherlands
Jo Pierson   Vrije Universiteit Brussel, Belgium
Marianna Obrist   University of Salzburg, Austria

Table of Contents

UcMedia 2009 – Session 3: Multimedia & User Experience

UcMedia 2009 – Session 4: Multimedia Search and Retrieval

UcMedia 2009 – Session 5: Interactive TV

UcMedia 2009 – Session 6: Content Delivery

UcMedia 2009 – Session 7: Security, Surveillance and Legal Aspects

PerMeD 2009 – Session 1

PerMeD 2009 – Session 2

TrustVWs 2009 – Session 1

TrustVWS 2009 – Session 2

NSA 2009

MinUCS 2009 – Session 1

MinUCS 2009 – Session 2

MinUCS 2009 – Session 3

MinUCS 2009 – Session 4

ExpDes 2009

UCMedia 2009

Keynote Speech

InterMedia: Towards Truly User-Centric Convergence of Multimedia

Nadia Magnenat-Thalmann, Niels A. Nijdam, Seunghyun Han, and Dimitris Protopsaltou

MIRALab university of Geneva, 1227 Carouge/Geneva, Switzerland
{thalmann,nijdam,han,protopsaltou}@miralab.unige.ch

Abstract. In this paper, we present an interactive media with personal networked devices (InterMedia) which aims to progress towards truly user-centric convergence of multimedia system. Our vision is to make the user as a multimedia central to access multimedia anywhere and anytime exploiting diverse devices. To realize the user-centric convergence, we defined three key challenging issues: dynamic distributed networking, mobile and wearable interfaces, and multimedia adaptation and handling. Each field of interest investigates a transparent access to diverse networks for seamless multimedia session migration, a flexible wearable platform that supports dynamic composition of diverse wearable devices and sensors, as well as the adaptation of diverse multimedia contents based on the user's personal and device contexts. To prove our goals, we prototyped our scenario, called Chloe@University, which included interactive 3D multimedia manipulation with seamless session mobility, modular wearable interfaces with DRM, contextual bookmark with mobile interfaces, and interactive surfaces and remote displays which aim to overcome the limited output capabilities of mobile devices.

Keywords: Interactive media, Dynamic networks, mobile and wearable interface, multimedia content adaptation.

1 Introduction

Recent advances in computing devices and networks equip users with rich and interactive environment where media can be accessed, visualized and interacted with. Considerable efforts have been made in the field of audio-video systems and applications convergence for especially smart home environments [1, 5, 12] but they still have spatial limitations to access multimedia. Mobile computing, on the other hand, has focused on device-centric approach which makes a mobile device as a point of multimedia convergence to support nomadic access to multimedia [2, 15]. This, however, is still limited to the usage of a single mobile device with the limited resources.

To overcome these limitations, we address how to move beyond the home and device-centric convergence towards truly user centric convergence of multimedia. Our vision is to make a user as a multimedia central which means that "a user is a point at which services and interaction with them (devices and interfaces) converge

P. Daras and O. Mayora (Eds.): UCMedia 2009, LNICST 40, pp. 3–10, 2010.

across spatial boundaries". To realize the user-centric convergence, we defined three key challenging issues: dynamic distributed networking, mobile and wearable interfaces, and multimedia adaptation and handling. Dynamic distributed networking layer mainly focuses on a transparent access to diverse networks for seamless multimedia session continuity which enables a user to switch among different devices and networks with minimal manual intervention from the user. Mobile and wearable interfaces layer provides dynamic composition of wearable devices and various mobile interfaces to access multimedia contents exploiting diverse devices nearby to users, which make users free from using specific devices to access multimedia contents. Multimedia adaptation and handling layer support multimedia contents to be presented to different devices for personal manipulation which requires adaptation of multimedia to device or personal context along with seamless presentation of the multimedia for different devices.

We prototyped our proposed system and experimented with a specific scenario, called Chloe@University. Chloe@University scenario is instantiated from InterMedia vision [13] and contains all research issues including interactive 3D multimedia manipulation with seamless session mobility, modular wearable interfaces with DRM, contextual bookmark with mobile interfaces, and interactive surfaces and remote displays which aim to overcome the limited output capabilities of mobile devices.

The remainder of this paper is organized as follows. In section 2, we present research challenges and architecture. In order to clearly show our vision, section 3 describes the scenario "Chloe@University" and its prototype implementations. Conclusion and future work are discussed in section 4.

2   InterMedia Research Challenges and Architecture

The concept of user-centric convergence liberates a nomadic user from carrying a range of mobile devices by providing personalized access to multimedia regardless of devices and networks. It is accomplished not only by removing spatial constraints but also by making multimedia contents adapted to diverse devices and networks in our daily activities. An overview of the three key challenging issues (dynamic distributed networking, mobile and wearable interfaces, and multimedia content adaptation and handling) is shown in Fig. 1 with their internal components.

Dynamic distributed networking layer mainly focuses on a transparent access to diverse networks for seamless multimedia session continuity which enables a user to switch among different devices and networks with minimal manual intervention from the user. We defined two frameworks: Personal Address (PA) Framework and Sensor Abstraction and Integration Layer (SAIL). PA framework [9] provides a cross-layer user-centric mobility framework that accounts for terminal handover and session migration. It associates network addresses to the users and their sessions. It exploits context information to automate the processes of terminal handover and session migration. SAIL [3] gathers data from heterogeneous context sources and exports context information as they came from "virtual sensors"; through high-level standard interfaces (for example, HTTP and UPnP).

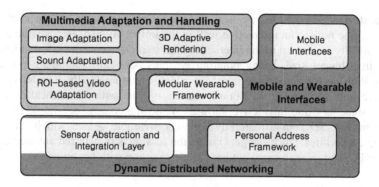

Fig. 1. InterMedia Research Challenges and Architecture

Mobile and wearable interfaces layer provides various interfaces to access multimedia contents exploiting diverse devices nearby to users which make users free from using specific devices to access multimedia contents. We use a modular approach towards a wearable interface in a sense that users do not have to decide between several garments according to their fixed respective functionalities, but they can rather select and attach modules (e.g., UI, storage, localization sensors, and communication protocols, etc.) that will suit their needs. Thus, it becomes entirely personalized as it depends on the selected wearable modules, user profile, and the available surrounding devices. To overcome the limited output capabilities of mobile devices, in particular mobile phones, new mobile interfaces are also developed such as display technologies like projector phones, interactive surfaces and remote displays. Using those interfaces, multimedia can be explored and shared in a collaborative manner using new interaction techniques.

Multimedia adaptation and handling layer support multimedia contents to be presented to different devices for personal manipulation which requires adaptation of multimedia to device or personal context along with seamless presentation of the multimedia for different devices. We provide diverse adaptation mechanisms according to multimedia types including video, image, sound, and 3D contents. We also provide a multimedia sharing architecture through dynamic networks with other users.

3 Prototype: Chloe@University

In this section, we describe our scenario, Chloe@University, and prototype implementations which contain research issues addressed in the previous section.

3.1 Scenario

Chloe attends a medical course at the university. All DRM-featured course materials are automatically downloaded to her wearable device without manual operations. On the way to the library after the class, she takes a picture of a poster on a bulletin board with her mobile device which results in a contextual bookmark. While doing so, her wearable device identifies nearby users and exchanges comments with them about the bookmark. She notices an interactive wall in the university hall. Her wearable device is automatically connected to the interactive wall in order to see the information of the concert in a bigger size. Using pointing gestures, she can manipulate multimedia contents with personalized user interface which is automatically transferred to the interactive wall. Arriving at the library, she can interact with 3D medical data with a group of students using interactive surface. While she comes back home by a tram, she can browse 3D course materials with her wearable device which has limited rendering capabilities. Once she gets home, the session is transferred automatically on her home PC, and she can carry out her learning session with her home PC.

3.2 Implementation

Chloe@University is divided into five applications: modular wearable interfaces with DRM, interactive 3D multimedia manipulation with seamless session mobility, contextual bookmark with mobile interfaces, controlling remote display with mobile devices, and interactive table with 3D content.

3.2.1 Modular Wearable Interface with DRM
We use a jacket as a wearable device interface with distinct modules attached to it in order to store multimedia content related to the profile of the user.

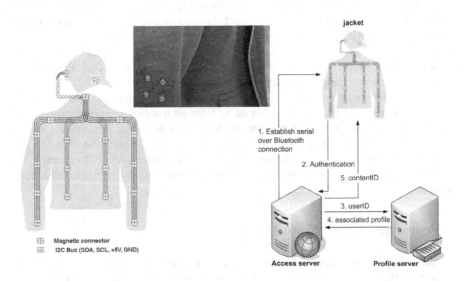

Fig. 2. The jacket with a modular I2C architecture (left), the process of the communication with the jacket (right)

To build a modular wearable interface framework [10] as shown in Fig. 2 (left), we use I^2C protocol [14], which has a built-in addressing scheme and requires only two data lines, because it is inexpensive, easy to upgrade, and convenient for monitoring attached modules. Users can select and attach multiple modules (e.g., UI, storage, localization sensors, and communication protocols, etc.) to the wearable jacket whereafter the modules are automatically identified and communicated to each other. The transfer and access of multimedia take licensed-content into consideration by supporting manipulation of the digital right management (DRM) client and DRM enforcement, in order to handle DRM-protected course contents as shown in Fig. 2 (right).

3.2.2 3D Adaptive Rendering with Seamless Session Mobility

To support users to interact with 3D contents seamlessly adapting to diverse devices and network situations, we implemented an adaptive rendering framework and personal address framework [9]. Adaptive rendering framework provides context-aware polymorphic visualization adaptation and dynamic interface adaptation. In order to provide perceptual real-time interaction, a micro-scale buffer based adaptation mechanism is introduced. It dynamically adjusts frame rate on a client device based on the current network situations and computational resources of the client. Data from a remote server is temporarily stored in a small buffer. If the client renderer continuously fails to consume data within a maximum tolerable latency, the client requests the remote server to decrease frame rate of the stream. Otherwise, the client requests the remote server to increase the frame rate. Interface adaptation mechanism dynamically binds operations provided by 3D application and user interfaces with pre-described device and application profiles. To manage seamless session, personal agent framework is used as shown in Fig. 3 (left). A Key characteristic of PA framework uses topologically-independent IP address based on the MIP infrastructure [8]. Session migration is managed at the application layer. The specific application includes the PA in its session context to be transferred on the new terminal. The old terminal stops the MIP operations, while the new terminal starts the MIP operations with the same IP address. From the Home Agent perspective, this appears like a standard terminal migration. No modifications are required at the MIP protocol. The signaling needed to move the session is application-specific. Fig. 3 (right) shows the snapshot of the prototype.

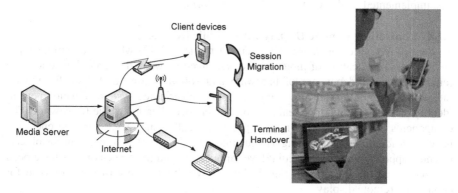

Fig. 3. 3D Adaptive Rendering with Seamless Session Mobility

3.2.3   Contextual Bookmark with Mobile Devices

The Contextual bookmarks [4] allow mobile interactions with physical objects using novel interaction techniques. Mobile devices are used to access information related to static objects such as posters and printed photos. Novel interaction techniques include pointing techniques, gestures, and Magic Lenses. It consists of the four components shown in Fig. 4 (left).

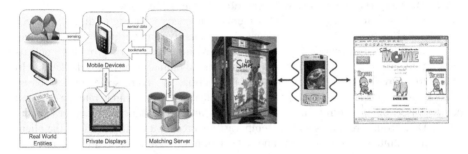

Fig. 4. The central components and their interaction within the contextual bookmark approach

Recognition and matching server extracts object descriptions from images of these objects and equips the object descriptions with meaningful information, such as names, descriptions and additional services. The used recognition algorithm is a speeded up version of Lowe's SIFT algorithm [6]. The application for the mobile phone is implemented in J2ME and runs on phones supporting MIDP-2.0 CLDC-1.1such as Nokia's S60 series. Communication between the mobile phone application and remote playback devices is facilitated via the Java backend of the framework for interactive services in ambient computing. We have prototyped different use-cases. For the recognition of printed photos inside photo-books, for example, we added a two-stage recognition and support for visual markers. For testing recognition using the live camera image's to provide the user with direct feedback the image analysis was implemented on the Windows Mobile platform.

3.2.4   Controlling Remote Display with Mobile Devices

To control remote display, we use a service framework [7] which uses virtual input devices for specification of input events exchanged between the input device and the remote service. Independent of hardware and software constraints, the framework integrates common approaches from the field of distributed systems development to identify all components to be realized for further specialization of the framework. The components for implementing remote procedure calls using XML for data encoding have been identified. Fig. 5 (right) illustrates the design of a solution in an object-oriented approach. The InterMediaPlayer application and the game controls have been realized based on this design. Fig. 5 (left) shows the snapshot of the prototype for controlling remote display.

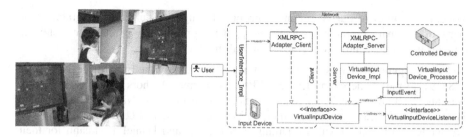

Fig. 5. Controlling remote displays (left) and the system architecture (right)

3.2.5 Interactive Table with 3D Content

To overcome the limited output and resource capabilities of mobile devices, we investigate to exploit diverse resource-rich devices, such as interactive table, if those are available nearby to user. We integrate interactive surfaces, 3D graphics library and NFC [11] as a communication tool in this prototype. The main concept is transferring data from a mobile device to the table and displaying on the table. We have implemented TUIO framework [16] which defines a protocol and API to simulate multiple pointing events on the single input system which have been integrated into OSG to replace its mouse-event handling functionalities with the multi-touch ones that are generated by interactive surface. With the integration of TUIO, OSG can translate the different interactions on the table surface to appropriate 3D manipulations of the 3D content. Also, interacting with the menu that is displayed along the model on the table is achieved through separate event handling mechanism that can detect 2D interaction apart from the 3D interaction.

Fig. 6. Mobile Interaction with Interactive Surfaces

4 Conclusion and Future Work

InterMedia project aims to support truly user-centric convergence of multimedia beyond device-centric convergence. It enables users to exploit diverse devices nearby for continuous consumption of multimedia anywhere and anytime. Advances in mobile devices, networks and sensors have lessened temporal and spatial constraints to access multimedia with diverse devices. In this paper, we introduced interactive media with personal networked devices and its preliminary results. Our goal is to provide technologies beyond the home network and device-centric media convergence enabling seamless multimedia services anywhere and anytime exploiting diverse devices nearby to users. To accomplish this goal, we challenged three main research issues: dynamic distributed networking, mobile and wearable interfaces, and multimedia content adaptation. As a case study, we prototyped Chloe@University system as a proof of our vision and concept. Although we currently integrated some parts of our main research modules, we plan to continuously extend our system to support truly user-centric convergence of multimedia.

Acknowledgement. The work presented was supported by the EU project InterMedia (38419), in the frame of the EU IST FP6 Programme. Authors would like to thank Susanne Boll, Raffaele Bolla, Fadi Chehimi, John Gialelis, Niels Henze, Gonzalo Huerta, Kyriaki Lambropoulou, Andreas Lorenz, Dongman Lee, Riccardo Rapuzzi, Xavier Righetti, Enrico Rukzio, Dimitrios Serpanos, and Daniel Thalmann for their collaboration in the InterMedia NoE.

References

1. Brumitt, B., Meyers, B., Krumm, J., Kern, A., Shafer, S.: EasyLiving: Technologies for intelligent environments. In: Thomas, P., Gellersen, H.-W. (eds.) HUC 2000. LNCS, vol. 1927, p. 12. Springer, Heidelberg (2000)
2. Capin, T., Pulli, K., Akenine-Moller, T.: The State of the Art in Mobile Graphics Research Computer Graphics and Applications. IEEE 28(4), 74–84 (2008)
3. Chessa, S., Furfari, F., Girolami, M., Lenzi, S.: SAIL: a Sensor Abstraction and Integration Layer for Context Aware Architectures. In: 34th EUROMICRO Conference on Software Engineering and Advanced Applications - Special Session on Software Architecture for Pervasive Systems (SAPS), Parma, Italy, September 3-5, pp. 374–381 (2008)
4. Henze, N., Boll, S.: Snap and share your photobooks. In: ACM Multimedia 2008, pp. 409–418 (2008)
5. Kidd, C.D., Orr, R., Abowd, G., Atkeson, C., Essa, I., MacIntyre, B., Mynatt, E., Starner, T., Newstetter, W.: The Aware Home: A Living Laboratory for Ubiquitous Computing Research. In: Streitz, N.A., Hartkopf, V. (eds.) CoBuild 1999. LNCS, vol. 1670, pp. 191–198. Springer, Heidelberg (1999)
6. Lowe, D.G.: Distinctive image features from scale-invariant keypoints. International Journal of Computer Vision 60(2), 91–110 (2004)
7. Lorenz, A., Fernandez De Castro, C., Rukzio, E.: Using Handheld Devices for Mobile Interaction with Displays in Home Environments. In: 11th International Conference on Human-Computer Interaction with Mobile Devices and Services (Mobile HCI 2009), Bonn, Germany, September 15-18 (2009) (accepted)
8. Perkins, C.: Mobility Support for IPv4, RFC 3220 (January 2002), http://www.rfc-editor.org/rfc/rfc3220.txt
9. Repetto, M., Mangialardi, S., Rapuzzi, R., Bolla, R.: Streaming multimedia contents to nomadic users in ubiquitous computing environments. In: Workshop on Mobile Video Delivery in conjunction with InfoCom 2009, Rio de Janeiro, Brazil (2009)
10. Righetti, X., Peternier, A., Hopmann, M.: Design and Implementation of a wearable, context-aware MR framework for the Chloe@University application. In: 13th IEEE International Conference on Emerging Technologies and Factory Automation, pp. 1362–1369 (2008)
11. Rukzio, E., Wetzstein, S., Schmidt, A.: Framework for Mobile Interactions with the Physical World. In: Proceedings of Wireless Personal Multimedia Communication (2005)
12. Schulzrinne, H., Wu, X., Sidiroglou, S., Berger, S.: Ubiquitous Computing in Home Networks. IEEE Communications Magazine 41(11), 128–135 (2003)
13. InterMedia project, http://intermedia.miralab.unige.ch
14. I2C Bus, http://www.i2c-bus.org/
15. Streamezzo, http://www.streamezzo.com
16. TUIO framework, http://www.tuio.org/

UCMedia 2009

Session 1: User Centric Multimedia

From Photos to Memories: A User-Centric Authoring Tool for Telling Stories with Your Photos

Fons Kuijk, Rodrigo Laiola Guimarães, Pablo Cesar, and Dick C.A. Bulterman

Centrum Wiskunde & Informatica,
Science Park 123, 1098 XG Amsterdam, Netherlands
{fons.kuijk,rlaiola,p.s.cesar,dick.bulterman}@cwi.nl

Abstract. Over the last years we have witnessed a rapid transformation on how people use digital media. Thanks to innovative interfaces, non-professional users are becoming active nodes in the content production chain by uploading, commenting, and sharing their media. As a result, people now use media for communication purposes, for sharing experiences, and for staying in touch. This paper introduces a user-centric authoring tool that enables common users to transform a static photo into a temporal presentation, or story, which can be shared with close friends and relatives. The most relevant characteristics of our approach is the use of a format-independent data model that can be easily imported and exported, the possibility of creating different storylines intended for different people, and the support of interactivity. As part of the activities carried out in the TA2 project, the system presented in this paper is a tool for end-users to nurture relationships.

Keywords: Animation, Content Enrichment, Multimedia Document, Pan and Zoom, Photo Sharing, SMIL, Storytelling, Togetherness.

1 Introduction

The Internet has been designed to facilitate exchange of information in an effortless way. Lately, it has turned into a social environment that facilitates communication, services, and interaction. Technology innovation has led to equipment and mechanisms to produce and distribute multimedia content efficiently and at a low cost. The World Wide Web enables easy access to this multimedia content. Web interfaces such as blogs, podcasts, video casts, image sharing, and instant messages have emerged, and their fast rising popularity indicates the need for facilities so that the 'common' user can become an active node in the content production chain. This trend, the user becoming a media producer and distributor, and the use of his media, is referred to as User Centric Media. In a current report from the EC [13], scientific challenges have been identified for research in the area of Future Media Internet. One of these challenges is the capability to actively support the common user in creating new content and sharing it with others. Professional artists have found their way to use 'technological' tools of the earliest kind to create media; the challenge now is to invent tools for the common user that focus on creation, rather than on technology.

P. Daras and O. Mayora (Eds.): UCMedia 2009, LNICST 40, pp. 13–20, 2010.
© Institute for Computer Sciences, Social-Informatics and Telecommunications Engineering 2010

2 Motivation and Contribution

In this paper we focus on sharing old memories – photos –with friends and family that do not live in the same household. Currently, the Web offers many interfaces (e.g., Flickr, Picasa, MySpace, and Facebook) that allow people to share photos. In some cases, users can turn their photos into movie-like presentations and slideshows, converting photo sets into an interesting class of temporal documents (Animoto, Stupeflix). Users may include metadata describing a photo, or parts of a photo, and add comments. *Fig. 1* shows a photo of Rio de Janeiro as present in Flickr, apparently published by a tourist that wants to share his experiences with others. Our tourist added notes: rectangular-shaped regions – indicating locations he visited and objects he observed – with associated descriptions. In addition, some of his friends have commented on the photo. This example represents one instance on how user-centric media can leverage togetherness between people living in different locations by providing users with easy-to-use interfaces to describe, comment, and tag their media.

Fig. 1. A typical Flickr annotated photo of Rio de Janeiro with its thread of comments

While digital photo sharing systems have been widely used for studying knowledge extraction [8], location mapping [6], and social networks [4][11], their restrictive rich media capabilities have not been challenged yet. Consider again the image in *Fig. 1*: regions of interest with metadata have been specified, and people have commented on the photo. The image itself, however, remains a static object, with an undefined focused area of interest. One way of improving the presentation of this image is to add a variant of the Ken Burns Effect, in which various descriptions associated with each region are presented as dynamic, synchronized pans and zooms across the image content. The presentation may start, for example, with the general description of the photo using some styling [2] and interactive options can be incorporated in the form of temporal and spatial hyperlinks to the comments. The presentation then continues with a number of sequenced pans and zooms across the content. The story becomes even more compelling if the resulting presentation can be customized for different friends and family members [7].

The contribution of this article is an authoring tool that provides users with an array of possibilities for creating meaningful and customizable stories as sketched above. The authoring tool uses an underlying declarative multimedia document format with the following characteristics:

- Regions of interest: users are able to identify regions of interest, which can be imported from and exported to other formats (eg., MPEG-7, SMIL);
- Descriptions and comments: users may associate annotations and comments to identified regions of interest;
- Path creation: users have the ability to animate transitions and zoom in on the regions of interest;
- Temporal synchronization: users can synchronize different media types (eg., audio commentary and animation within the photo);
- Base-Media integrity: users cannot alter the underlying based content. Regions, annotations, and comments are linked to media. They are not embedded into it;
- Customization: users may re-use regions/paths for the creation of customized stories using existing content control mechanisms;
- Accessibility: users are able to create accessible multimedia presentations; and
- Interactivity: users can create interactive presentations based on temporal and spatial hyperlinks.

The authoring tool for adding dynamic visual features to static images is targeted for the common user and focused on the creative aspect of storytelling. The tool generates a dynamic presentation of the author's story, in the form of a declarative multimedia document that can be shared with non-collocated peers.

An essential element of the multimedia document format is the SMIL MediaPanZoom module[1], an extension we have proposed to add to the SMIL 3.0 recommendation [3]. The extension is now standardized and supported by popular multimedia players such as the Ambulant Player[2] and the RealPlayer[3].

3 Related Work

Many photo management tools (e.g. iPhoto) and video production systems (such as iMovie or Photo Story) are available for creating digital artifacts from personal media to share experiences with others. Users can employ these applications to tell stories, but these applications typically package the resulting media in some encoded format (e.g. slideshow or video). This feature does not keep the image integrity or allow navigation based on timed enrichments, nor does it support selective viewing (since the annotations are hardcoded with the base photos).

---

[1] http://www.w3.org/TR/SMIL3/smil-extended-media-object.html
[2] http://www.ambulantplayer.org
[3] http://www.realnetworks.com

StillMotion[4] and Amara Photo Animation[5] are slideshow-authoring tools that enable users to create Flash presentations with sound, navigation, transitions, and pan and zoom functionality. These tools pack all media in a self-contained media file, so media integrity is not retained. Even though these tools produce Flash, they do not support navigation based on timed annotations or selective viewing. MemoryNet Viewer [12] is a Peer-to-Peer system for sharing and enriching photos (by adding voice and text comments) among people in personal networks. MemoryNet does not enable end-users to add timed enrichments, such as pan and zoom, nor to export to an open document suitable for other systems. StoryTrack [1] supports a touchscreen device for creating personal stories based on digital photos. Stories and metadata are stored in XML, which allows for translation to be shared with others who do not have a StoryTrack system. This tool does not allow users to zoom in to regions of interest and does not allow an audio track to go beyond the presentation of a single image. Flipper System [5] is a photo sharing application both for mobile and desktop environments. Users may add text comments to any image. It does not offer timed enrichments, and it is not possible to create a story based on a set of images. iTell [10] is a narrative composition tool that leads storytellers stepwise through the writing and media production processes. Forced to follow this stepwise approach, a user can specify sequencing of the imagery and voiceover. iTell departs from the typical timeline metaphor to the notion of associations in order to indicate relationships between narration script, voiceover and images. It exports stories to SMIL, but does not support timed annotations or hyperlinks to extra information or related stories. Web-based photo sharing services (e.g. Flickr) and community-sharing environments (e.g. Orkut) do allow users to share pictures with the ability to add notes and comments. End-users even can add spatial notes to third-party photos if they have been given editing rights to do so. Yet, annotations are a temporal and site-specific.

4 Authoring Process

To illustrate the creative process we envision for storytelling based on images, picture a family member that went on holiday to Rio de Janeiro, Brazil. He decides to tell the story of his trip based on the photo of *Fig. 1*, on which memorable locations are visible. Using our authoring tool, he identifies regions of interest within the photo (like Flickr's notes). He orders these regions on a timeline that is part of the user interface (see *Fig. 2*). Ordering regions defines temporal transitions: the default transition, inserted by the system automatically, is panning along a straight path from one region to the next. In this way, we obtain a presentation that starts by slowly zooming in on the first region of interest. Then by panning, following the connecting paths, a transition to the next region of interest is made, and so on, concluding with zooming out to reveal the entire image. A timeline representation of this presentation is shown in *Fig. 3*. The author can link recorded audio files (e.g., saying "On this mountain I saw....") and other information (textual descriptions and annotations) to regions as well as to transitions. Note that regions, transitions, and associated linked media are not embedded, thus assuring customizable and accessible capabilities.

---

[4] http://www.imagematics.com
[5] http://www.amarasoftware.com

We experimented with the system set-up and concluded that we could recognize two preferred modes of operation: operating on the basis of the visual aspects (the regions of interest), or operating on the basis of the audio (the narratives). Both modes of operation are supported by the authoring system shown in *Fig. 2*.

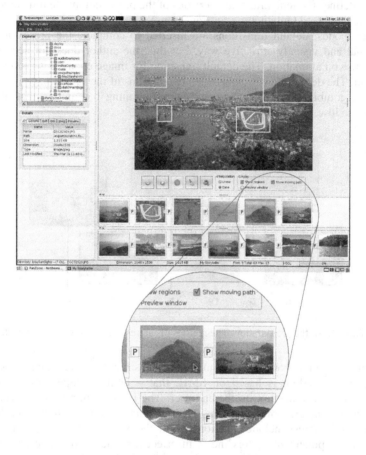

Fig. 2. User interface. On the left we see a *browser* for selecting images and an area to display *properties*. The large *work area* on the right of that serves for editing regions of interest. Below we see two timelines: the *image-level timeline* (bottom row) for ordering images, and the *region-level timeline* (upper row) for ordering regions.

In the *region driven* mode, the author selects images and annotates them with regions of interest. The author specifies the order in which images and regions are to be shown by ordering thumbnail representations on the image timeline or region-timeline. Clicking and dragging in the work area specify new regions of interest. To begin with, default duration is assigned to images and regions and default transitions are inserted automatically. The user can record narratives or specify audio files that go with an image, a region or a transition. Selecting an image, a region, or a transition, enables the record button. The duration of the audio fragment or recording defines the duration assigned to the selected item.

In the *audio driven* mode, the author starts recording the narratives and in the meantime makes up the visuals by selecting images and regions that relate to what is being told. As the recording and selection occurs in realtime, the time between selections defines the temporal characteristics of the presentation: the time images and regions are displayed (anticipating for transitions).

In both modes, default transitions are inserted to complete the presentation in a professional and visually attractive way. For this, the author does not need to have any expertise on animation. The default transition for image changes is a simple fade-in fade-out. The default transition between two regions of the same image is enhanced panning (a combination of panning and some zooming out and in that helps the viewer not to lose the context).

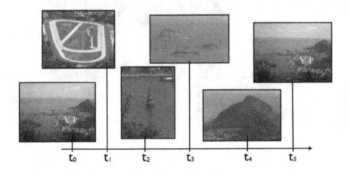

Fig. 3. Timeline of the presentation. The regions of interest and the order in which show up.

The author can export the presentation as a structured multimedia document. Currently the encoding is based on SMIL that – being an open format – offers navigation and customization to the end-user. The images are referred to via URI's, maintaining the integrity of the sources. The timed annotations for transitions between the regions of interest include functionality of the SMIL MediaPanZoom module[6]. The visual component of a storyline is in effect a sequence of static and dynamic panZoom components [9]. Support for interaction is obtained by using temporal and spatial hyperlinks. Specific storylines can be targeted to individual users, so that watching the presentation may become an interactive, personalized experience. Although there is a common ground, our tourist may want to convey a story to his family that differs from the one he will tell his close friends.

5   Implementation

The authoring system is implemented in Java. We use a format-agnostic data model (see *Fig. 4*), designed to accommodate the authoring process rather than the process of

---

[6] http://www.w3.org/TR/SMIL3/smil-extended-media-object.html

generating the structured multimedia document. An XML representation of the model is used for persistence and to cut, copy and paste a storyline or parts there of. This organization allows for import and export of functionality to manifold formats, such as MPEG-7 and SMIL.

A `DataManager` handles all access to the data model. A storyline is represented by `Path`, being a collection of `PathElements`. Each `PathElement` has an associated `TimeInterval` that specifies duration; it can be coupled with an audio fragment. We distinguish two types of `PathElements`: `Transition` and `Hold`. A `Path` of a typical storyline is a series of alternating `Holds` and `Transitions`. A `Hold` is linked with one `Region`. It represents the part of a storyline when a region of interest is highlighted (e.g. by zooming in on that region and playing the audio fragment that may be coupled to `TimeInterval`). A `Transition` is linked with two `Regions`. It represents the animated transition from one region of interest to another (e.g. by panning). Its `TimeInterval` may also be coupled with an audio fragment.

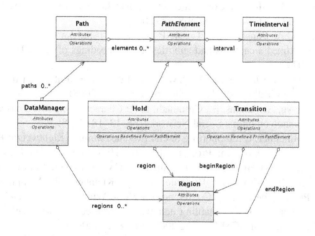

Fig. 4. Data model of the authoring system

6 Discussion

The EU-funded project TA2[7] studies how technology can help to nurture family-to-family relationships by sharing experiences, holidays, celebrations and moments of fun and laughter. TA2 media and communication support can be characterized by naturalness; clear, relaxed voice communication and intelligently edited video.

The authoring system we presented in this paper is part of this overall TA2 effort. Our family member can define a storyline in a natural way, simply by talking and clicking on images, much like browsing through a family photo album. Simplicity led to a set of requirements (cf. Section 2), on which we designed the structure of the underlying data model. Simplicity is key, especially to encourage elderly people (such as our grandparents) to produce and augment their stories – to help them to document

---

[7] http://www.ta2-project.eu/

their historic memories and experiences in general. The storyline is transferred into a SMIL document: a compact structured multimedia document that can be shared in an efficient manner with non-collocated peers, making use of URI's instead of transferring full image data wherever possible. It is important to highlight that SMIL allows the addition of metadata to any of the elements contained in the document specification. Moreover, authors can synchronize pan and zoom effects with audio or captions, and even generate customized presentations for distinct audiences.

Acknowledgments

This work was supported by the EU FP7-ICT project TA2. The NLnet Foundation supported the development of the Ambulant Player.

References

1. Balabanović, M., Chu, L.L., Wolff, G.J.: Storytelling with Digital Photographs. In: Proceedings of the SIGCHI Conference on Human factors in Computing Systems, CHI 2000, pp. 564–571. ACM Press, New York (2000)
2. Bulterman, D.C.A., Jansen, A.J., Cesar, P., Cruz-Lara, S.: An efficient, streamable text format for multimedia captions and subtitles. In: Proceedings of ACM DocEng 2007, pp. 101–110 (2007)
3. Bulterman, D.C.A., Rutledge, L.W.: SMIL 3.0: Interactive Multimedia for Web, Mobile Devices and Daisy Talking Books. Springer, New York (2008)
4. Cha, M., Mislove, A., Adams, B., Gummadi, K.P.: Characterizing social cascades in flickr. In: Proc. Workshop on online Social Networks, pp. 13–18 (2008)
5. Counts, S., Fellheimer, E.: Supporting Social Presence through Lightweight Photo Sharing On and Off the Desktop. In: Proc. of the SIGCHI Conference on Human factors in Computing Systems, CHI 2004, Vienna, Austria. ACM Press, New York (2004)
6. Crandall, D.J., Backstrom, L., Huttenlocher, D., Kleinberg, J.: Mapping the world's photos. In: Proc. of the 18th int. Conference on World Wide Web, pp. 761–770 (2009)
7. Jansen, J., Bulterman, D.C.: Enabling adaptive time-based web applications with SMIL state. In: Proceedings of ACM DocEng 2008, pp. 18–27 (2008)
8. Kennedy, L., Naaman, M., Ahern, S., Nair, R., Rattenbury, T.: How flickr helps us make sense of the world: context and content in community-contributed media collections. In: Proceedings of the 15th international Conference on Multimedia, pp. 631–640 (2007)
9. Kuijk, F., Guimarães, R.L., Cesar, P., Bulterman, D.C.A.: Adding Dynamic Visual Manipulations to Declarative Multimedia Documents. In: Proceedings of ACM DocEng 2009, München, Germany (2009)
10. Landry, B.M., Guzdial, M.: iTell: Supporting Retrospective Storytelling with Digital Photos. In: Proceedings of the 6th Conference on Designing Interactive Systems, DIS 2006, University Park, PA, USA, pp. 160–168. ACM Press, New York (2006)
11. Negoescu, R.A., Gatica-Perez, D.: Analyzing Flickr groups. In: Proceedings of the 2008 international Conference on Content-Based Image and Video Retrieval, pp. 417–426 (2008)
12. Rajani, R., Vorbau, A.: Viewing and Annotating Media with MemoryNet. Extended Abstracts on Human Factors in Computing Systems, CHI 2004, Vienna, Austria, pp. 1517–1520. ACM Press, New York (2004)
13. User Centric Media Cluster of FP6 projects: User Centric Media, Future and Challenges in European Research Luxembourg: Office for Official Publications of the European Communities, p.76 (2007) ISBN 978-92-79-06865-2

User-Centric Context-Aware Mobile Applications for Embodied Music Listening

Antonio Camurri[1,*], Gualtiero Volpe[1], Hugues Vinet[2], Roberto Bresin[3],
Marco Fabiani[3], Gaël Dubus[3], Esteban Maestre[4], Jordi Llop[4],
Jari Kleimola[5], Sami Oksanen[5], Vesa Välimäki[5], and Jarno Seppanen[6]

[1] Casa Paganini - InfoMus Lab, DIST - University of Genova, Genova, Italy
antonio.camurri@unige.it
www.sameproject.eu
[2] IRCAM, Paris, France
[3] KTH School of Computer Science and Communication, Stockholm, Sweden
[4] Music Technology Group, UPF - Universitat Pompeu Fabra, Barcelona, Spain
[5] TKK, Department of Signal Processing and Acoustics, Espoo, Finland
[6] Nokia Research Center, Helsinki, Finland
Viale Causa 13, I-16145 Genova, Italy

Abstract. This paper surveys a collection of sample applications for networked user-centric context-aware embodied music listening. The applications have been designed and developed in the framework of the EU-ICT Project SAME (www.sameproject.eu) and have been presented at Agora Festival (IRCAM, Paris, France) in June 2009. All of them address in different ways the concept of embodied, active listening to music, i.e., enabling listeners to interactively operate in real-time on the music content by means of their movements and gestures as captured by mobile devices. In the occasion of the Agora Festival the applications have also been evaluated by both expert and non-expert users.

1 Introduction

The concept of User-Centric Media entails the development of new technologies enabling an active, participative, personalized experience of media. Such technologies include, for example, innovative and intelligent real-time content processing techniques, new paradigms for natural multimodal interfaces, new devices, context-awareness. Moreover, since the strong emphasis on the user, technologies for User-Centric Media cannot avoid to take into account two major aspects of human interaction and communication: embodiment and the social dimension.

In this framework, music making and listening are an excellent test-bed for technologies for future User-Centric Media, since they are a clear example of human activities that are above all interactive and social.

The EU-ICT Project SAME (Sound And Music for Everyone Everyday Everywhere Every way, www.sameproject.eu), started in January 2008 and that recently reached half of its way, aims at developing mobile context-aware music

---

* Corresponding author.

P. Daras and O. Mayora (Eds.): UCMedia 2009, LNICST 40, pp. 21–30, 2010.

applications for active, embodied experience of music in cooperative social environments. The project is based on the concept of *active listening*, i.e., listeners are enabled to interactively operate on (pre-recorded) music content through their movement and gesture, by modifying and molding it in real-time while listening. This is obtained through the development of a networked end-to-end platform for mobile music applications enabling novel paradigms for natural, expressive/emotional multimodal interfaces, empowering the user to influence, interact, mould and shape the music content, by intervening actively and physically into the experience.

Active listening is the basic concept for a novel generation of interactive music applications, particularly addressed to a general public of beginners, naïve and inexperienced users, rather than to professional musicians. A particularly relevant aspect of active listening is its social, collaborative implication: active listening enables a social, collaborative, and context aware experience of music, allowing listeners to cooperate in the real-time manipulation and re-creation of music content.

Examples of the active listening paradigm are emerging. The *Orchestra Explorer* [1] enables users to explore a space populated by virtual instruments. *Mappe per Affetti Erranti* [2] introduces multiple levels of navigation: from navigation in a physical space up to emotional spaces populated by different expressive performances of the same music piece. Users can navigate such spaces by their expressive movement and gesture. *Mappe per Affetti Erranti* also addresses experience by multiple users encouraging social behavior. The virtual air guitars [3, 4] are examples of gesture-based mobile musical instruments. They can be made easier to play than conventional musical instruments, because user's gestures can be interpreted by the computer to produce the desired output sound.

This paper surveys a first set of such mobile context-aware music applications, presented by the SAME partners at the Agora Festival (IRCAM, Paris, France, June 2009) and representing the mid-term milestone of the project (Section 2). Prototype applications were evaluated by both expert and non-expert users visiting the festival. Results from such evaluation are also discussed (Section 3).

2 Sample Applications

In the following the sample applications presented at the AGORA Festival are shortly described. Applications can be grouped depending on how they address the concept of active listening: some of them implement active listening as an exploration of the music content, others put a particular focus on the possibility of molding the expressivity of a music piece, others adopt a game-like paradigm.

Three applications are built around the concept of exploration of a pre-recorded music piece by user's movement and gesture as captured by a mobile device. The *Audio Explorer*[1] is a mobile active-listening application allowing users to interactively de-mix commercial stereo recordings into different channels while being streamed to their mobile devices, also offering interactive re-mixing possibilities based on previously separated channels. Audio separation is carried out in a server by remotely exploring the panning position of different sources (or channels) in a stereo-mastered track. Separation parameters are controlled by means of either keypad buttons or processed

---

[1] Contributors: Esteban Maestre, Jordi Llop, Vassilis Pantazis – UPF; Alberto Massari – DIST.

accelerometer data gathered from the mobile phone device. The separation parameters are stored in the server, so that they are shared among users, who can access to them for using them in a re-mix context: users manipulate (gain by means of either keypad buttons or processed accelerometer data) the gain and panning position of each previously separated channel within the original recording, leading to an active listening experience. An overview of the system architecture is depicted in Figure 1. Original audio tracks reside on a database in the server side and are retrieved by the user. The EyesWeb XMI platform (v. 5.0.3.0) [5] is running on the server machine, giving support for audio streaming to the mobile device, audio processing through an extended and improved VST implementation of the audio separation algorithm described in [6], and application control protocol based on Open Sound Control (OSC). The mobile phone device (Nokia N85) runs the application control interface and offers visual feedback to the user through a user-friendly GUI. The processed audio stream is received from the server and played back locally. Real-time control of separation/remixing is performed remotely from the mobile device, which is in charge of gathering and processing accelerometer data, processing key pressing for retrieval of audio and separation preset files, and display of visual feedback to the user.

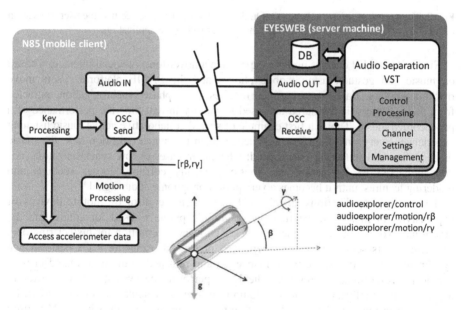

Fig. 1. System architecture of the *Audio Explorer* application

The *Mobile Orchestra Explorer*[2] is a mobile version of the former Orchestra Explorer [1]. Users can navigate a shared (physical or virtual) "orchestra space", populated by the sections or single instruments of a pre-recorded music (see Figure 2): a user can activate and listen to one or more sections of the music. The mobile phones

---

[2] Contributors: Antonio Camurri, Corrado Canepa, Paolo Coletta, Gualtiero Volpe, Alberto Massari, Maurizio Mancini – DIST; Markus Noisternig, Joseph Sanson, Olivier Warusfel – IRCAM for WFS extension.

are here used to detect the movement of the user, to activate and control the music sections, and to present, on the phone display, the user's position in the orchestra space. The music rendering is either based on 3D sound via loudspeakers (using WFS) or on the mobile phone using its headphones.

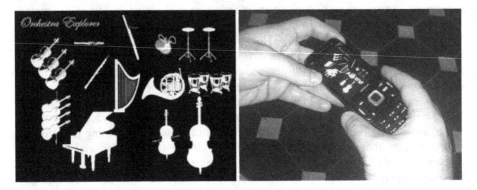

Fig. 2. The *Mobile Orchestra Explorer*. On the left the orchestra space that the user can see on the display of the mobile phone. On the right, an user trying the application.

Sync'n'Move[3] [7] enables users to experience a novel form of social interaction based on music and gesture, using mobile phones and the SAME platform. Users move rhythmically (e.g., dance) wearing their mobiles. Their phase synchronization, extracted from their gestures, is measured and used to modify in real-time the performance of a pre-recorded music. This is a first example of shared collaborative active music listening experience. Every time the users are successful in the synchronization task, the music orchestration and rendering is enhanced; whereas in cases of low synchronization, i.e., poor collaborative interaction, the music gradually corrupts, looses sections and rendering features, until it becomes a very poor monophonic audio signal.

Two sample applications are devoted to real-time control by mobile devices of expressivity in music. In the first one, a mobile phone is used for controlling the emotional expression of ringtones. The user chooses an emotion for his/her ringtone. The ringtone is sent to a server where it is processed using the KTH performance system for expressive music performance [8] and returned to the user's handset with the desired emotional expression. The KTH performance system controls different aspects of the performance, such as tempo, dynamics, articulation, orchestration, by associating pre-assigned values for each emotion[4]. In the second one, *pyDM* is used for expressive control of a piano performance[5]. A computer-controlled piano is connected to a computer running pyDM. This is a program for interactive control of expressivity in music performance, using the KTH rule system for music performance (see Figure 3). Again, each rule controls different aspects of the performance, such as tempo, dynamics, and articulation. Rule values can be adjusted separately, or mapped

---

[3] Contributors: Giovanna Varni, Paolo Coletta, Gualtiero Volpe, Antonio Camurri, Corrado Canepa, Maurizio Mancini, Barbara Mazzarino – DIST.
[4] Contributors: Roberto Bresin – KTH; Jarno Seppanen – Nokia.
[5] Contributors: Marco Fabiani, Roberto Bresin, Gaël Dubus – KTH.

to more intuitive control parameters, such as the Activity-Valence space, in which different basic emotions can be expressed (e.g., happiness, sadness, tenderness, anger). In pyDM, the Activity-Valence value is shown by a moving circle, whose color and dimension vary according to the expressed emotion (see Figure 3). The program can be controlled using a mobile phone graphical interface, or by tilting the phone, as well as by shaking it in different ways to express different emotions.

 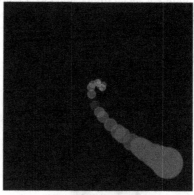

Fig. 3. On the left *PyDM*. On the right, representation of trajectories in the Activity-Valence space as moving circles, whose color and dimension vary according to the expressed emotion.

Context-awareness is particularly addressed in *Zagora*[6], a context-aware mobile music player, which detects the ambient situation using audio analysis and retrieves a playlist of suitable music. The Zagora player is doing advanced audio processing to differentiate between situations like *street, restaurant, car, office,* and *meeting,* and uses the situation information to filter down an online music catalog. The user can see the current audio analysis results, generate a playlist online, and start streaming music. Finally, all resulting playlists can be browsed for other similar online music.

Further sample applications adopt a game-like paradigm. The *Mobile Sonic Playground*[7] demonstrates examples of individual and collective sonic applications using mobile phones as musical instruments and sounding toys. The user interacts with the phone accelerometers and keypad keys, and generates control events that are captured, processed, and rendered to sound using the phone embedded Mobilophone framework. Several playing styles and synthesized sound selections are available.

The *Fishing Game*[8] illustrates novel technologies on gestural sound control and embodied active listening. The system makes use of gesture recognition and analysis, driving a sound engine. When users with their mobile phones mimic gestures such as pouring a glass or brushing teeth, they can listen to the sound associated to such gestures. This illustrates emerging uses of embedded sensors in mobile phones.

[6] Contributors: Antti Eronen, Jussi Leppänen, Jarno Seppänen – Nokia.
[7] Contributors: Jari Kleimola, Sami Oksanen, Vesa Välimäki – TKK.
[8] Contributors: Pierre Jodlowski, Baptiste Caramiaux, Grace Leslie, Norbert Schnell, Diemo Schwarz, Bruno Zamborlin, Frédéric Bevilacqua, Hugues Vinet, Olivier Warusfel – IRCAM.

Finally, the *Grain Stick* installation[9] offers a collaborative interactive experience featuring music by Pierre Jodlowski. One or two participants shake a virtual tube by means of two manual sensors that set off a waterfall of sound grains (like a rain stick) in a sound space spatialized with WFS technology. The sounds of the grains, generated by the corpus-based synthesis engine CataRT, overlap the surrounding soundscape and percussive sounds that are triggered by the users' movements. The virtual stick can be used by one person alone with both hands or by two users, thus including a social dimension (see Figure 4). Beyond technical feasibility, an important aspect of this work on the artistic side has been to experience a new kind of interactive musical form.

Fig. 4. Two users experiencing the *Grain Stick* installation

3   Evaluation

Evaluation has been carried out by asking participants to fill questionnaires. These included general questions, concerning the overall evaluation of the active listening concept, and questions for the sample applications. Moreover, the evaluation of some of the sample applications included specific questions especially devoted to them.

The general questions concerning all the prototypes are reported in Figure 5. The questions that participants answered for each sample application are reported in Figure 6. The questionnaire also included information about age, gender, musical skills and habits (e.g., preferred music genre, time spent in listening to music, etc.).

Evaluation involved 108 participants. 82 attended public sessions; 16 attended a special session dedicated to expert users (music professionals). Table 1 shows the results for the questions in Figure 6. Answers were collected using 11 point scale

---

[9] Contributors: Pierre Jodlowski, Grace Leslie, Markus Noisternig, Norbert Schnell, Joseph Sanson, Diemo Schwarz, Bruno Zamborlin, Frédéric Bevilacqua, Hugues Vinet, Olivier Warusfel – IRCAM.

from -5 to 5. The table include the number of subjects that answered the question for a specific application (N), and mean and standard deviation of the results.

Q1: What did you expect from this experience (check all that apply) ?
    ☐Have fun
    ☐Learn
    ☐New music experience
    ☐ Better communicate with peers
    Other _____

Q2: What was you first impression ?
    Very negative ☐☐☐☐☐☐☐☐☐☐ Very positive

Q2: The strength of your experience was:
    Very weak   ☐☐☐☐☐☐☐☐☐☐ Very strong

Q3: Which of the following areas do you think could benefit from the project (check all that apply) ?
    ☐New entertainment
    ☐New technology
    ☐New form of art
    ☐ Kinesthetic/motor abilities
    ☐Ability to communicate
    ☐For no good use at all
    Other _____

Fig. 5. Evaluation questionnaire: the general questions on the active listening experience

---

Q1: How easy is it to understand how the application works?
    Very difficult ☐☐☐☐☐☐☐☐☐☐ Very easy

Q2: How much do you feel in control of the application?
    Very little ☐☐☐☐☐☐☐☐☐☐ Very much

Q3: How do you find the level of interaction?
    Low ☐☐☐☐☐☐☐☐☐☐ High

Q4: What do you think about this application?
              Boring ☐☐☐☐☐☐☐☐☐☐ Funny
     Uninteresting   ☐☐☐☐☐☐☐☐☐☐ Ineresting
Nothing for the future☐☐☐☐☐☐☐☐☐☐ Something for the future
      Not engaging ☐☐☐☐☐☐☐☐☐☐ Engaging
     I did not enjoy it   ☐☐☐☐☐☐☐☐☐☐ I enjoyed it

Fig. 6. Evaluation questionnaire: questions for each sample application

Table 1. Results from user evaluation for the SAME sample application presented at the Agora Festival, Paris, France, June 2009. Questions Q1, Q2, Q3, and Q4 are reported in Figure 6. Answers were collected using a 11 point scale from -5 to 5. The number of participants (N), average values ($\bar{x}$) and standard deviations (σ) are reported for the both session types: i.e., with expert and non-expert users.

| | Audio Explorer | | | | | | Fishing Game | | | | | | Grain Stick | | | | | | Mobile Orchestra Explorer | | | | | | Sync'n'Move | | | | | |
| | Experts | | | Non-experts | | | Experts | | | Non-experts | | | Experts | | | Non-experts | | | Experts | | | Non-experts | | | Experts | | | Non-experts | | |
Question	N	$\bar{x}$	σ	N	$\bar{x}$	σ	N	$\bar{x}$	σ	N	$\bar{x}$	σ	N	$\bar{x}$	σ	N	$\bar{x}$	σ	N	$\bar{x}$	σ	N	$\bar{x}$	σ	N	$\bar{x}$	σ	N	$\bar{x}$	σ
Q1 Very difficult vs. Very easy	11	2.45	1.51	18	2.22	2.62	10	4.5	0.71	15	3.53	2.33	12	3.33	1.15	24	1.54	2.21	10	3.9	1.52	13	4.31	1.18	9	3.56	1.74	10	4.3	1.57
Q2 Very little vs. Very much	9	2.89	1.05	15	1.87	1.81	10	3.3	1.7	13	2.45	1.66	13	3.23	2.09	20	1.95	2.31	8	1	1.93	13	4.15	0.8	8	0.38	2.62	9	2.11	1.83
Q3 Low vs. High	11	1.91	1.87	16	2.5	1.71	10	3.5	1.18	14	2.85	2.44	13	3.54	1.61	22	3.86	1.67	9	0.11	3.06	13	2.54	1.98	9	0.78	3.03	10	1.7	1.77
Q4 a) Boring vs. Funny	11	1.18	2.14	16	2.38	1.82	10	3.9	1.1	15	3.67	1.91	13	3.69	1.18	24	3.29	1.73	9	0.11	2.62	14	2.14	2.28	10	-0.7	3.02	10	2	3.27
b) Uninteresting vs. Interesting	11	1.73	2.41	17	3.06	1.75	10	2.7	1.95	14	2.43	1.79	13	3.62	1.19	24	3.96	1.65	9	0.78	3.19	13	1.77	1.24	10	-1	3.13	10	0.7	2.54
c) Nothing for the future vs. Something for the future	11	1.27	3.17	16	2.63	2.16	10	2	3.16	14	2	2.66	13	3.69	1.49	24	3.79	1.86	9	1.33	3.39	14	2.86	1.75	10	-0.2	3.19	10	1.5	2.76
d) Not engaging vs. Very engaging	11	1.36	2.54	16	1.56	2.13	10	2	1.99	13	2.46	2.15	13	3.85	1.07	24	3.42	1.84	9	-0.33	3.12	13	1.92	1.89	10	0.1	3.03	10	3.9	2.33
e) did not enjoy it at all vs. I enjoyed it very much	11	0.82	2.04	16	2.38	1.71	10	2.1	1.79	14	3.07	1.94	13	3.77	1.24	24	3.54	2.11	9	0.11	2.98	13	2.36	1.5	10	-0.3	2.63	10	1.4	2.67

| | pyDM | | | | | | Emotional expression of ringtones | | | | | | Zegora | | | | | | Mobile Sonic Playground | | | | | |
| | Experts | | | Non-experts | | | Experts | | | Non-experts | | | Experts | | | Non-experts | | | Experts | | | Non-experts | | |
Question	N	$\bar{x}$	σ	N	$\bar{x}$	σ	N	$\bar{x}$	σ	N	$\bar{x}$	σ	N	$\bar{x}$	σ	N	$\bar{x}$	σ	N	$\bar{x}$	σ	N	$\bar{x}$	σ
Q1 Very difficult vs. Very easy	12	4.5	0.8	23	3.61	1.75	10	3.9	1.29	15	3.2	1.93	13	4.69	0.63	11	2.55	2.54	12	3.75	1.42	10	2.8	1.87
Q2 Very little vs. Very much	11	1.73	3.07	21	3.05	2.01	10	2.1	2.6	13	2.54	1.71	10	-0.7	2.41	10	1.5	2.51	11	0.91	2.74	8	2.63	1.41
Q3 Low vs. High	12	0.92	3.48	23	2.91	1.68	10	-0.3	3.23	14	1.36	1.98	13	-0.85	2.54	11	0.91	2.12	12	-0.92	2.97	9	2.67	1.73
Q4 a) Boring vs. Funny	12	1.75	3.05	23	3.26	1.79	10	-1.2	2.82	15	1.8	2.37	13	-0.92	2.5	11	1.73	3.17	12	-0.83	2.48	10	2.3	3.23
b) Uninteresting vs. Interesting	12	0.42	3.2	23	3.13	1.66	10	-1.3	2.67	15	1.53	2.07	13	-0.92	2.66	10	2	2.79	11	-0.27	2.1	10	2	3.02
c) Nothing for the future vs. Something for the future	12	1.67	2.67	23	2.35	2.04	10	-1.7	3.3	15	1.8	1.93	12	0.25	3.19	10	2.6	1.43	11	-0.27	2.69	10	2.6	2.37
d) Not engaging vs. Very engaging	12	0.75	3.14	22	2.36	1.5	10	-2	2.45	14	1.36	2.53	9	-0.58	2.78	9	1.22	2.54	12	-1.08	2.5	10	2.4	2.46
e) did not enjoy it at all vs. I enjoyed it very much	12	0.92	3.03	22	2.95	1.59	10	-1.5	3.27	14	1.79	1.85	10	-0.08	2.71	10	2	1.49	12	-0.58	2.27	10	3.3	2.5

Based on such feedback, we can infer that visitors generally liked the SAME applications. Feedback from both groups included valuable criticism, suggestions, and proposals for improvements. The demonstrated applications have a potential for the future, but more research and development work needs to be done. Indeed, some of the applications were finished and well refined installations (e.g., Grain Stick), whereas others were rather proof-of-concepts that still need to be further developed (e.g., Synch'n'Move). Such different development stages of the applications may have affected the average appreciation by participants.

Evaluation pointed out that the users are anxious to see social networking features implemented as a part of the applications. This is encouraging for future research on embodied social interaction envisaged in SAME. In general, the interaction between users and personal file sharing was appreciated. Merging of the applications or combining them into more versatile systems was also suggested.

Finally, if from the one hand, the applications received an overall positive feedback by both expert and non-expert users, on the other hand, non-expert users especially appreciated active listening both as a new way for listening to music and also as an educational tool for gaining a better understanding of how music is made, structured, and performed.

4 Conclusions

This paper presented a survey of the sample applications the SAME EU Project presented at Agora Festival (IRCAM, Paris, June 2009), as mid-term milestone of its research and development work. We believe that such applications represent a useful test-bed for future paradigms of active experience and User-Centric Media.

The applications have been evaluated by expert and non-expert participants, whose feedback will be used for refining the requirements of the project and for moving towards the final set of comprehensive prototypes of systems and applications for context-aware mobile social and active listening to music.

Acknowledgments

This work has been partially supported by the EU-ICT Project SAME. We thank all the contributors to the sample applications for their precious work.

References

1. Camurri, A., Canepa, C., Volpe, G.: Active listening to a virtual orchestra through an expressive gestural interface: The Orchestra Explorer. In: Proceedings 2007 Intl. Conference on New Interfaces for Musical Expression (NIME 2007), New York, USA (June 2007)
2. Camurri, A., Canepa, C., Coletta, P., Mazzarino, B., Volpe, G.: Mappe per Affetti Erranti: a Multimodal System for Social Active Listening and Expressive Performance. In: Proc 2008 Intl. Conference on New Interfaces for Musical Expression (NIME 2008), Genova (2008)

3. Karjalainen, M., Maki-Patola, T., Kanerva, A., Huovilainen, A.: Virtual air guitar. Journal of the Audio Engineering Society 54(10), 964–980 (2006)
4. Pakarinen, J., Puputti, T., Valimaki, V.: Virtual slide guitar. Computer Music Journal 32(3), 42–54 (Fall 2008)
5. Camurri, A., Coletta, P., Demurtas, M., Peri, M., Ricci, A., Sagoleo, R., Simonetti, M., Varni, G., Volpe, G.: A Platform for Real-Time Multimodal Processing. In: Proceedings International Conference Sound and Music Computing 2007 (SMC 2007), Lefkada, Greece (2007)
6. Vinyes, M., Bonada, J., Loscos, A.: Demixing Commercial Music Productions via Human-Assisted Time-Frequency Masking. In: 120th AES Convention, Paris (2006)
7. Varni, G., Mancini, M., Volpe, G.: Sync'n'Move: social interaction based on music and gesture. In: Proc. 1st International ICST Conference on User Centric Media, Venice (2009)
8. Friberg, A.: pDM: an expressive sequencer with real-time control of the KTH music performance rules movements. Computer Music Journal 30(1), 37–48 (2006)

Sync'n'Move: Social Interaction Based on Music and Gesture

Giovanna Varni, Maurizio Mancini, Gualtiero Volpe, and Antonio Camurri

InfoMus Lab, DIST - Università degli Studi di Genova, Italy
giovanna.varni@infomus.dist.unige.it, maurizio.mancini@dist.unige.it,
gualtiero.volpe@unige.it, antonio.camurri@unige.it

Abstract. In future User Centric Media the importance of the social dimension will likely increase. As social networks and Internet games show, the social dimension has a key role for active participation of the users in the overall media chain. In this paper, a first sample application for social active listening to music is presented. Sync'n'Move enables two users to explore a multi-channel pre-recorded music piece as the result of their social interaction. The application has been developed in the framework of the EU-ICT Project SAME (www.sameproject.eu) and has been presented for the first time at the Agora Festival (IRCAM, Paris, June 2009). In that occasion, Sync'n'Move has also been evaluated by both expert and non expert users.

Keywords: active music listening, social interaction, synchronization.

1 Introduction

In the Future Internet users will play a crucial role both as individuals and in the social dimension. The User Centric Media concept "implies that the user will become an active member of the overall media chain by generating, distributing and experiencing high-quality media content" [1]. The worldwide spreading of social networks and Internet games bears witness of the importance of the social dimension for such an active participation of the users in the overall media chain.

Nevertheless, many existing multimedia interactive systems and Internet applications are still intended for a single user and social interaction is often neglected. Social networks, indeed, are mainly based on sharing of static textual and audiovisual content, whereas realtime interaction between users, immersiveness, and sense of presence are far to be fully reached.

In the framework of the EU-ICT Project SAME (www.sameproject.eu), novel paradigms of active, embodied, and social listening to music in context-aware mobile applications are explored [5], i.e., paradigms enabling both single and groups of users to actively mould and reshape sound and music content, based on movement, gesture, and social interactions.

Sync'n'Move is a first sample application in the direction of social active listening to music. It enables two users to explore a multi-channel pre-recorded music piece as the result of their social interaction. Users interact through the

P. Daras and O. Mayora (Eds.): UCMedia 2009, LNICST 40, pp. 31–38, 2010.

movements they perform by handling a mobile device. A phase synchronization index is extracted from movement. High-level of synchronization enhances
music orchestration and rendering, by adding music sections and rendering features. Sync'n'Move has been presented for the first time at the Agora Festival
(IRCAM, Paris, June 2009), as one of the sample applications proposed by the
SAME Project. In that occasion, Sync'n'Move has also been evaluated by both
expert and non expert users.

The remainder of this paper is organized as follows. Section 2 gives an overview
of the theoretical background, with particular reference on how social interaction is addressed and measured. Section 3 describes the technical details of
Sync'n'Move. Finally, evaluation results are discussed in Section 4.

2   Background

Our approach to the complex phenomenon of synchronization in social interaction is based on analysis of Phase Synchronization (PS) [8] In our model each
user is represented with an n-dimensional state vector in which the n dimensions
are n features like, for example, trajectories of body segments, amount of motion, audio descriptors and so on. On this assumption, interaction is addressed
considering how the state vectors evolve together in time and extracting, from
this joint dynamics, indices of a *global* system's behavior.

Phase synchronization can be measured using techniques based on the recurrence property of dynamical systems [9]. Recurrence Plots (RPs) [6] and Recurrence Quantitative Analysis (RQA) [7], [11] are techniques providing qualitative
and quantitative information on systems' dynamics and their interrelations in
terms of trajectories in a chosen features space. More specifically, a RP is a time-
time binary colorimetric plot displaying all time instants in which recurrences
in the state of a system are observed, whereas RQA quantifies the graphical
patterns occuring in RP.

Let us consider two systems, identified by their state vectors x and y, respectively. RP for the first system (the same formula is also valid for the second
system by replacing x with y) is defined through the recurrence matrix:

$$R_{i,j}(\varepsilon) = \Theta\left(\varepsilon - \parallel x_i - x_j \parallel\right) \qquad i,j = 1...N \qquad (1)$$

where, $x_{i,j} \in \mathbb{R}^n$ are the system states at times i and j, N is the number
of the states, ε is a closeness threshold, $\parallel \parallel$ and Θ are a norm (e.g., euclidean
norm, minimum or maximum norms can be adopted) and the Heaviside function,
respectively. The elements $R_{i,j}$ are binary values (0 or 1) according to whether
x_i is *close* or not to x_j.

By applying RQA to a RP we compute the probability $\hat{p}(\varepsilon, \tau)$ that the system
recurs at a certain state after some time τ [10]. The estimate of this probability
can be written as:

$$\hat{p}(\varepsilon, \tau) = \frac{1}{N-\tau} \sum_{i=1}^{N-\tau} R_{i,i+\tau}(\varepsilon) = \frac{1}{N-\tau} \sum_{i=1}^{N-\tau} \Theta(\varepsilon - \parallel x_i - x_{i+\tau} \parallel) \qquad (2)$$

Again the same formula is valid for the second system by replacing x with y. Finally we compute the PS between the two systems with the Correlation Probability of Recurrence (CPR), defined as:

$$PS = CPR = \langle \bar{p}_x(\varepsilon, \tau) \bar{p}_y(\varepsilon, \tau) \rangle \qquad (3)$$

where $\bar{p}_{xy}(\varepsilon, \tau)$ are the functions $\hat{p}(\varepsilon, \tau)$ normalized to zero mean and unitary standard deviation.

In the Sync'n'Move application we present in this paper, we aim to study the realtime synchronization between two systems: two users that move two mobile phones in space. In this case x and y are 1-dimensional state vectors representing the acceleration measured by on-board accelerometers the users mobile phones are endowed with. High phase synchronization between such accelerations corresponds to enhanced audio rendering as output, otherwise the audio output is minimal.

3   Application Description

Our application is developed in the SAME networked platform for mobile, experience-centric, and context-aware active music listening. The SAME platform is an end-to-end framework (i.e., between clients of a mobile service, producers and consumers of content) for context-aware, experience-centric mobile music applications, enabling embodiment and control of music content by user behaviour. It facilitates rapid prototyping of context sensitive experience-centric embodied music applications and it enables intelligent, real-time, distributed processing of integrated music, video, and multimodal signals through a generalized plugin mechanism, aimed at open connectivity of mobile music systems and emerging media centers and industrial standards. The platform includes one or more servers running software environments, such as for example EyesWeb XMI [4] and MAX MSP [2]. Sync'n'Move runs on the SAME platform by allowing mobile phones to be connected and communicate to EyesWeb XMI.

Figure 1 sketches out how our application works. Two users freely move their mobile phones and their hand/body acceleration is detected and measured by the tri-axial accelerometer embedded in the mobile. An index of phase synchronization is extracted from their gesture: every time the users succeed in synchronizing (index is high) the music orchestration and rendering is enhanced; if instead users cannot synchronize (i.e., the phase synchronization index is low), the music gradually looses sections and rendering features.

Figure 1 shows the main modules composing our application: *the data acquisition module* and *the feature extraction & audio processing module*.

3.1   The Data Acquisition Module

This first module reported in Figure 2 acquires, calibrates, and computes the normalized acceleration captured by the mobile phones the users are moving. Each mobile runs a Python script that collects data from the accelerometer and creates an OSC packet in the form:

```
/synchronizer ax ay az
```

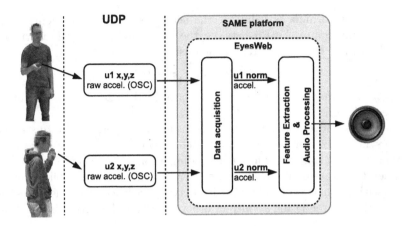

Fig. 1. Architecture of the Sync'n'Move application

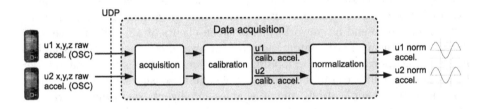

Fig. 2. Data acquisition module

The above packet is sent via UDP to the SAME platform running EyesWeb where the *raw accelerations* on the 3 axis are extracted from the OSC packet (*acquisition* box in Figure 2). The *calibration* and *normalization* blocks are necessary since every accelerometer has a different ground reference, that is, the *max* and *min* values of g change on every axis. Here we report the operations performed by the calibration block on the x axis, the same computation is necessary also on the other two axis:

$$A_{C_x} = A_{raw_x} - IRD_x; \quad IRD_x = \frac{gx^+ + gx^-}{2}; \quad (4)$$

in which:

- IRD_x is the *Inter Range Difference*, that is, the half of the difference between the maximum g measured on x^+ and x^-;
- A_{C_x} is the calibrated acceleration on x axis, that is, the acceleration obtained by subtracting the IRD_x from the raw acceleration on x, that is, A_{raw_x};

Finally the normalization block computes the absolute value of acceleration and nomalizes it in the range $[0, 1]$ subtracting g in order to ignore it:

$$A_{normalized} = \frac{\sqrt{A_{C_x}^2 + A_{C_y}^2 + A_{C_z}^2} - g}{A_{MAX}}; \quad (5)$$

where A_{MAX} is the maximal absolute value of acceleration detected by the phone.

3.2   The Feature Extraction and Audio Processing Module

Figure 3 shows the architecture of the feature extraction & audio processing module. This module is responsible for computing the phase synchronization index, which is used for controlling audio processing. From the normalized accelerations, we compute the probabilities of recurrence $\hat{p}_x(\varepsilon, \tau)$ and $\hat{p}_y(\varepsilon, \tau)$ and we normalize them to obtain the probabilities $\bar{p}_x(\varepsilon, \tau)$ and $\bar{p}_y(\varepsilon, \tau)$ having zero mean and unitary standard deviation. The next step is the computation of CPR. The final output of the application is an audio content produced by the *audio processing* block. This content changes according to the synchronization degree between the users. The following three cases can occur:

- no audio: the users are not interacting at all, that is they are not moving their mobile phones. In this case the *movement detection* block detects that the two accelerations are equal to zero and inhibit audio generation;
- metronomic audio: (i) only one user is moving or (ii) both are moving but they are not synchronized. In the first condition, the *movement detection* block detects that just one of the accelerations is different from zero and enables the generation of a metronomic section in the audio output, e.g., the charleston instrument. In the second condition, the CPR is computed but it is too low to allow the generation of the full audio output.
- full audio: the two users are moving in a synchronized way. The CPR assumes a high value (almost one) and the *synchronization timer* measures the time along which the two users keep synchronized. According to the duration of this time, new sections are added to the audio content: the longer is the synchronization time the larger is the number of the enabled instruments, e.g., drums, bass and guitar, voice.

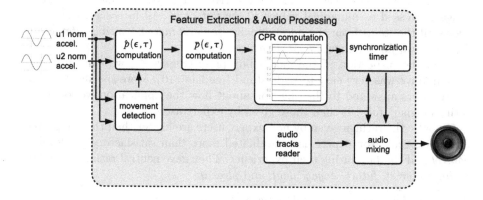

Fig. 3. The feature extraction & audio processing module

An example of how Sync'n'Move works can be found at:

www.sameproject.eu/Demos

4 Evaluation

Sync'n'Move was tested during the multidisciplinary encounter Agora Festival 2009 in Paris [3]. We gathered qualitative information on the application using an anonymous assessment questionnaire[1] filled up by 22 participants (19 male and 3 female from different european and extra-european countries) attending the Festival. Mean age of the participants was 33.7 years (range 18y-64y). All the participants were volunteered for this study and they were only asked to have a spontaneous behavior as much as possible. Before their performance, they are provided with a short demonstration of how the system works.

The participants were classified in two groups of users: expert and non-expert users. The experts'group was composed by 12 subjects having attested expertise in the musical field such as composers and professional musicians.

The questionnaire was composed by three parts: the first one, collecting general information about the participants such as gender, age, nationality and work; the second one, including general questions on the SAME applications presented at Agora Festival; the third one concerning the evaluation of each single application. This last part was different for each application. The analysis in this paper took into account data only from the first and the third part of the questionnaire. The third part of the questionnaire conceived for Sync'n'Move was composed by four questions about understanding, control, interaction level, fun, interest, future exploitation, engagement, and pleasure. Participants were asked to express their ratings on scales divided into eleven steps ranging from *not at all* to *very*. In the final part of the questionnaire, participants could write their comments and suggestions. We took into account partially filled up questionnaires also.

Figure 4 summarizes the results for both groups simultaneously, whereas Figure 5 shows the results obtained for the expert users (Left Panel) and for the non-expert users (Right Panel) respectively. The y-axis shows the ratings range expressed as numerical values (0 for *not at all*, 10 for *very*), the x-axis shows all the items from the questionnaire: *understanding, control, interaction, fun, interest, future, engagement, pleasure.*

From the inspection of the global box plots (Figure 4), we can infer that Sync'n'Move was very easy to understand (median=10). More specifically, non-expert users answered to the question about how the application works giving ratings higher (full 10) than those given by expert users.

Globally the ratings given by non-expert users are higher than those reported by the experts. Non-expert users evaluated more than satisfactory the level of *control* and the *fun* during their experience. They gave neutral ratings for *interaction, interest, future, engagement,* and *pleasure.*

---

[1] The questionnaire was designed by the SAME consortium
http://www.sameproject.eu

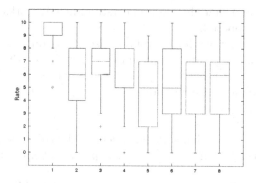

Fig. 4. The global box plot. The numbers on the x-axis stand for the items: 1-understanding, 2-control, 3-interaction, 4-fun, 5-interest, 6-future, 7-engagement, 8-pleasure.

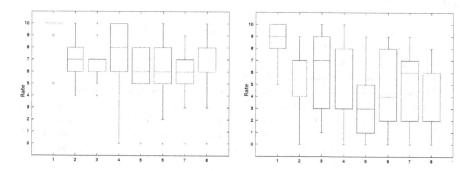

Fig. 5. Box plots of the ratings of the non-expert users (Left Panel) and of the expert users (Right Panel)

Differently, the ratings provided by expert users were more spread, for example the 50% of ranked ratings for *interaction* ranges from 3 to 9, whereas the 50% of ranked ratings for *engagement* ranges from 2 to 7. This may be due to the fact that expert users could perceive better than non-expert users that the audio tempo generated by Sync'n'Move was not matching in time with the tempo chosen by the users. This could also explain the low ratings obtained for *control*, *interest*, and *future*. Moreover, Sync'n'move exploits very simple music content and interaction mechanisms, whereas expert users probably expect more complex paradigms and content.

5   Conclusion

This paper presented an application on active music listening. In Sync'n'Move two users generate an audio content by syncronizing their movements using mobiles phones as a collaborative interface. Evaluation shows that users are engaged in the application and that they well-understood how it works and the interaction paradigm we proposed.

We would like to improve the control that users have on generation of the music sections, e.g., enabling audio tempo to vary depending on the users' movement frequency, and allowing users to play their preferred music in Sync'n'Move. In this last case, we plan to use an audio sections separation algorithm developed within the SAME project.

Acknowledgments. We thank Paolo Coletta, Alberto Massari, and Corrado Canepa. We also thank Carlo Chiorri for his precious help in the set up of the questionnaire.

This work is partially supported by the FP7 EU ICT SAME Project n. 215749 on active music listening (www.sameproject.eu).

References

1. Laso-Ballesteros, I., Daras, P. (eds.): User Centric Future Media Internet, EU Commission (September 2008)
2. http://www.cycling74.com
3. http://agora2009.ircam.fr/843.html
4. Camurri, A., Coletta, P., Varni, G., Ghisio, S.: Developing multimodal interactive systems with EyesWeb XMI. In: Proceedings of the 7th international conference on New interfaces for musical expression, pp. 305–308. ACM, New York (2007)
5. Camurri, A., Volpe, G.: Active and personalized experience of sound and music content. In: Proc. 12th IEEE International Symposium on Consumer Electronics. IEEE Press, Los Alamitos (2008)
6. Eckmann, J.P., Kamphorst, S.O., Ruelle, D.: Recurrence plots of dynamical system. Europhysics Letters 5, 973–977 (1987)
7. Marwan, N., Romano, M.C., Thiel, M., Kurths, J.: Recurrence plots for the analysis of complex systems. Physics Reports 438, 237–329 (2007)
8. Pikovsky, A., Rosenblum, M.G., Kurths, J.: Synchronisation: a Universal Concept in Nonlinear Sciences. Cambridge Nonlinear Science Series, vol. 12. Cambridge University Press, Cambridge (2001)
9. Poincaré, H.: Sur la probleme des trois corps et les équations de la dynamique. Acta Mathematica 13, 1–271 (1890)
10. Romano, M.C., Thiel, M., Kurths, J., Kiss, I.Z., Hudson, J.L.: Detection of synchronisation for non-phase coherent and non-stationarity data. Europhysics Letters 71(3), 466–472 (2005)
11. Zbilut, J., Webber Jr., C.L.: Embeddings and delays as derived from quantification of recurrence plots. Physics Letters A 5, 199–203 (1992)

UCMedia 2009

Session 2: Content in Media Communities

From Coach Potatoes to TV Prosumers: Community-Oriented Content Creation for IDTV

Oscar Mayora-Ibarra[1], Christian Fuhrhop[2], and Elizabeth Furtado[3]

[1] CREATE-NET
[2] Fraunhofer FOKUS
[3] University of Fortaleza

Abstract. The transition from mere content consumers to more proactive "prosumers" that transformed the web is finding now place in the context of idtv. In the close future it will be common that TV content will be empowered with interactive applications that will be created not only by professionals but also on the final users' side. This fact will change the traditional role and stereotype of TV users from "coach-potatoes" to active generators of content. This paper presents a step forward to this vision through the development of a content creation tool for enabling local communities to create and share idtv content. The paper presents the tool development process as a user-centered approach and its evaluation in a real context.

Keywords: idtv, content creation, prosumers, content management system.

1 Introduction

In the latest years there has been a major transformation in the way users consume and generate media content. In fact, it is clear that the new trend for users is to increase their participation in the development and share of content through online communities. This shift on users' involvement in media lifecycle has redefined their role of mere consumers of media to become "prosumers" [1]. In the present, this trend is evolving from the web domain to other spheres such as the mobile web [2] and more recently to other platforms such as the interactive digital television (idtv) [3]. In particular in this latest platform, the involvement of users has started to comprise the different stages of the production-consumption lifecycle in a similar way as happened some years ago within the web domain [4]. It is now an emerging reality, the possibility of creating virtual idtv communities dedicated to social networking [4][5]. In particular, the community-level dimension of idtv has a significant peculiarity when compared to traditional online web communities. An example of this is the coverage area of idtv that is limited to well define geographic areas making possible to cluster communities together with specific local content of interest. In other words, the communities built around idtv are intrinsically connected to particular territories and can be naturally associated to specific local services. This fact is particularly beneficial to specific communities and territories affected with low accessibility to the Internet (such as rural areas in developing countries), where idtv presents a good potential for providing interactive services focusing the local needs. Moreover, based

P. Daras and O. Mayora (Eds.): UCMedia 2009, LNICST 40, pp. 41–49, 2010.

on the strong penetration and ubiquity of TV in our society its use as a mean for social inclusion is a powerful concept that requires further exploration [6]. For example, the use of idtv in marginalized areas or addressing digital inclusion needs of certain sectors of the population (like elders) is of paramount interest [7][8].

This paper presents a content creation tool for enabling local communities to create and share idtv content. In order to provide good accessibility to the tool that is targeted to users with limited knowledge of technology, different users studies were performed. The results of the studies obtained during both, the requirement definition of the tool and the evaluation of its use are presented. These studies were of relevance from two different viewpoints: 1) as a tool for adequate identification of requirements for this specific work and 2) given the early stage of idtv in Latin America, as a preliminary, more general overview of requirements setting for the development of idtv in such region. The following sections describe both, the main characteristics and components of the content management system tool based on a user centric approach and the results of its evaluation.

2   User Centric Content Creation System

The community-oriented Content Management System (CMS) presented in this paper was elaborated within the framework of EU-SAMBA project [9]. In this project, the objective was to provide rural communities with means for creating and consuming relevant content, mostly impacting on digital inclusion for the target region in a sustainable way. In this context, a set of local institutions supporting the community as reference partners for improving societal inclusion and development were identified and involved. These institutions contributed as "Secondary Users" of the system, so those in charge of creating and sharing the community-oriented content with the rest of the population (Primary Users).

The CMS was designed following a user-centric approach by involving Secondary Users together with HCI experts. The analysis conducted by the experts was done based on information available on the Internet and documents existing at the local organizations (such as reports on educational levels, local economy and census, etc.) as well as from information gathered from users living or working in the town center. Some relevant requirements identified from the understanding of secondary and primary users experiences were related to the Content Management System (CMS) module and the data presentation and interactive services templates. In particular, several relevant functional features were identified related to Content Creation and Personalization as well as other non-functional aspects:

- Content Creation – The main requirements for content creation included the availability of content preview in the way the end user would see it, automatic tools for text verification (e.g. spell checker), manipulation of images within the CMS, creation of content reuse features (asset repositories, templates, galleries, etc.)
- Personalization – referred to the definition of different ways to deliver contents to primary users related to its format, language, and grouping. Examples of this are the existence of web templates, availability of multilingual content, visual themes, etc.

- Non-functional requirements – The main non-functional requirements identified together with Secondary Users included the system capacities (e.g. computer speed, operating systems, access to the web, etc.) and usability (focus on reusability and accessibility).

3 Content Management System

3.1 CMS in SAMBA System

The CMS in SAMBA project is part of the Platform Domain of the overall idtv system. A detailed description of the platform domain is presented in [10]. In particular, the CMS is responsible for the creation and management of the content and it is to provide the Web-database in which user data and content metadata are generated and stored. The CMS allows availability of information either through the Digital TV broadcasting channel or through a generic client to access to it through a simple HTML connection. To allow this, some preliminary operations are required in order to properly present these contents in a format suitable for the TV screen. In particular, the CMS communicates with the Playout Centre and provides it with a specific package consisting of the broadcasted application and the associated content.

The CMS in SAMBA generates DVB–J based applications which contain all the JAVA .class files referring to the different specific scenarios scheduled by the application itself. In addition, a XML file is needed in order to describe the broadcasting modalities of transmission of the application (see figure 1). This couple of files is transferred to the Playout Centre at specific pre-defined (and optionally periodic) instants. In this way, it is possible to refresh the contents of the application continuously in the time, satisfying the intrinsic constraint to have dynamic applications which can bring different kind of information (that is, text and photos) that can change with time. In fact, the Secondary Users have the possibility to insert/remove contents in/from the CMS at any time, and these modifications have to be visible to the Primary Users.

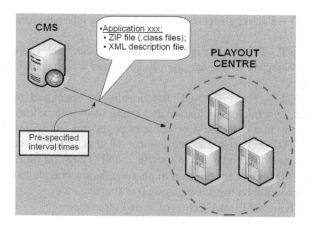

Fig. 1. Transmission of the applications from CMS to Playout Centre

3.2 CMS Functionalities

The CMS includes a set of pre-defined templates for simplifying the process of content creation (see figure 2). Indeed the motivation of a template-oriented approach was due to improvement of accessibility, reusability and personalization of the user interfaces.

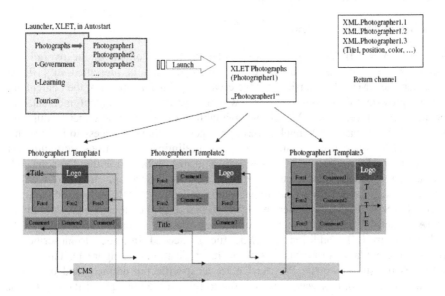

Fig. 2. Template-based interface design in CMS

These templates provide a set of different functionalities for creating different types of interactive content. The main functions of the templates include:

- **T-INFO** – Enabling secondary users to create info pages based on text and images.
- **F-ALERT** - The F-Alert is a functionality of the T-INFO application which has the objective to provide the primary users with means for creating alert messages
- **F-RANGE** - The F-RANGE functionality is composed by two different features: dynamic pages feature and rating feature for allowing respectively the display in a dynamic way of list of pages that can be reached through a specific T-INFO page and the number of clicks (of the single user of the entire community) related to a specific T-INFO page for ranking purposes.
- **T-PHOTO** – Used for creation of photo galleries with photo descriptions for exchange/commenting within a group of users.
- **T-VOTE** – Enable the creation of voting interfaces

- **T-SMS** – This application allows the primary users to send text messages (SMS) to mobile phones.
- **T-RSS** – To allow primary users to access through the DTV channel different kinds of content provided by several RSS feeds.

4 Community Content Production

The Community content production process involved three main steps; the CMS training process, the actual production of content and finally the system validation and usability tests.

4.1 CMS Training Process

In order to verify the system functionalities and usability, a set of tests were carried out. The objective of the tests was to utilize the system in a realistic scenario involving secondary users representative of a real community with digital inclusion needs. The community that was selected corresponded to the rural town of Barreirinhas in the North-East in Brazil. In this town three different institutions that were identified as potential Secondary Users were recruited:

- SEBRAE – That is national-wide agency supporting development for small-medium enterprises and community-oriented services in Brazil with a representation in Barreirinhas
- Secretary of Education – That is a government dependency responsible of education services
- TV-MIRANTE – That is a local TV broadcaster

A detailed training session was given to secondary users about the CMS functionalities and content creation. During this training he following techniques were applied: i) Workshop; ii) Direct Observations of CMS use; iii) Questionnaires. The aim was to validate the usability of CMS system by empowering secondary users to create interactive applications. Four secondary users (one man and three women) participated during the training for three full workdays. The artifact used during the training was a help manual, with a step-by-step description of the CMS with the objective to help the secondary users in the content creation process. The training involved usability experts and one of the system developers.

4.2 Content Creation

The CMS can work in both, a local or a remote server accessible through the internet that allows secondary users to create their own content and to reutilize other existing one through RSS feeds. During the training sessions, the recruited users were assigned the task of creating specific content of their interest. Then using the combination of CMS functionalities, the templates allowed the creation of idtv pages and the association among them by forming a navigational structure. Figure 3 illustrates in its left side an example of selected content, and on the right its visualization through the T-INFO application.

Fig. 3. Preview option of CMS for the T-INFO application

Table 1 summarizes the content and test services that were created:

Table 1. Contents created during secondary users training

CREATED PAGES	SERVICES
- Nutrition (cashew, fruit of the cashew)	Business
- Free educational courses	Education
- Fashion	Business
- Tourism	Tourism
- Crafts	Business
- Poll on SEBRAE courses	Education
- Alert on business opportunities	Business
- Alert on dissemination of business results	Business
- Noise pollution	Education
- Pro-Literacy in Barreirinhas	Education
- Information about specific local tours	Tourism
- Generic business information	Business

4.3 System Validation and Usability Tests

The results of evaluation of CMS with secondary users provided in general positive insight about the use of the system. SAMBA CMS system was perceived by users as a very relevant tool for supporting the local community. Such affirmation is related to the intrinsic limitations of the town related to access to digital services or other forms of interactive information sharing, few secondary schools, and few options of entertainment offered to the community. In addition their answers in the questionnaire advocate the fact the primary users would access local content rather than access to a nationally produced content .The summary of the results on the validation process and the usability inspections is presented in Table 2. After the validation and verification process, some useful suggestions improved the iTV applications before the final usability tests with target primary users.

Table 2. Evaluation Results

Validation and Verification activities	Description
General Descriptions	
Verification tests	Ad hoc verification in group and Heuristic evaluation
System validated in real context	CMS applications, T-Info and T-Photo Gallery
Methodology Used for the Validation	
Secondary Users	5 users of 3 local organizations
Activities	Creation of content, reuse of content and visualization of created content
Artifacts used	CMS tutorial, questionnaires, diary and consent forms
contextualization with users	Invitation by correspondence and by personal contact as well as via telephone, five meetings with gifts, snacks and certificates
usability tests	During three days, team at least of three (designer, evaluator and psychologist) and one technician. They observed the execution of the activities and applied the questionnaires.
Created Content	
Type of Content	Education, Business, Tourism , News
Main Results	
Users´ acceptance and utility of CMS applications	Users show interest in producing content for all types of people including the desire to reach users from more far away houses. They were very positive about the access of users to the local content.
Usability of CMS applications	Despite reporting an average degree of difficulty, users were able to create their contents using the applications and the CMS manual with occasional requests of assistance of the evaluators' team. Main problem: low internet access
Users´ acceptance and utility of idtv applications	Most users had positive experiences when using the applications. "I liked more the application of text. Because as a teacher I am part of that context." They were involved by the content and by the possibility of seeing what they refer as a "possibility" becoming "real". The utility for tackling digital divide was perceived by most users, according to the fact that TV is the most common communication mean in the city.
Usability of idtv Applications	Some difficulties appeared in function of the lack of familiarity with technology, legibility (in some TV screens, the applications were cut or had a black line) and the buttons on the remote control were in English. Suggestions were mainly related to have dynamic information.

5 Concluding Remarks and Future Work

The CMS proposed in this work is provided in the framework of the changing context of users' role from content consumers to prosumers. This phenomenon that was originated in the web domain is starting to migrate and impact in other digital domains such as idtv. . In fact, the main contribution of this paper is to present a content creation system for community-oriented idtv as a step forward to the vision of more proactiveness in the users side. The CMS tool presented here was tested and evaluated following a user-centric approach in a realistic environment for addressing digital inclusiveness in a small town in the Brazilian North-East. The ultimate outreach of this vision of users' involvement in the content creation process will change the traditional role and stereotype of users from TV "coach-potatoes" to active actors in the context of emerging idtv markets.

Regarding future work, the authors are aware of the need to extend this work towards the creation of relevant services based on the use of the CMS according to local opportunities and users needs. In fact, the users´ field studies approach presented here was used for the requirement definition of the tool and the evaluation of its use. Current work includes the investigation of how primary user attitudes could be transformed into new opportunities for the business and how these businesses can be implemented in a sustainable way.

Acknowledgements

The authors would like to thank Raju Vaidya and Marilia Mendes for their valuable contribution to the development of the CMS and the usability tests respectively.

References

1. Participative Web and User-Created Content: Web 2.0, Wikis and Social Networking. Source OCDE Science et technologies de l'information, 15 (September 2007); OECD Organization for Economic Co-operation and Development
2. Kramer, M.A.M., Reponen, E., Obrist, M.: MobiMundi: exploring the impact of user-generated mobile content – the participatory panopticon. In: Proceedings of MobileHCI 2008 (September 2008)
3. Cesar, P., Chorianopoulos, K.: Interactivity and user participation in the television lifecycle: creating, sharing, and controlling content. In: UXTV 2008: Proceeding of the 1st international conference on Designing interactive user experiences for TV and video (October 2008)
4. Mantzari, E., Lekakos, G., Vrechopoulos, A.: Social tv: introducing virtual socialization in the tv experience. In: UXTV 2008: Proceeding of the 1st international conference on Designing interactive user experiences for TV and video (October 2008)
5. Collini-Nocker, B.: Community TV: a new dimension for immersive social networking. In: MoMM 2008: Proceedings of the 6th International Conference on Advances in Mobile Computing and Multimedia (November 2008)

6. Squire, K., Johnson, C.: Supporting Distributed Communities of Practice with Interactive Television. Journal of Educational Technology Research and Development 48(1), 23–43 (2000)

7. Kurniawan, S.: Older women and digital TV: a case study. In: Proceedings of the 9th international ACM SIGACCESS conference on Computers and accessibility (October 2007)

8. Schibelsky, L., Piccolo, G., Baranauskas, M.C.: Understanding iDTV in a developing country and designing a T-gov application prototype. In: Proceedings of the 7th ACM conference on Designing interactive systems (February 2008)

9. Mayora, O., Costa, C.: The SAMBA Approach to Community-Based Interactive Digital Television. In: Proceedings of ChinaCom Conference, Shanghai, China (August 2007)

10. Martucci, M., Hirakawa, A., Jatoba, P.: SAMBA Project: A test bed for PLC application as a digital inclusion tool. In: Proceedings of IEEE ISPLC 2008 - International Symposium on Power Line Communications and Its Applications, Jeju, Korea (April 2008)

Innovation in Online Communities –
Towards Community-Centric Design

Petter Bae Brandtzæg[1], Asbjørn Følstad[1], Marianna Obrist[2], David Geerts[3], and Rüdiger Berg[4]

[1] SINTEF ICT, Forskningsveien 1, 0314, Oslo, Norway
[2] ICT&S Center, University of Salzburg, Sigmund-Haffner-Gasse 18, 5020 Salzburg, Austria
[3] CUO, IBBT / K.U.Leuven, Parkstraat 45 Bus 3605, 3000 Leuven, Belgium
[4] Netcontact GBR, Republikplatz 5, 52072 Aachen, Germany
pbb@sintef.no, asb@sintef.no, Marianna.Obrist@sbg.ac.at, david.geerts@soc.kuleuven.be, berg@itsberg.de

Abstract. Online communities are changing how companies and non-profits innovate, lower costs, tap talent, and realize new socio-economic opportunities. In this paper, online communities are predicted to be an important resource in open innovation processes, based on the emerging online community trend of sharing and collaboration among non-professional users. However, in order to take full advantage of online communities for innovation purposes, challenges to community interaction, commitment and co-creation have to be overcome. The paper describes these challenges and discusses how future developments in community-centric design can facilitate innovation. Different community-centric methods are suggested with the aim to provide a research direction for redesigning community collaboration and a new approach to innovation in the Future Media Internet.

Keywords: Online communities, innovation, community-centric design.

1 Introduction

According to Clay Shirky, group action gives human society its particular character, "and anything that changes the way groups get things done will affect society as a whole" (p. 23) [1]. One important factor for such change is the proliferation of online communities as they encourage people to find, contact, interact and co-create with an extended web of employees, customers, and stakeholders [2]. Online communities are changing how companies and non-profits innovate [3], lower costs, tap talent, and realize new opportunities [4]. Innovation is no longer necessarily conducted only within a company, but also through co-creation involving non-professional users and external enterprises [5]. Facebook, MySpace, and Twitter are all successful examples of how online communities facilitate innovation and idea generation among groups. Moreover, large companies are increasingly turning to online communities for innovation. Innovation initiatives that used to take months to coordinate and launch can often be started on a very short notice and include groups around the globe.

P. Daras and O. Mayora (Eds.): UCMedia 2009, LNICST 40, pp. 50–57, 2010.

Co-creation, the practice of product or service development that is collaboratively executed by developers and stakeholders together [11], changes how knowledge, services and applications are created. Co-creation in online communities has the potential to lower the costs and time of doing research and innovation, with a greater disruptive potential. Online community is therefore predicted to be a key enabler of novel innovation chains and networks, to emerge both in the industry [3] and in non-profit organizations and e-Government [6]. Despite these predictions, important challenges exist in (a) how we can change innovation processes to involve non-professionals into efficient and systematic innovation processes and (b) how we can design systems that lower the threshold for commitment, contribution and co-creation within an online community.

The overall objective of this paper is to present and discuss (a) the challenges and opportunities to innovation involving online communities and (b) how user centric methods can be adapted to community-centric design methods supporting online community commitment, contribution and co-creation. We hope that the resulting insights may benefit future design and development of open innovation solutions for the Future Internet communities.

The paper is based on lessons learned from the European research project CITIZEN MEDIA (www.ist-citizenmedia.org) and RECORD, a Norwegian national project (www.recordproject.org). Both projects explore how people interact and create user-generated content in online communities. Taken together, these projects draw on experiences from the user involvement of more than 10.000 people.

The next sections describe online communities, their benefits for social innovation, key challenges and community-centric methods that can overcome these challenges.

2 Background: Innovation in Online Communities

2.1 Online Communities

Online communities are groups of people with a purposeful interaction supported by technology, guided by norms and policies [7]. Online communities are distinguished by e.g. frequency of interaction, links within or outside the community, enabling technology, and the community member characteristics [8]. Like traditional communities, online communities have their own identities, norms, and goals, and several community objectives may be shared with one or more related communities.

An online community will not survive without lasting user motivation and user participation [2], and the outcome of the innovation is dependent upon characteristics related to the community or communities involved [5]. The characteristics shared by successful online communities are often poorly understood; yet, this is critical knowledge for designers and human factor engineers developing online communities for co-creation and innovation. For example, why individuals or groups help and co-create with strangers in online communities is not very well understood [9], and why this happens more or less in different types of communities is key to know more about this process. An efficient online community platform for co-creation will also be dependent upon ease of use [2].

2.2 Open Innovation, Co-creation, and User Innovation

In a globalised economy and networked world, the way innovation is conceived is changing. Both non-profit organisations and businesses are opening up their innovation processes, involving larger parts of the value chain in the innovation activities, and turning towards the users for inspiration. In particular, three theoretical perspectives seem to be dominating the way we think of innovation: Open innovation [10] co-creation [11], and user innovation [12]. The theory of open innovation target businesses' needs to elicit and use both external and internal input in their innovation processes and technology development. The theories of co-creation and user innovation seem to suggest approaches to how innovation processes may be opened up by targeting the vendor and customers' co-creation of experiences through dialogue, access and transparency [11], and by looking at lead users for innovative ideas and product modifications [10].

The value creation potential of online communities has been clearly demonstrated by consumer market successes such as MySpace and YouTube [1]. The novelty of utilizing communities as an arena for co-creation to be utilized in open innovation processes is that users are not only asked about their opinions, wants and needs; they are invited to contribute with their creativity and problem solving skills. As an example, Peugeot initiated an online design contest where nearly 2800 design enthusiasts from 90 countries proposed car designs on the theme of "Retrofuturism" (www.peugeot-avenue.com) [3]. Other examples, in addition to those mentioned in the introduction, include how enterprises and NPOs such as IBM, Adidas, Reebok, Mazda, Sun, Reef Ball Foundation, Creative Commons and the American Cancer Society use SecondLife to gain benefit from networked innovations [13]. Most of these companies involve users to contribute with ideas, but some even give away some of their control (e.g. source code, professional content) and enable others (individuals and communities) to further improve their product and extend their services. In addition social networking accelerator allows business professionals to monitor and analyze customers' conversations on social networking sites, and as a result, provides real-time status updates about their products and services.

Taken together, online communities have the potential to improve the ability for people to co-create with others for the purpose of innovation. This gives high expectations as to how we can involve these communities, and how they can achieve innovations in public organizations, enterprises and governments.

3 Challenges to Innovation in Online Communities

The development and introduction of online communities for innovation still face important challenges in how non-professional users and communities can be involved more effectively to enhance collaboration and creativity between people. According to Daras and Alvarez [14] the key for the Future Internet is to bring together professional and amateur media creators, which still are two separate worlds: the amateurs collaborate in the consumer oriented communities and the professionals compete in the commercial media world. The challenge will be to create a culture of sharing, in which information and knowledge is distributed without losing important content or trust in the process. On a general level it might be easier than ever to get *low commitment* from people and communities, but harder than ever to get *high commitment*. Therefore, designers should strive to make users participate with little effort by making their contributions a side-effect of something else they are already doing. Finally, organisations and companies must see the potential of transparency and openness to both trust and the promotion of collaboration between industry, governments and citizens.

Another major challenge is that the greater part of the European population is either not using online communities [18] or not being skilled in ICT, and is thus excluded to participate in such online innovation processes. This is an important gap, because community-driven innovations can lead to new solutions that create more benefits for non-professional users and wider uptake of services and applications. This gap is also important since the advancement of the ICT society is characterized by more complexity and rapid communication, which is an opportunity that will speed up group actions [1] as well as innovation, but might in addition increase the *digital divide*. ICT-competent people will more easily adapt to this fragmentation and speed, and even customize technology to their own personalised needs, while others don't. Furthermore, the social dynamics on the Internet put a greater demand upon the user to be productive in terms of a "do-it-yourself" movement [5], resulting in a divide between those who consume and those who produce: a *digital production divide* [15].

4   Possible Solutions: Community-Centric Design

As more and more people get involved in experimentation and community based innovation online, companies and researchers will also need to change their focus in education and training efforts for innovation. Instead of just interpret large volumes of data using a user-centric approach, companies will need to help to develop the skills to rapidly design provocative community-based experiments. Passive analysis will change towards active experimentation and *conversation*, using a community-centric design [16]. Therefore, this section recommends and discusses some applicable community-centric methods experienced in the projects CITIZEN MEDIA and RECORD that can enable co-creation and open innovation in communities, and tackles these challenges.

So far, media developments have been focusing on single users and user-centric design, while the design for co-creation and community interaction is still weak. However, collective action is different from individual action [1], and passive consumption is different from co-creation and collaboration. Light [17] explains that a traditional media perspective is a one-producer-to-many-recipients model, with little focus on user participation and co-creation. By contrast, a range of collaborative activities is open to users in online communities, which can stimulate non-professionals from being passive consumers to be the producers themselves of new services and new applications. A key question is how these models of behaviour can be combined to design systems, including trust, privacy, security and commitment to encourage co-creation and co-activity.

4.1 From User-Centric Design to Community-Centric Design

Evaluation and design of online communities requires sensitivity to utility and user experience issues rather than just usability, while at the same time user feedback methods should not overburden participants in community applications. The nature of these applications makes it important to focus on continuous evaluation of running services (a) in order to gain knowledge regarding the users' motivation over time, and (b) because community applications are dynamic entities that are subject to continuous change. This can be supported by introducing new ways to keep existing users involved as well as new members that are joining the community. Finally, different types of communities need different tools for very particular needs. These tools might not meet most previous criteria of successful design, but nevertheless function well because they are situated in the community that uses them [1].

Table 1. Comparison of user-centric and community-centric goals

	User-centric design	Community-centric design
Who are the users/community	Individual users, contributor	Community members, networks, groups, collaborators
Users/community goals	Specific tasks and task completion	Social exchange, co-creation, creativity in collaboration.
Users/community needs	Efficiency, satisfaction, experience, searching	Community interaction, co-experience, privacy, social belonging

Table 1 shows some key differences between a user-centric approach and a community-centric approach, and how communities should be analyzed based on type and characteristics. Analyzing requirements is not only needed for the individual community members, but also for the broader community needs and characteristics, which are not always visible or apparent to individual members. This will give more insight in the dynamics of the selected community and what tools and components can be created that support collaboration and co-creation within this community.

As user-centric design and community-centric design are asking the same questions but have a different focus, existing user-centric design methods need to be adapted in order to fit this new paradigm.

Table 2 below shows how user-centric design methods can be adapted for a community-centric approach, and which future research directions still need to be explored. Some of the methods in Table 2 should also in future research be combined with social network analysis to identify a social structure made of nodes referring to individuals and communities that are tied by one or more specific types of interdependency, such as values, visions, ideas, financial exchange, friendship etc.

Table 2. A selection and examples of user-centric vs. community-centric methods

Goals	User-centric	Community-centric	Research direction
In-depth knowledge of user experience and user requirements	Interviews. One interviewer speaking to one participant at a time, getting individual requirements	Blogs or wikis to extend the scope of an interview from a mere study of user requirements to a study of collaboration with participants via e.g. online apps such as blog, YouTube, Picasa, and Blogger	Explore: a) the use of community-centric methods in existing communities, where the participants are familiar, b) involvement of diverse stakeholders and communities e.g. in co-creative activities.
Large scale user analysis and system evaluations	Survey questionnaires, by e-mail or ordinary mail, often with questions targeting individual user analysis	Online surveys including measures of community requirements, community usage, social capital, sociability issues, group relationships, etc. Extend text based surveys by visualization tools – pictures, sketches or videos to test ideas in collaboration	Explore a) how online community survey can be more effective, easy and interesting to use from the perspective of community members, b) the use and efficiency of visualisation tools in questionnaire surveys
Information on user motivation and ideas	Focus groups and single in-depth interviews	Instead of physically present focus groups, online focus groups can be used to facilitates participation among geographically distributed participants	Explore a) the potential of online discussion rounds, and how different form of structured or unstructured processes is doing, b) how different tools and contexts could be more or less useful for different innovations.
Observing users context and social interaction	Often lab, but also field studies and ethnographic studies are also used, measuring individual tasks	Online ethnography is emerging and adapts ethnography to the study of online communities and cultures	Explore a) how researchers can be immersed in the life of an online community or be engaged in the culture as a member, b) ethical and privacy concerns
Exploring users' contextualized experiences, needs, ideas	Cultural probes. Explorative contexts through tools in a probe package	Online probes are getting popular and can make use of various social networking tools and the participants' audiovisual equipment	Explore a) the right combination of social probing tools for each community interaction, b) innovation in real time

4.2 Living Labs as a New Approach to Community-Centric Design

In addition to community related adaptations of user-centric methods, new approaches to community-centric design are emerging: Living Labs, and in particular online Living Labs, where the users are actively involved in the innovation process, are assumed to enable the industry to meet innovation challenges, has generated a great deal of interest within the field of information and communication technology (ICT). This is in particular seen in the explosive growth in the European Network of Living Labs (ENOLL), now counting more than one hundred Living Labs across Europe.

According to Følstad [18], Living Labs are defined as "environments for innovation and development where users are exposed to new ICT solutions in (semi)realistic contexts, as part of medium- or long-term studies targeting evaluation of new ICT solutions and discovery of innovation opportunities". But, the concept of Living Labs is still evolving. An important emerging trend is to see Living Labs as a way to tap into the creative potential of communities [18]. Instead of being recipients of the outcome of innovation and development, users may be engaged in co-creative innovation processes of a Living Lab, which is recognized by ENOLL as being a key aspect of Living Labs. An online Living lab has also the potential of running co-creative experiments with high speed and test ideas right away.

Within the RECORD project, an online Living Lab has been established on the basis of two main components: a participant panel of >3000 persons meant to be representative of Internet users of a given geographical area (Norway), and an online environment enabling presentation of ICT concepts or prototypes and feedback from the participants. Survey and social media functionality are combined in order to enable (a) discussions between the participants, designers, and developers, (b) uploading of the participants' own redesign suggestions, and (c) analysis of the participants' feedback based on the participants' characteristics. This online environment allows the establishment of project-oriented online communities to participate in product and service development and innovation. Early experiences with the RECORD Living Lab indicate that online Living Labs may represent a relevant approach to the involvement of online communities in innovation processes.

The use of Living Labs is an interesting approach to community-centric design and open innovation, but we are only at the beginning of exploring how Living Lab participants may be involved in targeted and beneficial co-creation processes.

5 Conclusion

Internet usage is shifting from passive consumption of information services to active and collaborative group actions in online communities, which offer new opportunities for innovation. This paper has discussed challenges related to innovation in online communities, and proposed possible solutions. Still, community-centric methods are immature, and existing user-centric design methods need to be adapted in order to fit a new paradigm for innovation. New approaches such as Living Labs are promising, but further development of associated methods and processes is needed.

The different approaches and related research directions described in this paper will contribute to enhancing open innovation in communities and giving them new tools to facilitate new forms of group actions, collaboration and innovation among both different user groups and communities in the Future Media Internet. We hope that the need for a future focus on such a community-centric design approach is being recognized, as this will help companies and non-profit organizations to tap into the enormous potential co-creation with communities offer for creating new applications and services on the Future Media Internet. This approach can also promote the development of services with social benefits and thus closing the digital divide.

Acknowledgments. The research leading to these results received funding from the CITIZEN MEDIA project (038312) FP6-2005-IST, and the RECORD-project, supported by the VERDIKT programme in the Research Council of Norway.

References

1. Shirky, C.: Here comes everybody. The power of organizing without organizations. Penguin books, New York (2008)
2. Brandtzæg, P.B., Heim, J.: User Loyalty and Online Communities: Why Members of Online Communities are not faithful. In: INTETAIN, ICST Second International Conference on Intelligent Technologies for Interactive Entertainment. ACM, Cancun (2008)
3. Füller, J., Bartl, M., Holger, E., Mühlbacher, H.: Community based innovation: How to integrate members of virtual communities into new product development. Electronic Commerce Research 6, 57–73 (2006)
4. Weber, L.: Marketing to the Social Web. How digital customers communities build your business. Wiley, New Jersey (2008)
5. Obrist, M.: Finding individuality in the technological complexity: Why people do it themseves. The International Journal of Interdisciplinary Social Sciences 2, 203–211 (2007)
6. Brandtzaeg, P.B., Lüders, M.: eCitizen 2.0: the Ordinary Citizen as a Supplier of Public-Sector Information. SINTEF, Oslo, 89 (2008)
7. Preece, J.: Online communities: designing usability, supporting sociability. Wiley, Chichester (2000)
8. Maric, J.: Web Communities as a Tool for the Social Integration of Immigrants. In: WebSci 2009: Society On-Line, Athens, Greece (2009)
9. Wasko, M., Faraj, S.: Why should I share? Examining social capital and knowledge contribution in electronic networks of practice. MIS Quarterly 29, 35–57 (2005)
10. Chesborough, H.W.: Open Innovation: The New Imperative for Creating and Profiting from Technology. Harvard Business School Press, Cambridge (2006)
11. Prahalad, C.K., Ramaswamy, V.: The Future of Competition: Co-Creating Value with Customers. Harvard Business School Press, Boston (2004)
12. Von Hippel, E.: Lead users: Management Science 32, 791–805 (1986)
13. Wikipedia: Businesses and organizations in Second Life, Vol. 2009. Wikipedia (2009)
14. Daras, P., Alvarez, F.: A future perspective on the 3D Media Internet. In: Tselentis, G.e.a. (ed.) Towards the Future Internet - A European Perspective, pp. 303–312. IOS Press, Amsterdam (2009)
15. Brandtzæg, P.B.: The Innovators in the New Media landscape. In: The Cost 298 Conference - The Good, the Bad and the Unexpected, Moscow, Russia (2007)
16. Brynjolfsson, E., Scharge, M.: The New, Faster Face of Innovation. The Wall Street Journal (2009)
17. Light, A.: Design: Interacting with Networked Media. Interactions 11, 60–63 (2004)
18. Følstad, A.: Towards a Living Lab For Development of Online Community Services. Electronic Journal of Virtual Organisations 10, 47–58 (2008)

Multimedia Source Management
for Remote and Local Edition Environments

Alejandro González, Ana Cerezo, David Jiménez, and José Manuel Menéndez

Grupo de Aplicación de Telecomunicaciones Visuales, Universidad Politécnica de Madrid,
Avda. Complutense. 30, 28040 Madrid, Spain
{agn,ace,djb,jmm}@gatv.ssr.upm.es

Abstract. This paper aims to detail an innovative multimedia edition system. A special functionality provides this solution with different ways of managing audiovisual sources. It has been specially designed for media centralized environments dealing with large files and groups of users that access contents simultaneously for editing and composing. Mass media headquarters or user communities can take advantage of the two working modes, which allow online and offline workflows. The application provides a user edition interface, audiovisual processing and encoding based on GPL tools, communication via SOAP between client and server, independent and portable edition capabilities and easy adaptable source handling system for different technologies.

Keywords: Source Management, Multimedia Edition, Centralized Data, Multiple Users, SOAP, XML.

1 Introduction

The full networking development of work environments leads to the necessity of content digitalization and ubiquitous access to it. The combination of both aspects reveals new requirements. Large capacity for data storage is immediately recognized as one of the most urgent necessities, so new resources management models for optimization are needed. This requirement is a key factor in environments that need to manage large amount of data, like mass media, publishing and advertising companies.

Daily, these companies are constantly generating multimedia data and its associated metadata with pending edition works. Even if that information is not used in shortly, it could be necessary to stored it for a possible future use.

Furthermore, the storage of these contents is often centralized in data servers and it has to be easily accessible and editable from each user terminal. This exchange of files involves a great computational load that could saturate both the edition terminals (which capacities are meaningful lower than servers) and the servers (if there are many file requests simultaneously), so this implies a new problem.

Consequently, it is necessary to change the way the audiovisual content is edited. This work expounds a software solution based on a client-server model for multimedia edition avoiding long download times and freeing resources for a more efficient system.

P. Daras and O. Mayora (Eds.): UCMedia 2009, LNICST 40, pp. 58–66, 2010.

2 Functionalities

The main object of this work is to define a source management model for multimedia content within a new application for multimedia edition based on tools under a GPL license. The features covered by this kind of software are related to multimedia basic editing and encoding. The system functionalities can be classified in two categories: source management and multimedia edition.

2.1 Source Management

The application is based on a client-server model. It allows multimedia edition on files not necessarily stored in the client computer, but also in a server. Two source management work modes are allowed, according to the user necessities:

Offline Mode. It is the standard working mode. The importing module downloads low quality sources from the server catalog when they are requested. The multimedia files are stored in the client computer and no connection is required between client and server for the editing process. When the edition is over, its parameters are sent to the server in an XML[1] file. The edition is done using low resolution files within the client and the server repeats it with the high quality media once the process is over.

Online Mode. This mode is aimed to permanently connected environments, such as local networks. The importing module does not download any source file. Instead of this, it saves a reference to each imported multimedia event. This mode has a stronger bandwidth requirement, as the server core has to generate previews that are sent back to the client via streaming during the whole edition process.

2.2 Sources Origin and Prosumer Capabilities

The system supports several kinds of sources attending to their origin.

FTP Catalog. An online multimedia files set which is actually the main source provider for the editor. The catalog can be either private or public.

HTTP. Multimedia files available through the web can be imported to the projects directly, and referenced to the server as well in order to include external content.

Personal Content. It is also possible for the user to upload personal content to the FTP server. The user becomes then a "prosumer", content consumer and producer at the same time. The access to the uploaded content is also configurable, so the system orientation allows many different points of view, including editor communities for content sharing.

2.3 Multimedia Edition

As the main target of the application is to solve multimedia source management, its audiovisual edition capabilities are supported by some tools under a GPL license. Two external applications are used in order to satisfy these functionalities: AviSynth and MEncoder. AviSynth [2] is a *frameserver* capable of processing local multimedia files in real time. The user develops custom scripts using basic functions in order to edit the source content. In the other hand, MEncoder [3] provides the final solution with its encoding capabilities. It manages several codecs and containers, and accepts AviSynth scripts as input stream. MEncoder's software package also includes MPlayer, a multimedia player that is used in this application for the preview function.

A custom Java Swing user interface has been developed to allow easy multimedia edition. It generates project structures that will be translated either into an XML file or into an AviSynth script. Thus, the XML file can be sent to the server for high quality processing and the script can be executed for local previewing of the project. The complete list of its functionalities is explained in section 3.3.

3 Architecture

In order to support the functionalities previously explained, different structures have been defined and created. In this chapter, these structures and the communication between them are explained.

A general scheme of the work is shown in Fig. 1. As the diagram shows, the process implies several connections between client and server. The server, or an external media server, contains two versions of the sources: the master in original quality and another in low quality. The importing module either downloads or references (depending on the work mode) the desired multimedia events from the light sources catalog. This can be done using several communication protocols, such as http or ftp.

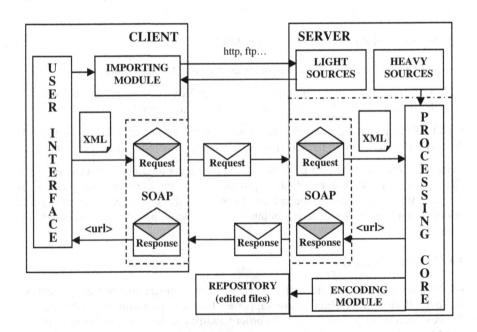

Fig. 1. General architecture scheme

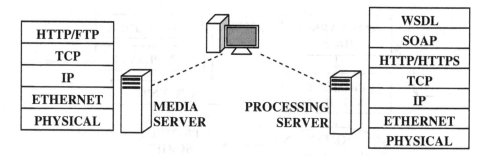

Fig. 2. Protocol stacks for media download and edition processing connections

An edition project can be built graphically by the user using the interface. Once the project is completed, the resulting XML edition file is enveloped following the SOAP standard [4] and sent as a Request message to the server. The server extracts the XML and processes its instructions using the original quality videos. An internal Java application automatically builds an AviSynth script following the directions established by the client in the XML file. Afterwards, this script's output stream is encoded by MEncoder with the extracted parameters. The resulting stream is encoded and stored in a repository that could be a public site, a private shared folder or any other accessible target. A SOAP Response message containing the URL to the generated file is sent back to the client.

SOAP is used due to its portability. In conjunction with the project structure definition, it allows alternative client solutions to work properly even if they are implemented in different programming languages. In addition, SOAP is based on XML, as the edition instructions file, so the application is easily adapted to this standard. The system protocol stacks are shown in Fig. 2.

3.1 The Processing Core

The multimedia processing core is in charge of extracting the XML parameters and building a valid AviSynth script that could be played or encoded. The whole work has been implemented using Java. The schema in Fig. 3 shows how it works.

The core's input is an XML file with the editing and encoding parameters. These parameters are extracted using JDOM [5], a Java library for handling XML structures.

The project builder uses the extracted parameters to build a Java structure of classes that defines each aspect of the edition. The project structure stands by itself and it is independent from AviSynth or any other external application used in this work. This means that a technological change on the multimedia processing within the application does not imply any change in the elements on the figure's left column.

The project structure is processed in order to write an AviSynth script that is ready to be played or encoded. Depending on how the processing core is accessed to, MPlayer or MEncoder is launched using the parameters specified in the XML file.

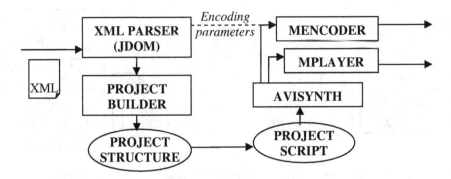

Fig. 3. The multimedia processing

3.2 The Project Structure

In this section, the structure used to define an edition project is described. It is based in parental relations easily exportable to an XML document. These relations are shown in Fig. 4 and they are shared by the implemented XML and Java structure.

The main node is *Project*, which contains the essential video and audio parameters that will be inherited by every node. These parameters are: project name, frames per second, video resolution, audio sampling rate and color system. Each new event added to the project will be treated in order to match these parameters. This treatment is included in the AviSynth script generation process and is supported by customized scripts. The most notable custom script is the *Autofit* function that, if used, matches different video sizes taking up the most of the available screen blank space without affecting on quality or aspect relation. A *Project* is made up of audio and video

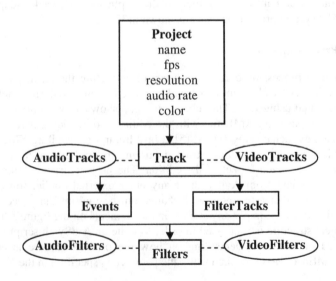

Fig. 4. The project structure

Tracks, which are independently configurable. Each *Track* is filled with *Events*, i.e., audio and video fragments distributed through time. These fragments can be modified trimming its content, changing its position in time or applying independent filters. These *Filters* were previously explained in this paper and can be applied directly to events, to certain tracks, to the final output, or by using *FilterTracks*, which allow the user to modify filter parameters through time.

As we introduced at the beginning of this section, the project structure is always exportable to an XML document or to an AviSynth script. The XML feature makes possible to implement an easy and fast *save* system, to recover a previously started project; and an *undo* system, in order to fix some undesired changes during the edition process. In the other hand, the AviSynth feature makes possible the Preview functionality, which is based in playing the resulting script.

3.3 The User Interface

The user interface's design and functionalities are based on professional edition tools, and provides the user with friendly commands that support a main part of the editing needs.

It may be important to emphasize the necessity for a friendly user interface. The final user is not going to be a computers' expert, nor an engineer or a scientist. Actually, the typical user will be an editor with experience using this kind of software and, for that reason, it is important to preserve the way of working developed by the industry standard edition tools, and avoid compromising it because of the different source management technology used by this system.

The interface includes the following features:

File Importing Module. An embedded window allows the user to explore the accessible sources and select the desired multimedia event to be included in the project. Accessible sources include web content, private and public FTP content and personal sources. Another tab in the module shows the already added sources with a snapshot preview.

Multi-track System. Multimedia events (video and audio clips) are represented in the interface by bars. They can be freely moved through a customizable set of video and audio tracks in order to build the final stream. The bars' width is proportional to the events duration, and there is a time ruler that helps the user distributing and ordering events during the editing process. Each track has its own parameters; so many different configurations can coexist in the same project. For example, each audio track has independent equalization, stereo balance and gain controls.

Filter Manager. Filters and effects are common tools in multimedia editing. They change the video and audio properties of the output stream. For example, a filter could modify video brightness or normalize an audio event. The application offers two ways of applying filters. Firstly, a single event or track can be assigned a filter in order to change its properties without affecting the rest of events. Secondly, filter tracks can be created to apply filters or arrays of filters to a certain range of a track or a project.

Some of the filters that are supported by the developed application are: *sharp* (blur adjustment), *bricon* (brightness and contrast adjustment), *levels* (black and white levels, and gamma correction), *crop*, *space denoiser*, *temporal denoiser*...

Preview Module. Video editing strongly requires previewing the result frequently. For this purpose, the application includes a module for playing the edited stream. It is based on the GPL tool MPlayer, which accepts AviSynth scripts as input. It is important to notice that big projects and complex editing features could require some parts to be rendered, this means previously encoded. This limitation is shared with professional editors, because it mainly depends on the machine where the application is running.

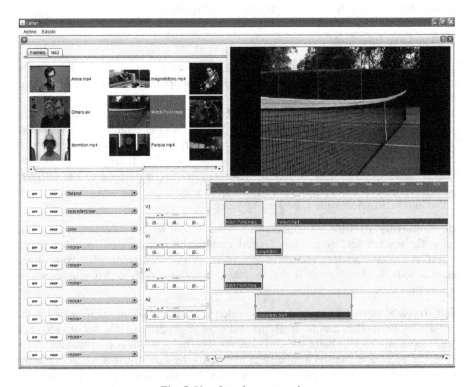

Fig. 5. User Interface screenshot

4   Test and Evaluation

This chapter shows how this system can tangibly bring some improvements into a multimedia edition environment. Several variables influence on the time spent on an entire edition process. Depending on the situation, different values can be found for sources bit rate, low quality sources bit rate, available bandwidth, sources duration, percentage of the sources' duration used in the final output stream or time spent by the user working in the edition. These values will determine the savings implied by the use of the solution.

An example should help on quantifying these savings. The high quality bit rate in this example will be 4 Mbps, which is less than the average bit rate for a DVD, but it is enough for a standard SD contribution; so it is not a very high quality value. The low quality bit rate will be 512 Kbps, for both downloading and streaming, which is enough for editing in the most of the situations. The percentage of the sources' length actually used for the project will be 40%.

It is necessary to study the work modes separately.

Offline Mode. Obviously, the main time saving is due to the difference between bit rates on high quality and low quality sources. In our example, a 4 Mbps high quality bit rate and a 512 Kbps low quality bit rate imply an 87.5% download time saving. As the resulting video has not to be uploaded again, a new time and bandwidth saving appears. In the example, the final file's length is the 40% of the sum of every source length. If we had to upload this file in high quality, time saving using the application's offline mode becomes then a 91.08%. Fig. 6 shows the difference implied by the use of the application's offline mode, using a 3 Mbps symmetric connection.

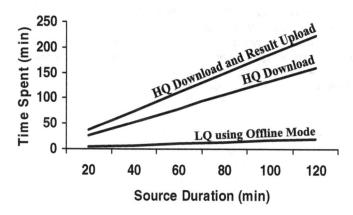

Fig. 6. Comparison between using the system with no improvement and using the offline mode

Online Mode. In this mode, the most notable advantage is that no previous downloads are needed. This means that the only spent time in an edition process is that invested by the user. The disadvantage is the necessity of a large bandwidth in order to give access simultaneously to a certain number of users. Anyway, this mode can be more efficient than the original way of working. Downloading the sources and uploading the result implies the same load than previewing the entire project more than ten times.

5   Conclusion

The explained application provides multiple user environments with a necessary source management system, flexible, and configurable. It defines as well a portable edition structure and coordinates different multimedia processing tools to work together, without making the final solution conditional to the selected tools.

The existence of two work modes reveals several uses for this application. For example, the online mode is useful for press offices, edition classrooms or multimedia file directories while the offline mode is useful for foreign correspondents, online production and edition educational applications or common edition workspaces, like replays on TV live shows or sports.

The improvements over a traditional edition system are notable. Less data load implies less spent time, and this implies the work to be done earlier and easier. This source management system may be a very useful tool for networking on content production and sharing not only for a professional use, but also for personal issues. If applied to more powerful multimedia edition software, it may result in a standalone application for online content edition and production.

References

1. Extensible Markup Language (XML) 1.0 (Fifth Edition),
   http://www.w3.org/TR/REC-xml/
2. AviSynth, http://avisynth.org/mediawiki/Main_Page
3. MEncoder/MPlayer, http://www.mplayerhq.hu/design7/news.html
4. SOAP Specifications, http://www.w3.org/TR/soap/
5. JDOM Project, http://www.jdom.org/

UCMedia 2009

Session 3: Multimedia and User Experience

In Search of the Uncanny Valley

Frank E. Pollick

Department of Psychology, University of Glasgow,
Glasgow, UK
frank@psy.gla.ac.uk

Abstract. Recent advances in computer animation and robotics have lead to greater and greater realism of human appearance to be obtained both on screen and in physical devices. A particular issue that has arisen in this pursuit is whether increases in realism necessarily lead to increases in acceptance. The concept of the uncanny valley suggests that high, though not perfect, levels of realism will result in poor acceptance. We review this concept and its psychological basis.

Keywords: uncanny valley, animation, robot.

1 Introduction

This essay reviews the concept of the uncanny valley which clearly states that increased realism does not necessarily imply acceptance. This review is not alone and related writings on the uncanny valley can be found elsewhere [1-4], as well as discussion of its prominent role in considerations of robots which are human-like in appearance [1, 5-7]. Thus, any review runs the risk of repeating what has already been said. To attempt to avoid this I have tried to focus on the psychology of the uncanny valley and to discuss what psychological principles might underlie its existence. With this in place we can look at falling into the uncanny valley not from the usual perspective of ever more realistic artifacts, but instead from the viewpoint of how normal human activity might be modulated to fall into the same uncanny valley.

The body of this essay is contained in four parts: The first three discuss, in turn, the history of the uncanny valley, evidence for its existence and theoretical arguments for its plausibility. The final section provides an operational definition of the uncanny valley that is examined in the context of human behavior, and the shortcomings which arise are discussed.

2 History

In 1970 Dr. Masahiro Mori, a Professor of Engineering at Tokyo Institute of Technology, put forth the following thought experiment: Assume we could make a robot more and more similar to a human in form, would our affinity to this robot steadily increase as realism increased or would there be dips in the relationship between affinity and realism [8]. Mori put forth the proposition that the latter would be the case - as the robot became more human-like there would first be an increase in its acceptability

P. Daras and O. Mayora (Eds.): UCMedia 2009, LNICST 40, pp. 69–78, 2010.

and then as it approached a nearly human state there would be a dramatic decrease in acceptance. He termed this precipitous drop "bukimi no tani" and the translation of "bukimi no tani" into "uncanny valley" has become popularized. The hypothesized shape of the uncanny valley revealed in the relationship between affinity and realism is shown in Figure 1.

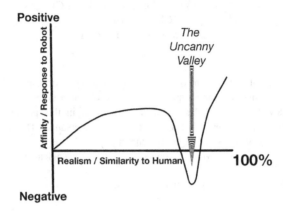

Fig. 1. Simplified diagram showing the hypothesized relationship between affinity and realism with the uncanny valley appearing as a negative response as one approaches total realism

Mori also appreciated that robots are not defined by the single dimension of form and considered the effect of robot motion. Here he proposed that motion and form together will produce a different function of acceptability versus realism. For example, motion could deepen the valley since form sets up expectations in an observer and if other factors such as motion do not match these expectations then there is further rejection of the entity. Mori tread a little further into the realm of thought experimentation and illustrated this phenomenon with the example of viewing a corpse. Certainly a corpse has to be very similar in form to an actual human yet we find viewing a corpse unpleasant, and if the corpse would suddenly move to stand up we would be terrified. Motion could also be used to circumvent a fall into the uncanny valley as Mori illustrated with the example of Japanese bunraku puppets. These puppets are somewhat basic in form and are accompanied by their black-cloaked puppeteers as they appear on stage, however their lifelike motion leads them to be accepted as nearly human as one follows the action on stage.

The practical significance, and lingering influence of Mori's proposal of the uncanny valley is found in his suggestion to designers of robots and other related artifacts. He proposed that the first peak in acceptability is an effective target for design. Here there are moderately high values of acceptance and a safe distance from the uncanny valley. Striving for realism will only lead to the risk of catastrophic tumbling into the uncanny valley. The imaginative example of zombies is transformed into advice for designers of artificial limbs that if they cannot get the motion, texture and temperature of an artificial hand to be correct then having it look perfect could lead to awkward situations when the artificiality is discovered in a social interaction like a handshake. A characteristic of Mori's advice to designers is that he did not provide precise definitions of realism or affinity and thus the concept of an uncanny valley has proven to be a broadly applicable guidepost to designers in a variety of domains.

Before leaving a discussion of historical aspects of the uncanny valley it is worth briefly considering the term "uncanny valley" itself and its translation from the Japanese "bukimi no tani". Here "tani" is quite directly "valley" and "no" is a connecting particle, while "bukimi" has several translations including "eery", "strange" and "uncanny". How "uncanny" was chosen for "bukimi" is an interesting question. The first appearance of this translation appears to be in the book "Robots: Fact, Fiction and Prediction" [9]. We can speculate that "uncanny" was chosen due to its psychological resonances with the 1919 essay of Sigmund Freud entitled "Das Unheimlich" which was translated to "The Uncanny" [10]. While "unheimlich" appears to be one of those problematic words which defy a simple translation, Freud specified the uncanny as that class of the frightening, which leads back to what is known of old and long familiar. This sense of the word seems particularly appropriate to the phenomenon Mori described as the valley appears as we approach the familiar.

3 Evidence for Existence

The original writing of Mori in Japanese referring to "bukimi no tani" and its translation into English as the "uncanny valley" both occurred during the 1970s. However, the concept seems to have laid dormant for almost 30 years and resurfaced as technology inched towards increasing levels of sophistication in computer graphics and robotics. This increase in sophistication makes it possible for greater and greater realism to be attained. However, increased realism has not necessarily equated with increased acceptance by the public and the existence of an uncanny valley has been called on to describe this phenomenon.

The uncanny valley entered the popular lexicon not long after the full-length feature film Final Fantasy appeared. This film consisted entirely of characters generated by computer graphics and used high levels of realism. The audience response was lukewarm and a general consensus began to evolve that it failed due to falling into the uncanny valley. Andy Jones, the Final Fantasy animation director gives a telling quote in Wired magazine when he says "it can get eerie. As you push further and further, it begins to get grotesque. You start to feel like you're puppeteering a corpse" [10, 11]. Film critic Roger Ebert continues this thread in his column in the Chicago Times when discussing the role of Andy Serkis in portraying the character Gollum in the Lord of the Rings: Return of the King, he says "If Serkis brought Gollum to life, other artists fine-tuned the balance with the uncanny valley" [12]. A final discussion point is a comparison of the computer-animated films Polar Express and The Incredibles which both opened in fall 2004. Polar Express featured realistic animation while The Incredibles was more stylistic. The initial reluctance of audiences to view the Polar Express began to be accounted for by its tumble into the uncanny valley while The Incredibles avoided this fate [13]. These examples are compelling in suggesting that the uncanny valley exists, however, they still do little to explain what the uncanny valley is and what conditions are critical for its occurrence.

Recently the field of robotics has moved towards robots designed to leave the assembly room floor and work alongside humans [14-16]. The most ambitious of such designs are humanoid robots that are modeled upon human structure and androids that strive for greater similarity to human form and function. Justifications for mimicking human form include that teaching the robot by demonstration might be facilitated by the teacher and robot having the same structure. Additionally, the robot will be able to function in spaces designed for humans and with human tools and that this would eliminate the need for special design considerations for the robots. A survey of such humanoid robots as the Honda Asimo, Sony Qrio and Toyota's partner robot reveal that all present a distinctly artificial appearance. At least for Qrio, this appearance is intentional as revealed in an interview with Toshitada Doi about Qrio presented on the Sony web pages. When asked – "What do you think about the "character" of robots? – he answered "Take QRIO as an example. We suggested the idea of an "eight year-old space life form" to the designer -- we didn't want to make it too similar to a human. In the background, as well, lay an idea passed down from the man whose work forms the foundation of the Japanese robot industry, Masahiro Mori: "The Valley of Eeriness". If your design is too close to human form, at a certain point it becomes just too . . . uncanny. So, while we created QRIO in a human image, we also wanted to give it little bit of a "spaceman" feel." [17].

While much of the evidence to support the uncanny valley, like that above, is anecdotal there have been limited attempts to experimentally confirm its existence. The primary evidence to support its existence comes from research by MacDorman and Ishiguro [1] that explored observers reactions to facial morphs from a mechanistic robot - to a human looking robot – to an actual human. What they found was that at the boundary of the mechanistic robot and the human looking robot there was a rise in judgments of the eeriness of the display that was consistent with judgments of the morph being seen as less human. However, using the same technique of morphing and identical stimuli as the bases for the morphing space, David Hanson has asked the question of whether falling into the uncanny valley is inevitable [18]. What he did first was to replicate the findings of MacDorman and Ishiguro to find a peak in eeriness judgments. What he did next was to "tune" the different morphs so that they would appear more attractive. What he found was that the eeriness ratings were a flat line although there still was a distinct transition of ratings from human to nonhuman. This clearly indicates that for the case of the single cue of appearance, uncanny reactions can be circumvented by skillful manipulation. It is possible that such a process can be extended to multiple cues if their complex interactions did not make the tuning process intractable. Another experiment investigating the basis of the uncanny valley has shown that the inanimate features of human-like robots which denote death could instill responses consistent with a fear of death in observers [1]. Finally, results from Ishiguro [19] have shown that an android robot undergoing small movements equivalent to postural adjustments could be viewed for 2 seconds without an observer detecting that they were viewing an artificial agent. Without motion observers were much more likely to detect that the agent was not human. This indicates that for very brief encounters the uncanny valley can be avoided without difficulty.

The preceding discussion brings into focus the current situation regarding information about the existence of the uncanny valley. It can be seen that the examples from feature films indicate that the uncanny valley exists. Moreover, many robot designers, animators and game designers appear sufficiently respectful of the concept that they design away from the uncanny valley. However, it can be argued regarding feature film that the uncanny valley is being used as a catch-all phrase when a realistic animation fails. Moreover, the limited empirical evidence both restricts extensive conclusions being drawn and further suggest that falling into the uncanny valley is not inevitable.

4 Psychological Plausibility

If the uncanny valley exists then it should be possible to explain why it exists and possibly to mitigate its effects. Such an explanation doesn't yet fully exist, however we can examine various proposals and related research findings to estimate the plausibility of the phenomenon and explanation. There are at least four descriptions of relevant psychological processes that could predict the uncanny valley and they will be discussed in the following paragraphs.

One common explanation is related to the perceptual issue that increased realism seems inextricably linked to increased information and thus if there are errors in our approximations to realism then they might simply become more evident as more information is provided. One issue with this explanation is that it begs the question as to *why* the errors would become more apparent. It would be just as easy to predict that with the greater and greater amounts of generally correct information being presented that any incorrect information would be drowned out. If this doesn't happen then there must be a peculiar sensitivity to the information which is incorrect. Thus, while this explanation has a ring of truth, it does not appear to be a complete explanation.

A cognitive issue noted by Ramey [20] is that although the uncanny valley is modeled to lie along a continuum of realism, the appreciation of what is being viewed lies at a categorical boundary between humans and machines. Since processes of event and object *categorization* are obligatory, the uncanny valley then is predicted once a lack of genuineness is discovered, the clever animation or robot seems not to fit solidly into either the living or non-living category. This inability to categorize will then lead to a state of dissonance. It appears that this cognitive issue of classification cannot be avoided, however since category boundaries are not necessarily static, the possibility then arises that increasing exposure will lead to a third category being developed which resolves the dilemma.

Another possible explanation for how the uncanny valley might come about is a refinement of the first proposal and inspired by the observation of Mori that motion could exacerbate an uncanny situation already existing in form. The generalization of this idea is that human actions consist of a wealth of different sensory cues. If these cues are not mutually consistent then reconciling the differences among cues might lead to a state of unease or at least uncertainty about what is being observed. The case of form and motion are interesting since various research leads to the view, consistent with the observation of Mori, that there are separate visual pathways that initially process form and motion information and then at a later stage integrate this information in the process of representing human actions [21]. One implication of this is that form and motion might contain different cues to human activity, a view supported by experimental results which indicate that the recognition of affect and emotion from human movement is represented by dimensions of activity and valence (positivity/negativity). The dimension of activity is supported by the speed of a movement and the valence dimension appears to be related to structural relations among the body parts [22]. These arguments point towards the question of whether the uncanny valley could arise out of mismatches between sensory cues where the subtle inconsistencies between cues or missing inputs might lead to finding an experience unpleasant. Certainly, one testable claim about Mori's presentation of the uncanny valley would be to find an uncanny form and to see if motion can be used to modulate the experience.

The previous paragraphs took the position that human activity forms a multidimensional signal and that an uncanny valley might come about in artificial systems either due to a subtle disorganization of the information carried along these dimensions or the subsequent difficulties on categorizing an event that falls on a category boundary. These explanations are not mutually exclusive but the emphasis on the available information and its categorization leaves out one potentially important aspect that has gained increasing interest in the field of neuroscience. This is that our sensitivity to particular information and its subsequent classification is driven by social (and survival) needs to communicate and react to the individuals around us. Related research is asking the parallel question of what brain processes are involved during observation of another social agent (another human) or a non-social agent (a robot). At present the results are inconclusive, some researchers find different responses to humans versus robots at the brain (Tai *et al.*, 2004) as well as behavioral levels [23, 24]. Other results find that both robot and human movements elicit automatic imitation [25], and finally some find mixed results in comparing responses to human versus artificial agents [26]. It is early days and this research has yet to delve into the uncanny valley but it is asking a key question regarding how a "social" brain evaluates its environment. That is to say that the critical issue might not be the logical problem of evaluating human versus nonhuman or confusion over a mismatch of perceptual cues. Rather, the issue might be how the social brain evaluates these perceptual cues and cognitive scenario. Support for this view can be seen in studies showing differences in the acceptance of robots by different cultures [27] and across the lifespan [28].

The purpose of this section was to review principles from psychology that might lead to the prediction of an uncanny valley. Several concepts were presented which suggest that from the standpoint of psychological theory it is plausible that an uncanny valley would exist. However, an explanation of the uncanny valley did not appear to be the providence of any one unique concept.

5   The Human Side of the Uncanny Valley

Thus far I have discussed the original thought experiment of Mori that introduced the uncanny valley, described evidence for its existence including widespread acceptance as a design principle in robotics and computer animation, and put forth psychological concepts that argue for its plausibility. This suggests a definition of the uncanny valley as a phenomenon that exists in the stimulus space around normal human activity and is triggered from either perceptual mismatches or categorical effects, but that the critical level of evaluation might be social. This definition avoids an obvious and important question about how to precisely characterize the dimensions of realism and affinity used in the plots of the uncanny valley. However, this issue is possibly best addressed only through empirical investigations. What I want to examine now is the question of how exclusive is this definition. In particular, if instead of starting at a cute robot toy and moving towards the uncanny valley, what if we start with human activity and move towards the uncanny valley. To do this I will briefly examine three phenomena which seem to fit different criteria of the proposed definition. These phenomena include dubbed speech in cinema, fear of clowns and Capgras syndrome.

The first example of dubbed speech satisfies the property of being in the vicinity of normal human activity since it combines an actual human movement with an actual auditory signal from a different language which is not entirely congruent. This would lead to both perceptual mismatches as well as the possibility for categorical effects of which language is being spoken. Recent evidence has described the importance of audiovisual processes in understanding speech [29] and shown that even very young infants are skilled at appreciating audiovisual congruence [30]. So the question should be then why doesn't an uncanny valley exist for dubbed speech? Perhaps it does. Evidence from a 1988 survey of British television viewers revealed that of those 32% viewers who prefer subtitling to dubbing, 42% did so because they dislike dubbed programs [31]. Moreover, a study of young children shown a subtitled and dubbed version of the same program preferred the subtitled version even though they would not have had advanced skill in reading [32]. Possibly dubbed speech is an example of where habituation, particularly in media markets which make frequent use of dubbing, can overcome a natural tendency to find the experience unpleasant.

The next example of clowns, while somewhat lighthearted, still seems a useful case. It can also be used to make the serious point, that although more seems to be written about the uncanny valley there is about as much empirical evidence to support clown phobia as the uncanny valley. Consulting the limited published report [33], the internet and informal interview does however reveal that those who hate clowns are not alone. Clearly a clown is just a human with some facepaint and funny clothes so they satisfy the condition that they are close to a normal human stimuli. Moreover, any categorical issues should be resolved by the category itself of "clowns" that makes it clear what kind of agent is being encountered. Possible perceptual inconsistencies include that the facial expression painted on the face is not consistent with the actions. Perhaps this incongruence might be appreciated more on a social level, that the clown with a painted smile ought not to always look so happy for all its actions.

The final example of Capgras syndrome suggest that an uncanny situation could arise without any perceptual mismatches and for normal human activity. Capgras syndrome is a relatively rare condition where the sufferer believes that people, or in some instances things, have been replaced with duplicates [34]. These duplicates are rationally accepted to be identical in physical properties but the irrational belief is held that the "true" entity has been replaced with something else. Ellis and Lewis

[35] describe the recent situation of a man who after a car accident believed that his wife had died in the accident, and the woman he currently lived with (his wife) was a duplicate. Naturally, he found this situation to be uncomfortable. Some sufferers of Capgras syndrome have even claimed that the duplicate is a robot and these cases would seem to perfectly match the uncanny valley. Ellis and Lewis [35] argue that the syndrome arises from an intact system for overt recognition coupled with a damaged system for covert recognition that leads to conflict over an individual being identifiable but not familiar in any emotional sense. This example provides support for a view that the uncanny valley could arise from issues of categorical perception that are particular to the specific way that the social brain processes information.

What this section has attempted to demonstrate is that there are sufficient possibilities for deviation from normal behavior and normal recognition to lead to scenarios consistent with a definition of the uncanny valley. Perhaps a more precise definition could avoid this multiplicity of ways into the uncanny valley, though this has the danger of throwing the proverbial baby out with the bathwater. Thus, we seem left with the situation that the obstacle to interpreting increased realism might not be one great uncanny valley but rather a multitude of uncanny potholes.

6 Conclusions

The goal of this essay was to review the uncanny valley and not to either refute or describe its essential mechanisms. The hope was that a description of its history, context and psychological plausibility would inform what questions are important to pursue. One essential question to ask is just whether there is enough evidence to say that the uncanny valley exists? Surprisingly, the answer is equivocal. It is clear from practitioners that more realism does not always equate with greater acceptance by audiences and there is a wealth of anecdotal evidence to support this view. Moreover, from first principles of psychology one can build a case that something like an uncanny valley would exist. However, there is a dearth of empirical evidence on the topic and certainly no study that outlines essential properties that can be manipulated to navigate into and out of the uncanny valley. Thus, it would seem some care is needed in the evaluation of claims about the uncanny valley until a more rigorous understanding is reached.

The attempt made here to come up with an operational definition of the uncanny valley ran into difficulties with its assumption that the uncanny valley should occur in the vicinity of natural human actions. Namely, this difficulty was that from the perspective of psychology there doesn't seem to be a shortage of situations where actual human actions can be transformed into the uncanny. What would be helpful to resolve this problem is further specification, by those animators and roboticists pushing into the uncanny valley, of what bit of the human response to these artifacts is the essential aspect of its uncanny nature. This does not seem a simple task since the uncanny region is at the cutting edge of technology and can be achieved only with substantial resources and talent and these are typically devoted to avoiding the uncanny valley.

References

1. MacDorman, K.F., Ishiguro, H.: The uncanny advantage of using androids in social and cognitive science research. Interaction Studies 7(3) (2006)
2. Brenton, H., et al.: The Uncanny Valley: does it exist? In: The 11th International Conference on Human-Computer Interaction. Lawrence Erlbaum Associates, Las Vegas (2005)
3. Gee, F., Browne, W., Kawamura, K.: Uncanny Valley Revisited. In: 2005 IEEE International Workshop on Robots and Interactive Communication. IEEE Press, Nashville (2005)
4. Hanson, D., et al.: Upending the Uncanny Valley (2006)
5. Kosloff, S., Greenberg, J.: Android science by all means, but let's be canny about it! Interaction Studies 7(3) (2006)
6. Chaminade, T., Hodgins, J.: Artificial agents in the social cognitive sciences. Interaction Studies 7(3) (2006)
7. Canamero, L.: Did Garbo care about the uncanny valley? Interaction Studies 7(3) (2006)
8. Mori, M.: Bukimi No Tani (the Uncanny Valley). Energy 7, 33–35 (1970)
9. Reichardt, I.: Robots: Fact, Fiction and Prediction, p. 166. Thames and Hudson, London (1978)
10. Freud, S.: The Uncanny. The Standard Edition of the Complete Psychological Works of Sigmund Freud, vol. 17, pp. 219–252. The Hogarth Press, London (1960)
11. Weschler, L.: Why is this man smiling? Digital animators are closing in on the complex system that makes a face come alive. Wired, 2002(10.06)
12. Ebert, R.: Gollum stuck in 'Uncanny Valley' of the 'Rings', in Chicago Sun Times, Chicago (2004)
13. Horneman, J.: The Incredibles, Polar Express, the Uncanny Valley, Pixar (2004)
14. Atkeson, C., et al.: Using humanoid robots to study human behavior. IEEE Intelligent Systems 15, 46–56 (2000)
15. Coradeschi, S., et al.: Human-Inspired Robots. IEEE Intelligent Systems 21(4), 74–85 (2006)
16. Hale, J., Pollick, F.: Sticky Hands: learning and generalization for cooperative physical interactions with a humanoid robot. IEEE Transactions on Systems, Man, and Cybernetics, Part C 35(4), 512–521 (2005)
17. Sony, Interview with Toshitada Doi: Personal robots make the 21st century more fun (2006)
18. Hanson, D.: Expanding the Aesthetic Possibilities for Humanoid Robots (2005)
19. Ishiguro, H.: Android science: conscious and subconscious recognition Connection Science 18(3) (2006)
20. Ramey, C.: The uncanny valley of similarities concerning abortion, baldness, heaps of sand, and humanlike robots. In: Proceedings of the Views of teh Uncanny Valley: workshop, IEEE-RAS International Conference on Humanoid Robots, Tsukuba, Japan (2005)
21. Giese, M., Poggio, T.: Neural Mechanisms for the Recognition of Biological Movements. Nature Neuroscience Review 4, 179–192 (2003)
22. Pollick, F., et al.: Perceiving Affect from Arm Movement. Cognition 82, B51–B61 (2001)
23. Kilner, J.M., Paulignan, Y., Blakemore, S.J.: An interference effect of observed biological movement on action. Current Biology 13(6), 522–525 (2003)
24. Castiello, U.: Understanding Other People's Actions: Intention and Attention. Journal of Experimental Psychology: Human Perception and Performance 29(2), 416–430 (2003)

25. Press, C., et al.: Robotic movement elicits automatic imitation. Cognitive Brain Research 25(3), 632–640 (2005)
26. Pelphrey, K.A., et al.: Brain activity evoked by the perception of human walking: Controlling for meaningful coherent motion. Journal of Neuroscience 23(17), 6819–6825 (2003)
27. Kaplan, F.: Who is afraid of the humanoid? Investigating cultural differences in the acceptance of robots. International Journal of Humanoid Robotics 1(3), 465–480 (2004)
28. Turkle, S., et al.: Relational artifacts with children and elders: the complexities of cybercompanionship (2006)
29. Munhall, K., Vatikiotis-Bateson, E.: Spatial and Temporal Constraints on Audiovisual Speech Perception. In: Calvert, G.A., Spence, C., Stein, B.E. (eds.) Handbook of Multisensory Processes, pp. 177–188. MIT Press, Cambridge (2004)
30. Hollich, G., Newman, R., Jusczyk: Infants' use of synchronized visual information to separate streams of speech. Child Development 76(3), 598–613 (2005)
31. Kilborn, R.: 'Speak my language': current attitudes to television subtitling and dubbing. Media Culture Society 15(4), 641–660 (1993)
32. Koolstra, C.M., Peeters, A.L., Spinhof, H.: The Pros and Cons of Dubbing and Subtitling. European Journal of Communication 17(3), 325–354 (2002)
33. Austin, R., McCann, U.: Ballatrophobia: When clowns aren't funny. Anxiety 2, 305 (1996)
34. Ellis, H., et al.: Delusional misidentification: The three original papers on the Capgras, Frégoli and intermetamorphosis delusions. History of Psychiatry 5, 117–118 (1994)
35. Ellis, H., Lewis, M.: Capgras delusion: a window on face recognition. Trends in Cognitive Science 5(4), 149–156 (2001)

Investigating the Use of the Experience Clip Method

Jeroen Vanattenhoven and Dries De Roeck

Centre for User Experience Research, IBBT / KU Leuven,
Parkstraat 45 Bus 3605, 3000 Leuven, Belgium
{jeroen.vanattenhoven,dries.deroeck}@soc.kuleuven.be

Abstract. The focus of this paper is on the use of the Experience Clip method for evaluating mobile applications. In this method one participant uses the to-be-evaluated mobile application while the other participant films whatever happens during the field test. Afterwards, participants are interviewed by the researcher(s) to clarify the recorded situations. Investigating the use of the Experience Clip method was achieved by evaluating two mobile multimedia geocaching applications. The method was found very suitable for capturing rich, emotional user experiences and uncovering usability issues. Furthermore, digital photo cameras, used by participants for filming, offer great image quality, ease-of-use, sufficient storage capacity, and attract little attention of bystanders. However, there is one concern that when participants film each other, another experience is added on top of the experience with the mobile application. It is unclear how the filming experience affects actual use and the validity of the collected data.

Keywords: User Experience, Mobile Computing, Geocaching, Evaluation, Experience Clip, Tangible Interaction.

1 Introduction

Evaluating mobile applications still involves several challenges. Currently, different methods exist to evaluate mobile applications such as diaries, questionnaires, mobile labs, logging, ESM, shadowing, etc [1][3]. Recently, Experience Clips was presented as a new way to evaluate mobile applications [2][3]. This method involves two participants who preferably know each other. One participant uses the application, while the other one shoots video clips – "Experience Clips" – to document the user's experiences and emotions. Halfway in each field test the user takes over the role of the shooter and vice versa, so both participants have used the application, and both participants have created Experience Clips. Afterwards both participants are invited to talk about their experiences using the videos they recorded. The main benefits are that the captured experiences are more natural because the researcher is not following or shadowing the participants, and that the participants already know each other. Furthermore, the elicitation afterwards is helped by the fact that one relies less on memory because the video captures the activity as it unfolds.

P. Daras and O. Mayora (Eds.): UCMedia 2009, LNICST 40, pp. 79–86, 2010.
© Institute for Computer Sciences, Social-Informatics and Telecommunications Engineering 2010

Geocaching "is the practice of hiding a container in a particular location, then publishing the latitude and longitude coordinates of the location on a geocaching web site for other geocachers to find using a GPS device" [4]. Some caches involve intermediate locations before getting to the cache, and involve puzzles that have to be solved in order to get to the end result. The evaluated applications have made use of the possibilities of mobile technology (GPS, pictures, video, audio and music), to augment such a geocaching experience.

In this paper we first describe the applications and their features. We continue by presenting the participants and the procedure, followed by the analysis and results of the user experience and usability evaluation. Finally, we discuss the most significant issues in using the Experience Clip method, formulate the main conclusions and identify future work.

2 User Experience Evaluation

2.1 Applications

The application referred to as 'Fluisterdingen' was developed as part of the ITEA SmartTouch project by Alcatel-Lucent Bell Labs and globally allows users to share media via NFC technology through RFID tags. Fluisterdingen consists of an online editor for PC and a mobile phone application. Only the latter was included in the user study. The online editor was used by the research team to create a route through the city centre of Leuven which consisted of five points of interest. Each point of interest was marked with a cardboard indicator that had an RFID tag on it (Fig. 1). Users had to find this tag and physically touch it with the phone in order to receive a description and a picture that contained hints about their next location. This application ran on a Nokia 6131 mobile phone (Fig. 1).

The 'Mediacaching' application was developed by researchers of the University of Cologne in the CITIZEN MEDIA research project, and is a multimedia geocaching application that incorporates Semacodes, GPS, audio, pictures, video and 3D models. Users can create their own geocaches with the online editor (http://www.mediacaching. org). We used this editor to create a route for the user experience evaluation in the city centre of Leuven. After the application started, the user would receive a hint to their first location. This hint could either be composed of text, pictures, video, audio, and a GPS-driven location indicator. When the user arrived at the location, an assignment would have to be carried out. One of the assignments in our study was answering a question (multiple choice, multiple selections) and providing evidence to that question using the camera on the phone. Another assignment was that the participants had to create a short clip that could serve as an item on the evening news. After the assignment they would receive the hint to the next location. When participants completed the tour, they received a "treasure"; in our case, a fireworks movie. This application ran on a Nokia N95 mobile phone (Fig. 1).

Fig. 1. RFID tag, Fluisterdingen application, and Mediacaching application

2.2 Participants and Procedure

Participants were recruited via a mailing list. Six student couples participated; one male couple, one mixed couple and four female couples. All couples knew each other or were friends. An incentive was provided in the form of a coupon of 20€ for a multimedia store in Belgium.

Participant couples were invited to the research centre where they were briefed about the study which included an explanation of how to use the applications and the digital photo camera, how to shoot experience clips and what to film, and were to go [2]. One of the participants carried the application while the other carried a digital photo camera (Panasonic Lumix with a 2GB SD Card). They were also asked to switch half-way during the test so both participants would have used the application and would have created experience clips.

Both applications were evaluated; for each application we foresaw a briefing of approximately 5-10 minutes, a field test using Experience Clips for 30 minutes, and an interview afterwards about participants' experience clips, making the total evaluation time one hour. Hence, the evaluation of two applications by one couple took approximately two hours. Three of the couples started with the Fluisterdingen application; the other three couples evaluated Mediacaching application first.

2.3 Analysis and Results

When investigating the use of the method, it is important to look at how the method is able to fulfill its purpose: evaluating user experience and usability of mobile applications. Therefore, in this section the analysis is described, followed by the results of the usability and user experience evaluation. More specifically, we first present the most significant usability issues and participants' experiences with the evaluated applications; then we discuss the evaluation of the devices itself.

The Experience Clips are transcribed into an "action-table" (Table 2) [2][3]. The first two columns contain the transcription of the dialogue between the two participants; in the third column a description is given about what is shown on the video, and what is happening in the field. We incorporated two more columns to add the transcriptions of the interviews about the experience clips with the participants after they returned from their field test. What a participant and researcher said in the interview, was placed exactly at the same line of the respective moment in the action table of the experience clip. In this paper however, we will not be able to present five columns because of a lack of horizontal space. This provided us with an overview on what happened during the field test and what was said during the interview afterwards. These action-tables formed the basis for the analysis, together with the video data. The interviews were semi-structured and inquired about what they liked and disliked about the application, the device, and the filming of the experience clips.

We collected 123 videos in total (see Table 1); a total duration of 36 minutes and 49 seconds for the Mediacaching application, and 15 minutes and 12 seconds for the Fluisterdingen application. Average length per video was 28 seconds for Mediacaching and 20 seconds for Fluisterdingen, while the median length was 22.5 seconds for Mediacaching and 12 for Fluisterdingen. The shortest videos for Mediacaching and Fluisterdingen were one second long, the longest Mediacaching video was two minutes and 51 seconds long whereas the longest Fluisterdingen video was 58 seconds long.

Table 1. Number of videos per application

Participant couple	1	2	3	4	5	6	Total
Mediacaching	17	21	9	6	12	13	78
Fluisterdingen	6	7	17	5	4	6	45
						Total	123

There were several usability issues with the Mediacaching application; only one problem was found with the Fluisterdingen application - the RFID tag did not always register immediately upon touching it with the mobile phone. One of the main problems in the Mediacaching application was the GPS-driven location indicator. Participants were confused by the fact that the indicated direction in which they were supposed to walk changed quite often, and did not know that they did not had to walk until the meter counter (that counted down towards the next destination) reached zero. The longest video, two minutes and 51 seconds, was a result of these problems, and all participant couples experienced this problem. To add to the problem, the message "walk faster" was shown simultaneously, because the direction based on GPS could only be calculated if users were moving. However, users were not informed that the direction based on GPS only worked this way. Another usability problem with the Mediacaching application was the screen in which participants has to select one of the multiple choice options. They would move the cursor down to their preferred answer and then press the right softkey on the mobile phone which would then process the answer ("OK" in the application). The problem was that none of the participant couples first selected the answer using the central button on the mobile phone. Most participant couples later found out what they did wrong, but were always surprised at first when the application informed them that their answer was wrong. In the end only two couples completed the entire route of the Mediacaching application.

The videos illustrated the user experiences to a great extent. However, to clarify the participants' meaning of the videos to the researcher(s), the interviews afterwards were really necessary. Frustration was one of the emotions we saw a lot on the clips mainly due to usability problems. Furthermore, enjoyment was also seen several times. One good example is illustrated by the experience clip illustrated in Table 2 and Figure 2, where participants first have some difficulties in registering the RFID tag, but then succeed in finding the final location. For the Fluisterdingen application participants liked the fact that they had to touch the RFID tags, and the trembling of the mobile phone that followed, in other words, the tangible interaction. This was illustrated in some experience clips, and mentioned by all participants in the interviews. The Mediacaching application incorporated many different features such as multiple choice questions, GPS guidance, video and photo assignments etc. This variety was very much appreciated by participants in comparison with Fluisterdingen because with the latter the only interaction took place at each point-of-interest - no interaction was provided in-between. The fact that participants were active more frequently with the Mediacaching application, and were confronted with more usability problems, form the main reasons for the difference in number and length of experience clips, as indicated during the interviews.

Table 2. Experience clip illustrating enjoyment

Application user	Video shooter	Description
Yeey!		The user has found the final RFID tag. The shooter films the user from behind while walking toward the RFID tag.
	Now what? Now what? Now what?	The shooter is very excited and impatient when she asks the user to explain what is happening now.
It doesn't do anything.		The user tries to put the mobile phone against the tag; but the application doesn't register the tag immediately.
	Did it tremble?	The shooter asks whether the mobile phone trembled, something that happens when the tag is registered.
Yes, but it doesn't show anything		The shooter goes to the user and grabs the mobile phone to inspect what is happening.
	Yes but you did "thingy". It's busy.	
Bullseye! You have completed this route.		The user reads the final message out loud.
	Perform a "round dance".	The shooter ask the use to dance around. Then the shooter also dances around.

Fig. 2. Participant performing a "round dance"

We asked participants' opinions on the two different mobile used in the user study. There was a small issue with the Fluisterdingen application in that it did not always register the tag immediately. This could be seen on the Experience Clips when participants tried several positions and interactions to get the application to register the tag. There was a more severe issue with the Mediacaching device, the Nokia N95. When this device is held horizontally, the display changes from portrait to landscape. For all participants this happened quite often, and especially very unexpectedly.

3   Discussion on the Use of the Method

In this section we will discuss the use of the method. First, we look at how the method is able to achieve its main objective: evaluating the user experience and usability of mobile applications. The previous section on the results of the user study illustrates that the applying the method provides us with an adequate view on the usability issues, and on participants' experiences. The user study generated rich data by means of video clips, which illustrated what happened during the field study, and the interviews afterwards, in which participants clarified the recorded situations to the researchers who were not present, and therefore missed some context.

Moreover, when we compare the user experience of both applications our findings illustrate some temporal aspects of user experience. The tangible interactions with the RFID tags of the Fluisterdingen application were found very enjoyable. But our participants preferred to be entertained between the locations as well. The variety of technologies available in current mobile devices, such as GPS, RFID, Semacodes, and features such as multimedia questions and assignments in the Mediacaching application provided participants with sufficient interaction between the points of interest, in other words, a more continuous user experience.

An essential part of the Experience Clip method is the fact that participants have to walk and film each other in public. Therefore, we asked participants how they felt about this during the interview afterwards. All participants did not mind filming each other, although not all of them were completely comfortable, mainly because sometimes they felt being watched. We also asked if they would film using a mobile phone, as in the original work [2], or using a digital video camera. None of the participants would film using a digital video camera because they expected that this device would be harder to use and control, and because of its size, the device would attract even more attention, making it very uncomfortable. Participants would not mind using a mobile phone camera to film, but noted that, based on experience with their personal mobiles, video quality on mobile phone cameras is insufficient.

There is one important methodological issue to consider: does the filming experience influence the application experience? Participants were free to film as they saw fit; one couple was acting (performing) in front of the camera and created mini plays, presenting their experiences to the researchers. The other couples mostly logged the behavior of the user, while engaging in dialogue with them at certain points. The researchers that originally developed the method are aware of this performance aspect, but discussed it mainly as an opportunity for participatory design [2][3]. What the impact of the filming experience is on the use of the application, and the validity of the results, remains unclear. One important strength of this method, and an important issue in evaluation mobile applications, is that rich data can be gathered from users in the field, with minimal intrusion from researchers. If one considers the creation of mini plays by participants, the possibility that this is done to seek approval from the researchers might impact the strength of the method.

Because two people are involved in this method, it is not straightforward to capture individual experiences and opinions. On the other hand, this might not be that important since the tested applications were designed to be used by multiple people, as "social walking" is one of the motivations for geocaching [4].

The audio of the videos participants made, was acceptable most of the time, but due to noises in the city environment i.e. cars passing by or the strong wind at the time of the study, at times we could not understand what participants were saying. The inaudible parts on the videos were very limited however, so it did not impact the results that much. Moreover, since the videos helped the participants' elicitation afterwards during the interviews, they were sometimes able to recall what they were saying during the inaudible parts.

The use of the digital photo cameras was found appropriate for filming each other. Using this kind of device instead of a mobile phone camera (as in the original studies [2][3]) or a digital video camera offered advantages for both participants (easy-to-use, does not attract too much attention from bystanders, sufficient storage capacity), and researcher (good video quality).

4   Conclusion and Future Work

The objective of this paper was to investigate the use of the Experience Clip method. We found that applying this method resulted in a rich view on participants experiences' and uncovered several significant usability issues. The resulting video is rich, and helped participants' elicitation in the interviews afterwards, which served to clarify the recorded situations to the researchers. The results of the user study allows us to improve the design of the application, and compare both applications with each other.

More research should be carried out to investigate the differences between the devices used for filming the experience clips more in depth, since we only provided participants with a digital photo camera, and asked their opinion about the use of a mobile phone camera and a digital video camera. In other words, what people think about certain uses is not always identical to the actual use situation.

Most participants felt comfortable filming each other in the city centre. However, since we recruited students, it is uncertain to what extent other possible user profiles feel the same way about creating video clips in the presence of bystanders.

We believe that when one participant is filmed by the other, and they engage in dialogue to clarify to each other what is happening in each situation, a new experience is created. Future research should therefore investigate the impact of the filming experience on the actual use of the to-be-evaluated application and on the validity of the data.

Acknowledgments. SmartTouch is an ITEA-IWT (ITEA No 05024) project which ran from 2006 until 2008. The research described in this paper has been made possible by close collaboration with Alcatel-Lucent Bell Labs Belgium. The work for Mediacaching was supported by the CITIZEN MEDIA (www.ist-citizenmedia.org) research project funded by FP6-2005-IST-41. Special thanks to Thomas van Reimersdahl of the University of Cologne for providing fast technical feedback concerning the application, and colleague Christof van Nimwegen for feedback on this paper.

References

1. Carter, S., Mankoff, J.: When participants do the capturing: the role of media in diary studies. In: 23rd SIGCHI conference on Human factors in computing systems, pp. 899–908. ACM, New York (2005)
2. Isomursu, M., Kuutti, K., Väinämö, S.: Experience clip: method for user participation and evaluation of mobile concepts. In: 8th conference on Participatory design, vol. 1, pp. 83–92. ACM, New York (2004)
3. Isomursu, M., Tähti, M., Väinämö, S., Kuutti, K.: Experimental evaluation of five methods for collecting emotions in field settings with mobile applications. International Journal of Human-Computer Studies 65, 404–418 (2007)
4. O'Hara, K.: Understanding geocaching practices and motivations. In: 26th SIGCHI conference on Human factors in computing systems, pp. 1177–1186. ACM, New York (2008)

Natural-Language-Based Conversion of Images to Mobile Multimedia Experiences

Bernhard Reiterer[1], Cyril Concolato[2], and Hermann Hellwagner[1]

[1] Klagenfurt University, Universitaetsstr. 65-67, 9020 Klagenfurt, Austria
firstname.lastname@uni-klu.ac.at
http://www.uni-klu.ac.at
[2] TELECOM ParisTech, 46, Rue Barrault, 75013 Paris, France
cyril.concolato@telecom-paristech.fr
http://www.telecom-paristech.fr

Abstract. We describe an approach for viewing any large, detail-rich picture on a small display by generating a video from the image, as taken by a virtual camera moving across it at varying distance. Our main innovation is the ability to build the virtual camera's motion from a textual description of a picture, e.g., a museum caption, so that relevance and ordering of image regions are determined by co-analyzing image annotations and natural language text. Furthermore, our system arranges the resulting presentation such that it is synchronized with an audio track generated from the text by use of a text-to-speech system.

Keywords: image adaptation, text analysis, image annotation, digital cultural heritage, computer animation.

1 Introduction

Images on the Web, from personal photography or any other source very often have resolutions higher than the displays they are to be shown on. For mobile devices, we commonly find image width and height each about ten times larger than the display. Such cases are usually dealt with in one of two disadvantageous ways: (1) creating a smaller image, either by scaling, cropping or seam carving [1], which is a more recent technique, all of which discard a lot of pixel data; or (2) navigating the image manually, e.g., as provided by the iPhone[1] or certain websites[2], which tends to be tedious for images at high scaling factors.

An emerging alternative family of approaches for viewing images on mobile devices is the automatic transmoding of images to videos, resulting from moving a virtual camera over the image at varying zoom levels, so that the most relevant regions are clearly visible one after the other. Such approaches, using different detection techniques for finding interesting regions, are shown in [2] and [3].

We argue that the steering of the virtual camera could consider more information than what can be extracted from the image. The main idea underlying our

---

[1] http://www.apple.com/iphone/, accessed on 28 September 2009.
[2] E.g., http://memorabilia.hardrock.com/, accessed on 28 September 2009.

P. Daras and O. Mayora (Eds.): UCMedia 2009, LNICST 40, pp. 87–90, 2010.
© Institute for Computer Sciences, Social-Informatics and Telecommunications Engineering 2010

approach is that a natural language description of an image, such as a painting's caption in a museum, can be a valuable source for determining the relevance of image regions. Furthermore, if presented to the user in an appropriate way, image and text can augment each other: the text explains the image, while the image illustrates the text. So, by transforming the image to a video and presenting the text in a synchronized way, either spoken or as subtitles, we seek to enhance the user experience in comparison to alternative methods.

2   Natural-Language-Based Image Transmoding

2.1   System Overview

Figure 1 shows a high-level overview of our Natural-Language-based Image Transmoding Engine. The core step, called *Virtual Camera Control* (VCC), is responsible for generating the script for the virtual camera, instructing the camera when to show which rectangle of the image. This step takes image annotations, text, and various constraints as its input. Internally, VCC creates *Sync Points*, tuples that link positions to be synchronized across different media, from the input data and refines them until all relevant aspects are considered.

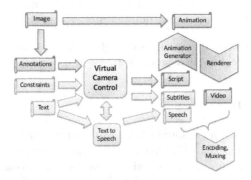

Fig. 1. Overview of the Natural-Language-based Image Transmoding Engine

Image annotations, either created manually or from computer vision systems, consist of shapes of regions of interest (ROIs) annotated with keywords. Constraints to be considered include the display size, the maximum video duration, and limits ensuring a pleasurable camera motion. A text-to-speech system, currently a MARY[3] server, produces audio and timing data from text.

From the script, an SVG[4] animation is generated. In the finalization steps, which, unless configured otherwise, delegate to software of GPAC [4] and FFmpeg[5],

---

[3] http://mary.dfki.de/, accessed on 28 September 2009.
[4] http://www.w3.org/Graphics/SVG/, accessed on 28 September 2009.
[5] http://ffmpeg.org, accessed on 28 September 2009.

the animation is rendered to a video, encoded (by far the biggest part of the execution time) and multiplexed with the encoded synthetic speech (and, if desired, subtitles), yielding an audiovisual file ready for playback.

We provide a growing set of example inputs and results on our project website[6], illustrating the functionality and rich configurability of the system so far.

2.2 Matching and Synchronization

The initial implementation of VCC (in Java) is realized by looking up keywords from ROI annotations in the text. For each match between an ROI's keyword and the text, a Sync Point is generated. This approach will be enhanced by more powerful text analysis in order to allow for annotations on a lower level and thus to allow for more automatic analysis, e.g., by interpreting descriptions of visual features (mainly shapes and colours) or spatial relations (such as "left of" or "in the center").

For making the text perceivable to the user, options are displaying it on the screen, either as subtitles or as running text, or transforming it to synthetic speech. The latter has the advantage of not using any display space, the resource whose limitation gives the main motivation for our work. The others are fallback solutions for cases in which the user cannot or prefers not to play audio, or for supporting languages for which the system is not able to generate speech.

We use text-to-speech preprocessing results for assigning a time value to each position in the text, thus enriching the previously generated Sync Points by precise timing. Unpleasantly short or long durations for motions or stays are resolved by slightly delaying Sync Points, discarding less relevant ROIs (e.g., repeated ones) or by adding unvisited ROIs, respectively.

3   Typical Application

As one possible field of application, our system could be incorporated into the Website of a museum for helping users decide which parts of the museum to see: the museum staff annotates an image once, and different texts for multiple target groups (language, age, expert level or interests, e.g., artistic style, historical background) are provided for each image. On the Website, users give the criteria for selecting among those texts, along with their choice of potentially interesting parts of the museum and the time they want to spend for watching the videos. With the users' approval, the server automatically receives information about the terminal's relevant properties, such as display resolution and capabilities for decoding and playback. The users then receive individually generated videos presenting the highlight paintings of the selected museum wings.

Other potential applications include maps and directions, social Websites, slide shows for collections of personal photography, which are currently often generated manually, and also image viewing as a side activity, where a user's visual attention and navigation capabilities are hampered.

---

[6] http://www.itec.uni-klu.ac.at/~reiterer/dawnlite/showcase.php,   accessed on 28 September 2009.

4 Conclusion and Future Work

We presented our approach for transforming a high-resolution image into audio-visual content for constrained devices by generating a video synchronized with synthesized speech after matching a descriptive text with image annotations. The thus achieved preservation or even enhancement of the users' experience will be evaluated by user studies at a later stage.

To get the most benefit from our approach, we will put the main focus of our future work on its most innovative component, the matching of text to image regions and the underlying (currently rather limited) text analysis, reducing the necessity of manual image annotations. However, any image analysis system (e.g., detection of persons, faces or certain objects) could be used with our system already, leading to functionality analogue to related work.

Along with linguistic improvements, the system should be enabled to rearrange the speech (e.g., inserting breaks, moving or erasing phrases) in order to even the temporal distribution of ROIs. Furthermore, if available, pre-existing audio should be usable, either by speech recognition or via textual transcripts.

The application of the text-based virtual camera control to 3D content instead of images seems feasible with manageable effort, since the software used for rendering now is already capable of handling 3D worlds. Example use cases would be illustrating text from natural sciences as well as 3D navigation systems for virtual or real places. A main challenge in this extension is expected in controlling the camera's motion (supporting different movement types, e.g., walking, following roads, flying) along with viewing angle and zoom in an enjoyable way.

Acknowledgements

This work is supported by the NoE INTERMEDIA funded by the European Commission (NoE 038419).

References

1. Shamir, A., Avidan, S.: Seam Carving for Media Retargeting. Commun. ACM 52(1), 77–85 (2009)
2. Pinho, P., Baltazar, J., Pereira, F.: Integrating Low-Level and Semantic Visual Cues for Improved Image-to-Video Experiences. In: Campilho, A., Kamel, M.S. (eds.) ICIAR 2006. LNCS, vol. 4142, pp. 832–843. Springer, Heidelberg (2006)
3. Megino, F.B., Martínez Sánchez, J.M., López, V.V.: José M. Martínez Sánchez. In: WIAMIS 2008: Proceedings of the 2008 Ninth International Workshop on Image Analysis for Multimedia Interactive Services, pp. 223–226. IEEE Computer Society, Los Alamitos (2008)
4. Concolato, C., Le Feuvre, J., Moissinac, J.-C.: Design of an efficient scalable vector graphics player for constrained devices. IEEE Transactions on Consumer Electronics 54, 895–903 (2008)

UCMedia 2009

Session 4: Multimedia Search and Retrieval

Provision of Multimedia Content Search and Retrieval Services to Users on the Move

Dimitrios Giakoumis[1], Dimitrios Tzovaras[1], Petros Daras[1], and George Hassapis[2]

[1] Centre for Research and Technology Hellas, Informatics and Telematics Institute,
6th Km Charilaou-Thermi Road Rd., 57001, Thermi, Thessaloniki, Greece
{dgiakoum,tzovaras,daras}@iti.gr
[2] Aristotle University of Thessaloniki, Department of Electrical and Computer Engineering
ghass@auth.gr

Abstract. In this paper we present a mobile application for the seamless and effective provision of networked multimedia content search and delivery services to users on the move. The VICTORY project had as an outcome a framework enabling users to search and retrieve multimedia objects over a distributed Peer to Peer - based network. Building upon the VICTORY framework, a mobile user agent was developed, which by the use of appropriate Web Services has the ability to connect to different P2P communities in order to search for and retrieve multimedia content. The content delivered follows the notion of the "MultiPedia" object, defined for the purpose of seamless and effective content search and delivery within the VICTORY Framework.

Keywords: 3D Content Search and Retrieval, PDA Application, Web Services, Multipedia Content.

1 Introduction

Mobile devices play a key role in our everyday life. Their advanced networking and processing capabilities offer us the opportunity to use them at "any-place and any-time", in a wide application spectrum. Search and retrieval of networked multimedia content is an application area of great importance and is expected to be of even greater in the years to come. However, for each different network or infrastructure offering this kind of functionality, usually a different client application is needed from the end users in order to effectively utilize the offered capabilities.

Focusing on 3D content, one easily may identify that only a few 3D search engines are currently available worldwide, most of them providing their functionality through a dedicated web site: The "Princeton 3D Model Search Engine" [1], developed in the Princeton University, supports 2D-sketch-based search, 3D-based search and combination of them with text-based search. The "3DTrue Search Engine" [2] is a commercial tool that supports only text-based search over huge repositories. The "ITI Search Engine" [3] allows for searching with existing model as a query only. The "MeshNose" [4] search engine supports only text-based search, however it has indexed many 3D repositories. Finally, the "Geolus" search engine, a product of the

P. Daras and O. Mayora (Eds.): UCMedia 2009, LNICST 40, pp. 93–100, 2010.
© Institute for Computer Sciences, Social-Informatics and Telecommunications Engineering 2010

UGS Corporation, is an integrated 3D search component of the commercial software Partfinder [5], product of the Solid Edge Company. Searches for similar parts are based on geometrical descriptors, extracted from the 3D objects.

Even though users currently have the ability to search for multimedia content within various different repositories, in order to do so, they have to use different user agents to access each of them. The proper integration of content repositories like the above would significantly benefit the quality of the search and retrieval procedure, since it would allow for the seamless search of multimedia content within different sources. In order for the different infrastructures that offer 3D content search and delivery capabilities to converge in the future, three pre-requisites can be identified:

- Definition of common generic data formats and standards allowing for the seamless search and delivery of multimedia content.
- Deployment of architectures allowing for diverse end user agents to utilize the networks.
- Development of innovative and adaptive client applications, which will finally provide the required functionality to end users.

Towards this direction, we built upon the VICTORY framework described in Section 2, which provides the two first pre-requisites. In this context, a mobile application was developed, capable to connect to different networks and communities in order to seamlessly search for and retrieve multimedia content. Our developed client application can be considered to be a very helpful tool for a wide range of user groups dealing with multimedia and especially 3D content, like 3D game designers, engineers working in design departments of the automotive industry etc, since the demanding task of 3D and multimedia content search and retrieval, often necessary for their daily tasks is simplified and furthermore, has become "mobile".

In the following, a description of the VICTORY MultiPedia object concept, used within our mobile application is initially provided, followed by an overview of the VICTORY service-oriented framework's architecture that allowed for the development of our mobile user agent. Finally, we describe our developed mobile application which delivers all the required functionality to the end users.

2   An Overview of the VICTORY Framework

The VICTORY project [6] has developed a framework enabling users to search and retrieve multimedia objects over a distributed P2P-based network [7][8]. By defining the notion of the "MultiPedia" object and a three-layer, service oriented architecture; it provided the means for the deployment of diverse and adaptive user agents.

2.1   The VICTORY "MultiPedia Object" Concept

In order to effectively utilize the VICTORY P2P-based infrastructure's capabilities, our developed PDA application operates on the basis of the VICTORY framework's "MultiPedia object" concept, where visual information is described as a 3D object along with its accompanied information (2D views, text, audio, and video).

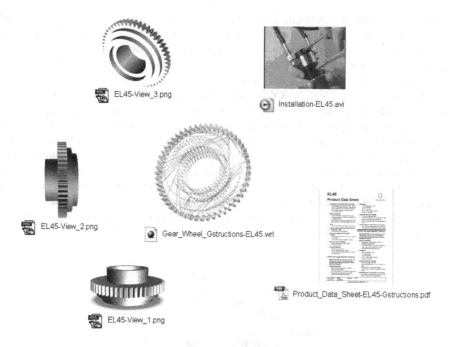

Fig. 1. An example of MultiPedia content

An example of MultiPedia content is given in Fig. 1, where the following content types are identified:

- A 3D object: it is available in VRML format (.wrl file). However, multiple other 3D file formats can be supported such as X3D, OFF, OBJ, 3DS, and so on.
- Multiple views of the 3D object: these are 2D images in any of the common image file formats.
- Textual information: it is available in several text file formats (.pdf, .doc, .txt, etc.) and it may include product information (data sheet), installation instructions, description of the 3D object or any other type of information that can be provided in text.
- Video: it is available in several common video file formats (.avi, .wmv, etc.) and may include additional information, such as demos, installation instructions, etc.

2.2 VICTORY Service-Oriented Architecture

VICTORY developed a framework for P2P-based search and retrieval of multimedia objects. The P2P middleware that was developed for VICTORY allows for generic content sharing among peers in the network. However, mobile peers, due to their limited resources, are not capable of running a fully featured P2P communication platform, neither are they capable of performing other resources consuming tasks locally. Therefore, it was decided that mobile devices would access the Internet-based segments of the VICTORY network through a dedicated Gateway. The services running on the Gateway perform various resource consuming tasks on behalf of the mobile clients, and provide the means for device and application-independent access to the overall functionality offered by the network.

The concept of the VICTORY framework's service-oriented approach is depicted in Fig. 2. VICTORY has proposed an architecture that consists of three layers. The "Backend Service Layer" provides all the required core functionality for the search and retrieval of MultiPedia objects. It consists of: a P2P Service module, providing a Networking Stack for seamless and efficient content delivery, a Search Engine (SE) Service module that provides the desired Indexing and Search functionality, a Low Level (LL) Feature Extraction Service providing the system with advanced descriptor extraction capabilities, and Remote Rendering Servers that offer the means for a "Remote Rendering Visualization" service [9] when remote rendering visualization is applicable.

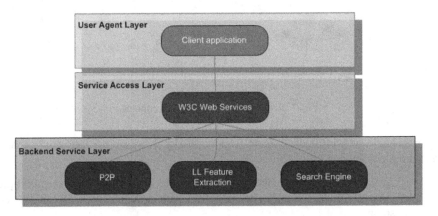

Fig. 2. The VICTORY service-oriented architecture

The "Service Access Layer" plays the key role in our case, by providing all the required functionality offered by the "Backend Service Layer" to the mobile client application in a device and application-independent form. The basic services offered by the VICTORY Mobile Gateway are the following:

- Extraction of low-level features from sketch, 2D and/or 3D objects. For the extraction the file needs to be uploaded to the mobile gateway first. The low level features are then extracted and returned to the mobile device.
- MultiPedia content search. This service is used in the combination with low-level features extraction service. Once the client compiles a search query (including low-level features), this query is sent to the search service which then propagates it to the P2P search engine.
- Retrieval of MultiPedia content from the P2P network. The content that is retrieved from the P2P network on behalf of a mobile device is stored on the gateway and can be retrieved by the mobile device using ordinary HTTP based file transfer.

The Service Access Layer was implemented on the basis of the WS-I Basic Profile Version 1.1 [10] and the Apache Axis2 Framework [11]. All this "wrapped" functionality is finally provided to end users through the "User Agent Layer", which can be implemented as any mobile client application that conforms to the standards defined within the Service Access Layer. Our developed VICTORY PDA application is such a mobile end user agent application.

3 The "VICTORY PDA" Mobile Application

In order to deliver all the functionality derived from the service-oriented approach described above to end users on the move, we developed "VICTORY PDA", a mobile user agent application for PDA devices, based on the .NET Compact Framework [12] v3.5. This application is the result of the integration of several modules, each of them dedicated to provide a specific functionality. These modules have the ability to communicate with each other, in order to provide seamless interaction to the end users. The main building blocks of the VICTORY PDA end user application are:

- A module allowing users to select 2D images or 3D models as input for the search of 3D objects.
- A module enabling users to draw their own sketches and use them as input for the search of 3D models.
- A module enabling the search for 3D models and the retrieval of the results.
- A module enabling users to form queries by using high and low-level features
- A 3D-Visualization module.
- A module enabling the invocation of the appropriate Web Services.

The VICTORY PDA application enables users to connect to any of the VICTORY framework's communities and: select the desired inputs in order to form a search query, submit it to the search engine, retrieve any of the search results and visualize it.

3.1 A "VICTORY PDA" Use Case

Prior to the submission of a search query, the user has the ability to form it, through the User Interface depicted in Fig. 3 (a) and (b). The User Interfaces shown in these figures have the same functionality, however, due to the fact that in these two cases the user has connected the application to different communities, their graphics differ in order to inform the user about the community s/he is connected.

The user can form the query by defining various high-level attributes like a keyword, the maximum allowed file size of the results retrieved, categories, and so on. Furthermore, the user has also the ability to "use a file in search", which means that s/he can either select a 3D model that already exists within the device's storage memory, a 2D image, or a new sketch drawn, get the selected file's low-level descriptors and use them as input for the search query. Fig. 3 (c) shows the UI enabling the user to select an existing 2D image, which can be a photo taken from the PDA device, or in general, any image file stored in it. The user draws the contour of the object depicted that s/he wants to search for (Fig. 3 (c)), and by selecting to "use it in search", the application contacts the Gateway's Web Service responsible for the extraction of Low Level Features, by providing as input the user's selection. When the service finishes the feature extraction procedure, VICTORY PDA retrieves the extracted features and populates the user's query with them. As a result, the user is shown again the UI depicted in Fig. 3 (a) or (b), with the difference that it now also shows the user's selection that will be used for the query (Fig. 3 (d)).

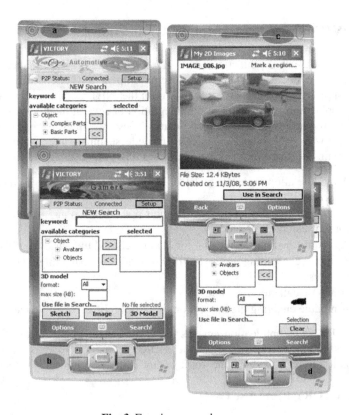

Fig. 3. Forming a search query

The sketch drawing functionality offered to the users is similar to the functionality described above, except from the fact that in this case the user draws on a "blank" (white) background. Regarding the 3D model selection procedure, the user is initially shown a list with the 3D model-files stored in the device. After the selection of a model, the Feature Extraction Gateway Web Service is invoked with the specific model as input; its low level features are retrieved and they finally populate the user's search query in a similar procedure like the one described above.

After the user finishes with the forming of a query, s/he can search for MultiPedia content within the VICTORY network. For this purpose, the PDA application invokes the "MultiPedia content search" Web Service provided by the VICTORY Mobile Gateway, responsible to forward the search request to the VICTORY P2P-network's search engine. As a result, the search engine searches within the network and if appropriate results are found, a list with details over them is returned back to the search web service. Finally, the service responds to the VICTORY PDA application with the list of the results found.

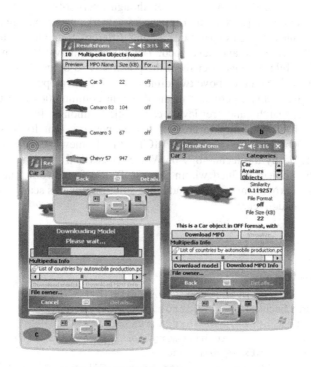

Fig. 4. MultiPedia content retrieval

This list is then presented to the end user by the means of the User Interface depicted in Fig. 4 (a). As a next step, the user can select to view the details of each MultiPedia object found (Fig. 4 (b)), and if desired, s/he can initialize the downloading procedure (Fig. 4 (c)). This involves the invocation of the "retrieval of MultiPedia content from the P2P network" Gateway Web Service. The file is retrieved from the network and stored in the device's storage memory. The user is now able to "visualize" the retrieved model. This functionality involves either the use of a 3D remote rendering-based visualization module, or the use of a "local 3D rendering" application for PDA devices.

A beta version of the system was initially tested on a PDA device with a 400 Mhz processor and 128 MB RAM, connected to VICTORY through a typical 8Mbps ADSL Internet connection. The response time for the feature extraction procedure was found in average around 6 seconds for 2D images, while for 3D models was higher, depending on the size of the 3D model file uploaded to the Gateway. For a relatively small file (22 kB), this response time was around 8 seconds in average. For the search process, the system's response time was found to be about 6 seconds in average.

4 Conclusions

Search and retrieval of networked multimedia content in a seamless fashion is one great challenge for the years to come. Even though internet-based search engines of multimedia content already exist, appropriate frameworks and client applications that will provide seamless search capabilities among different communities and networks have to be further research and developed. The advanced networking and processing capabilities of mobile devices offer them the possibility to be used in a wide area of applications. Focusing on the above two dimensions, our developed mobile user agent provides users the ability to connect to the different communities of a P2P network and utilize its search and retrieval capabilities among diverse repositories. Furthermore, it has the ability to connect to any network that follows the service oriented approach proposed by the VICTORY framework. By utilizing the capabilities offered by the technology of Web Services and by using the concept of the "MultiPedia" object defined within the VICTORY framework, our mobile user agent manages to provide an effective adaptive solution towards seamless multimedia content search and delivery.

References

1. Princeton 3D Model Search Engine,
   `http://shape.cs.princeton.edu/search.html`
2. 3DTrue Search Engine, `http://3dtrue.com/vrml/4.html`
3. ITI Search Engine, `http://3d-search.iti.gr`
4. MeshNose, `http://web.archive.org/web/20030622052315/`,
   `http://www.deepfx.com/meshnose/`
5. Partfinder, `http://www.smap3d.com/en-cad/CAD-partfinder.html`
6. VICTORY project official website,
   `http://www.victory-eu.org:8080/victory`
7. Project JXTA official web page, `https://jxta.dev.java.net/` (accessed on 22/11/2008)
8. Traversat, B., Arora, A., Abdelaziz, M., Duigou, M., Haywood, C., Hugly, J.-C., Pouyoul, E., Yeager, B.: Project JXTA 2.0 Super-Peer Virtual Network (2003)
9. Lamberti, F., Sanna, A.: A Streaming-Based Solution for Remote Visualization of 3D Graphics on Mobile Devices. IEEE Transaction on Visualization and Computer Graphics 13(2), 247–260 (2007)
10. WS-I Basic Profile Version 1.1,
    `http://www.ws-i.org/Profiles/BasicProfile-1.1.html`
11. Apache Axis2 official web page, `http://ws.apache.org/axis2/`
12. .NET Compact Framework,
    `http://msdn.microsoft.com/en-us/netframework/aa497273.aspx`

Ground-Truth-Less Comparison of Selected Content-Based Image Retrieval Measures

Rafał Fraczek[1], Michał Grega[1], Nicolas Liebau[2], Mikołaj Leszczuk[1],
Andree Luedtke[3], Lucjan Janowski[1], and Zdzisław Papir[1]

[1] AGH University of Science and Technology
[2] Technische Universitaet Darmstadt
[3] Universitaet Bremen

Abstract. The paper addresses the issue of finding the best content-based image retrieval measures. First, the authors describe several image descriptors that have been used for image feature extraction. Then, the detailed description of a query by example psycho-physical experiment is presented. The paper concludes with the analysis of the results obtained.

1 Introduction

Several content-based image QbE (Query by Example) techniques have been presented during the last years. While developing real QbE applications, a question arises: which image retrieval method should be applied? The available benchmarks are commonly incomplete in terms of the number of image similarity measures. Furthermore, re-executing a benchmark usually imposes possession of a well-annotated database of images (ground-truth). In this paper the authors present a ground-truth-less comparison of several content-based image retrieval measures. Results of the comparison are applicable for several usage scenarios. Below, one of them is briefly presented: a QbE search system for a Web portal being a gateway to archives of media art.

QbE systems are based, in most cases, on features extracted from the media. Sets of features are generally referred to as "descriptors" and their instances are called "descriptor values". The descriptor values are the meta-data of the media. Some of the descriptor extraction methods are standardised in the MPEG-7 (Moving Picture Experts Group) standard [6].

The experiments described in this paper have been carried out in the context of the European project GAMA (Gateway to Archives of Media Art). The main goal of this project is to give public and multidimensional access to European collections of media art. One of the objectives within the GAMA project is to "provide sophisticated multilingual query performance and implement advanced search functionality". Obviously today we cannot imagine advanced search without QbE functionality. Nevertheless, there is not standard answer to the question what best results mean in QbE, neither for images nor video sequences. Moreover, we cannot find a ground truth since asking the question of the definition of similarity to different users we will get different answers. Therefore, we tried to find out what kind of distance metric is the best in the most general case.

P. Daras and O. Mayora (Eds.): UCMedia 2009, LNICST 40, pp. 101–108, 2010.

The approach to the problem was to perform a set of psycho-physical experiments in order to allow the subjects to vote for the best QbE method. "Best" is here understood as the most satisfying and closest to the users expectations. A set of QbE retrieval methods was designed and implemented and the results were presented to the subjects. The experiment was designed to be as simple and uncomplicated as possible in order to allow it to be performed on inexperienced subjects. The task of the subject was just to choose from a set of available result images the one most similar to the presented query image.

The QbE techniques are nowadays utilised by numerous applications and widely researched. The VICTORY project can be referenced [8] as a good example of advanced research in the area of QbE – in this case focusing on search for 3D objects. As mentioned before, executing an image retrieval benchmark usually imposes possession of an image ground-truth. Most of available databases contain a rather limited number of images (e.g. [7]). Another problem with a ground-truth databases is related to the way the images are described. Some databases contain free-text descriptions only, meaning that there are no straightforward metrics allowing for measuring computational similarity between two annotated images (e.g. TRECVid [10]). Finally, for vast image databases that are annotated by the community, the quality of annotations is moderate (e.g. Flickr). Consequently, the authors decided to carry out an experiment that would not require the possession of a ground-truth database. The image database used in the experiment consists of more than 33.000 images downloaded from the Flickr service with the accompanying user-generated keywords.

There is evidence that some subjective measures such as, for example, AAMRR (Average Normalised Modified Retrieval Rate) [3] coincide linearly with the subjective evaluation results [9]. The authors however decided to create and perform their own subjective psycho-physical experiment in order to avoid any error introduced by objective measures.

The rest of the paper is organised as follows. Section 2 presents the similarity measures included in the subjective experiment. Section 3 describes the methodology of the subjective experiments and the results are presented and discussed in Section 4. Section 5 concludes the paper and gives and insight into the further work on the topic.

2   Similarity Measures

The experiment considered several image retrieval metrics. The predominant group were metrics based on MPEG-7 visual descriptors. Cross-combinations of MPEG-7 descriptors have been considered as well. Furthermore, the Picture-Finder content-based image retrieval algorithm has been applied. Moreover, the authors included a "Tag Metric", specifying the image-to-image distance, based on the degree of overlapping tags. Finally, a virtual "Random Metric" has been used in order to see how the real metrics actually differ from the totally random selections. The metrics (and the corresponding MPEG-7 descriptors, if applicable) have been described in detail below.

2.1   VS (PictureFinder)

VS [5] is a fast image retrieval library software for image-to-image matching. Similarity is measured based on spatial distribution of colour and texture features. It is especially optimised for fast matching within large data-sets.

VS applies a hierarchical grid-based approach with overlapping grid cells on different scales. For every grid cell up to 3 colours (in a quantised 10-bit representation in CIELab[1] colour space) and a texture energy feature are stored in the descriptor.

2.2   Metrics Based on MPEG-7 Visual Descriptors

MPEG-7 is an ISO/IEC standard for multimedia description defining a set of descriptors that are designed to extract specific information from the given content. This description allows for efficient multimedia content indexing and searching. In the experiment, the following MPEG-7 descriptors have been used.

DC (Dominant Colour) addresses the issue of finding major colours in the image. The descriptor quantises all colours present in the image and then the percentage of each quantised colour is calculated correspondingly.

SC (Scalable Colour) describes an image in terms of a colour histogram in HSV (Hue, Saturation, and Value) space. The descriptor representation is scalable in terms of both, bit representation accuracy and bin number. This feature makes the descriptor convenient for image-to-image matching.

CL (Colour Layout) is designed to efficiently represent the spatial colour distribution. The descriptor clusters the image into 64 (8×8) cells and the average colour of each block is derived. Finally, a DCT (Discrete Cosine Transform) is applied. The representation is very compact.

CS (Colour Structure) captures both, colour and structure information. The algorithm retrieves colour structure by analysing all colours in an 8×8 window that slides over the image. In consequence, the descriptor is able to distinguish between two images in which a given colour is present in identical amounts but where the structure of the groups of pixels having that colour is different.

EH (Edge Histogram) represents the spatial distribution of edges present in the image. These are four directional edges (vertical, horizontal, 45°, 135° and one non-directional edge. Then the image is divided into 16 (4×4) blocks and a five-bin histogram for each block is generated.

---

[1] Comission Internationale de l'Eclairage Lab.

2.3 TAG (Tag Metric)

The question "what does a similar image mean?" is difficult and very subjective. Nevertheless, instead of similarity definition one can ask what does a user see in the image. On the basis of such description obtained for two different images the similarity can be approximated since if the descriptions are similar the images should be similar too. Therefore, we decided to use the TAG, i.e. a metric based on a set of tags in order to know if other metrics have similar accuracy.

The TAG is not a perfect one mainly for two reasons. The first reason is the requirement of having the images tagged. Since we used images from the Flickr service our images were accompanied by the user-provided tags. The entered tags are far from perfect but it is almost impossible to obtain a database that is large and correctly described. The second problem is how to compute a distance between two different sets of tags. We decided to use the Jaccard similarity [2] denoted $\mathcal{J}_s(A, B)$, and defined as:

$$\mathcal{J}_s(A, B) = \frac{|A \cap B|}{|A \cup B|} \qquad (1)$$

where, A and B are sets of tags of images a and b respectively and $|A|$ is the cardinality of A set.

Jaccard similarity has an interesting property that will be explained by an example. Let us assume we have three images. The first one a has 10 tags with "tree" tag in it. The second b has only 3 tags with "tree" tag also. We are wondering which of a or b image is closer to c image with only one tag "tree"? Jaccard similarity will show b image as closer since $\mathcal{J}_s(C, B) = 1/3$ and $\mathcal{J}_s(C, A) = 1/10$. The obtained result is correct since in A tree is one of 10 objects and in B one of three.

The TAG is not perfect since synonymous tags change the obtained results and different people can tag different numbers of objects in the same picture. Nevertheless, our main goal was to examine whether the TAG is more accurate than other considered metrics.

3 Description of Psycho-Physical Experiments

We performed tree different experiments on different groups of people and with two different scenarios. All results obtained were very similar. Therefore, we are presenting the final result without a detailed description of each experiment.

The final result presented in this paper has been obtained by analysis of two last experiments. The first experiment had a different scenario and was used to calibrate the user interface. Since the two other experiments where slightly different we are presenting results obtained in the last two experiments.

3.1 General Assumptions

QbE interfaces show some results, i.e. n images that are the most similar to the query image, where n is determined by the user interface since a subject has to be able to see the results. The results order is determined by the comparing metric. We make two assumptions. The first one is that if a subject cannot find a (subjectively) similar image in the first n of them than the metric results are not correct. The second one is that $n = 10$, chosen based on our experience with existing QbE interfaces. Therefore, as a metric result we considered the set of the 10 most similar images. An important fact is that we did not consider if a picture was first or tenth, the only important property was to be in the first 10.

We analysed 12 different metrics:

1. 5 metrics based on MPEG-7 descriptors, introduced in Section 2.2
2. VS described in Section 2.1
3. TAG described in Section 2.3
4. Sum of ranks[2] obtained for EH and each MPEG-7-based metric (without EH) (4 different combinations)
5. Sum of logarithm of ranks obtained for each MPEG-7-based metric

The 13th metric was a random metric (i.e. a random image). We add it just to check subjects' reliability. On the other hand, we present only 7 images at a time since such an interface was the best in terms of the visual layout at all screen resolutions. Therefore, a subject can choose one of 7 different images (see Section 3.2). Each presented image is a result of two draws. First, the metric is drawn (for example the EH-based one) than from 10 images marked as 10 the most similar images one is drawn (in this case it is a random image from 10 the most similar images obtained for the EH-based metric). Since we considered 13 different metrics and each time we show 7 imagines not all of them were visible at once. Nevertheless, more than 2500 queries where answered therefore all possible combinations where properly represented.

Note that it is possible that there is no similar image in the set of 7 presented images. Therefore, we added an image that a subject can click if he/she cannot find a similar image among the presented. We added it as an image to make this answer identical to a similar image answer.

3.2 Experiment Setup

The user interface was implemented as a Web page in a form of a PHP (PHP: Hypertext Preprocessor) script. This allows for the execution of the experiment outside of the laboratory via Internet and gave access to a larger and more varied in the terms of age, occupation and nationality, group of subjects. The subjects were given a URL (Uniform Resource Locator) which directed them to the experiment.

3.3 Experiment Execution

The experiment was performed at two stages. First the subject was given the instructions and the general purpose of the experiment was explained. Information about the subject was collected, such as age, gender and nickname. Also

---

[2] Rank is a metric giving 1 to the most similar image, 2 to the second one etc.

Fig. 1. The web-based interface for the psycho-physical experiments

a 6-step colour blindness test was performed in order to identify colour-blind subjects.

The second stage was the experiment itself. 7 randomly chosen images were presented to the subject (Fig. 1 and Section 3.1). The subject's task was to choose from the small images the one most similar to the middle, large one. Subjects could also chose "no similarity" answer (see Fig. 1). Step five was repeated 300 times, but the subject was free to end the experiment at any time.

The middle image was the query and the surrounding images were the query results obtained with the different QbE techniques. So, the subjects were performing a vote for the QbE method that was most satisfying.

4   Analysis of Results

As the experiment could be terminated by a subject any time, we obtained a different number of answers from each subject. Therefore, for each subject a distribution of answers (i.e. probability of choosing any metric) was computed and analysed. Additionally we removed all subjects answering less than 50 queries. We collected 31 results and computed confidence intervals with $\alpha = 0.05$ [1]. The results obtained are shown in Figure 2.

The probability of the "no similarity" answer was the highest and reached 31%. Note that if lots of answers are "no similarity" probably the database was too small to have a similar picture. Since we are interested in comparing different descriptors and metrics quality and not the database quality we are not showing this value in the plot.

We considered two additional metrics. The first one is random metric i.e. a random image was shown. The random metric enables to conclude if the other metrics are better than random. The second one is the TAG which enables to

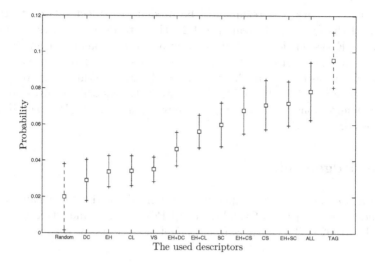

Fig. 2. The obtained results with the confidence intervals, solid lines — descriptors' combinations; dashed lines — the random metric and the TAG

compare MPEG-7-based metrics with descriptive metric used in classic search systems.

We analysed all MPEG-7-based metrics with the random metric and TAG by t-test [1] with $\alpha = 0.05$. The results show that metrics based on EH, DC, CL descriptors as well as the metric based on VS, are not statistically different from the randomly chosen image. On the other hand, only one metric (based on all MPEG-7 descriptors) is not statistically different from the TAG. Therefore we chose it as the best and implemented in GAMA project. However, CS is just slightly worse. Moreover, the CS value is computed on the basis of just one descriptor and thus computationally cheaper.

5   Conclusions

The paper addressed the issue of finding the best content-based image retrieval measures. First, the authors described several image descriptors that have been used for image feature extraction. Then, a detailed description of query by example psycho-physical experiment has been presented. The results show that the metric based on MPEG-7 CS Descriptor is the most commonly chosen among metrics based on single descriptors. Nevertheless, it is outperformed by the metric being a combination of various MPEG-7 descriptors as well as by manual tagging.

In the GAMA project the metric based on CS descriptor will be used as a pre-filter for the image and video QbE systems. This will allow for fast and accurate search in the vast repository of media art.

As a further work, the results will be used in a project related to the application of the QbE media search in P2P (Peer-to-Peer) overlays. In order to create a QbE system for the P2P overlay a decision has to be made on the selection of a descriptor or a set of the descriptors. It is planned then to implement the QbE mechanism in the unstructured and structured P2P overlays on the example of the Gnutella and CAN (Content Addressable Network) overlays. The implementations will be done in the environment of the PeerfactSim.KOM simulator [4].

Acknowledgements

The work presented in this paper was supported by the European Commission, under the projects: "CONTENT" (FP6-0384239) and "GAMA" (ECP-2006-DILI-510029). The authors thank panel of subjects for their efforts.

References

1. Nist/sematech e-handbook of statistical methods
2. Arasu, A., Ganti, V., Kaushik, R.: Efficient exact set-similarity joins. In: Proceedings of VLDB 2006 (2006)
3. Manjunath, B.S., Salembier, P., Sikora, T.: Introduction to MPEG-7: Multimedia Content Description Interface. John Wiley and Sons Ltd., Chichester (2002)
4. Graffi, K., Kovacevic, A., Steinmetz, R.: Towards an information and efficiency management architecture for peer-to-peer systems based on structured overlays. Technical report, Multimedia Communications Lab KOM, Technische Universitaet Darmstadt (2008)
5. Hermes, T., Miene, A., Herzog, O.: Graphical Search for Images by PictureFinder. Multimedia Tools and Applications. Special Issue on Multimedia Retrieval Algorithmics (2005)
6. ISO/IEC. Information technology – multimedia content description interface. ISO/IEC 15938
7. Li, Y.: Object and concept recognition for content-based image retrieval. PhD thesis, University of Washington (2005)
8. Mademlis, A., Daras, P., Tzovaras, D., Strintzis, M.G.: 3d volume watermarking using 3d krawtchouk moments. In: VISAPP (1), pp. 280–283 (2007)
9. Ndjiki-Nya, P., Restat, J., Meiers, T., Ohm, J.R., Seyferth, A., Sniehotta, R.: Subjective evaluation of the MPEG-7 retrieval accuracy measure (ANMRR). Technical report, ISO/ WG11 MPEG Meeting, 200
10. Smeaton, A.F., Over, P., Kraaij, W.: Evaluation campaigns and trecvid. In: MIR 2006: Proceedings of the 8th ACM international workshop on Multimedia information retrieval (2006)

Exploiting Complementary Resources for Cross-Discipline Multimedia Indexing and Retrieval

Virginia Fernandez, Krishna Chandramouli, and Ebroul Izquierdo

Multimedia and Vision Research Group,
School of Electronic Engineering and Computer Science,
Queen Mary, University of London, Mile End Road, London, E1 4NS, UK
{firstname.lastname}@elec.qmul.ac.uk

Abstract. In recent times, the exponential growth of multimedia retrieval techniques has stimulated interest in the application of these techniques to other alien disciplines. Addressing the challenges raised by such cross-discpline multimedia retrieval engines, in this paper we present a multi-user framework in which complementary resources are exploited to model visual semantics expressed by users. The cross-discpline areas include history of technology and news archives. In the framework presented the query terms generated by historians are first analysed and the extraction of corresponding complementary resources are used to index the multimedia news archives. The experimental evaluation is presented on three semantic queries namely wind mills, solar energy and tidal energy.

Keywords: Cross-discipline retrieval, multimedia indexing, complementary resource analysis, particle swarm optimisation.

1 Introduction

With the advances in information retrieval (IR) techniques the application of multimedia information retrieval (MIR) tools for alien domains has attracted researchers from various domains. Addressing this challenge, the doman of application considered in this paper is "history of technology'. In this field, the users typically also referred to as historians are required to study the evolution of technology. Typical examples include historians composing a documentary regarding the evolution of windmills from 1960's to 2000's. In particular the content for such reports are searched through news archives. Therefore, creating a critical necessity for semantic indexing of news archives with concepts obtained from modeling the domain of history of technology. In this paper, three concepts are chosen to be "wind mill", "solar energy" and "tidal energy". Understandably the news archives content includes multimodal sources which includes video, textual descriptions and images associated with the depiction of different concepts.

P. Daras and O. Mayora (Eds.): UCMedia 2009, LNICST 40, pp. 109–116, 2010.

Although the prototype developed will process all the information, the analysis of video data is considered out of scope of this paper. Therefore it is assumed that, the video data is analysed to extract keyframes and is only available for further processing.

For semantic indexing of the keyframes, it is critical to have training data to construct visual models for individual concepts. However, due to the limited availability of content, complementary resources are exploited to enable historians construct visual models which are used further for enhancing the performing the retrieval system. In order to build efficient retrieval engines different machine learning techniques have been studied and in particular research in developing new kernel methods for SVM [2] classification and ranking and biologically inspired systems [3] have shown to improve the performance of the retrieval systems.

The remainder of the paper is organised as follows. In Section 2, an overview of the proposed framework is presented, followed by the complementary resource analysis in 3. In Section 4, a brief overview of the visual indexing classifier is presented. Section 5 presents the preliminary experimental evaluation of the framework followed by conclusions and future work in 6.

2   Proposed Framework

In Fig. 1, an overview of the multimedia indexing framework exploiting complementary resource is presented. The framework consists of the user (in our case a historian[1]) who provides a query to the system (for example, wind mill) and the query is processed to extract corresponding complementary resources from online. Thus extracted information is further processed by extracting MPEG - 7 low-level features [4]. On the other hand, the database of news archives is considered to contain video material, which are already pre-processed by shot boundary detection module and key frame extraction module. The complementary resources from the web are used to automatically index the new archives and also, the system implements a Relevance Feedback mechanism through which the user correct the automatically corrected indexing schemes.

In Fig. 2, a multi-user environment is considered in which a set of individual users assume ownership of the individual research databases. The database schemes closely follow the MPEG - 7 content access definitions and in addition also, provides copyright protection for the ownership historians. The system infrastructure is configured in such a way that, it will enable a new query to be searched on the proprietary databases. However, the access to the content is limited only to the title. If the user prefers to have access to the content, then the user can make special request to the individual owners for content. Alternatively, if the owner has assigned the content to be public, then the content is freely available within the network.

---

[1] The term user and historian are interchangeably used in this paper.

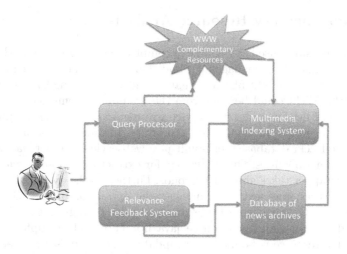

Fig. 1. Complementary Resource Analysis and Multimedia Indexing Framework

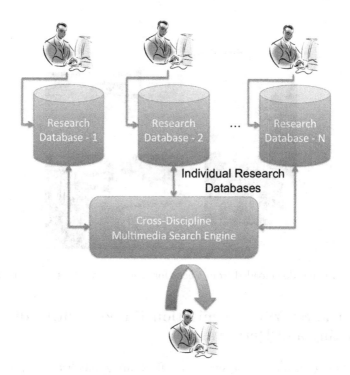

Fig. 2. Multi-User Relevance Feedback Framework from distributed Databases

3   Complementary Resource Analysis

The Flickr[2] website is used as a source of complementary resources in this paper. Flickr is an online photo management and sharing application which allows the user to upload photos, edit photos, organise and share among friends. The popularity of the Flickr could be seen from the 3.2 million items geotagged in one month[3]. Also, Flickr provides webservice wrappers for accessing the photo content directly using an authentication procedure. In order to obtain customised access to the Flickr database a pre-built java webservice wrapper was used with customised functionalities. On an average for extracting 100 images for a given query, the wrapper takes about 10 seconds. Further to the availability of direct access to pictures, the next steps involves the extraction of MPEG - 7 visual features namely Colour Layout Descriptor and Edge Histogram Descriptor. This is achieved by using an open source java library namely Caliph and Emir[4]. As the online resource is bound to get updated, the pictures and therefore the corresponding features are not temporarily stored in any internal databases. In Fig. 3 and Fig. 4, couple of examples for different queries are presented.

Fig. 3. Images downloaded from Flickr for Query: windMill (case sensitive)

4   Particle Swarm Optimisation Based Multimedia Indexing and Retrieval

In the PSO algorithm [5], the birds in a flock are symbolically represented as particles. These particles are considered to be "flying" through the problem space searching for optimal solution [6]. A particle's location in the multidimensional

---

[2] http://www.flickr.com/

[3] As of 2009-07-14.

[4] http://sourceforge.net/projects/caliph-emir/

Query: wind mill

Fig. 4. Images downloaded from Flickr for Query: wind mill (case sensitive)

problem space represents one solution for the problem. When a particle moves to a new location, a different solution to the problem is generated. This solution is evaluated by a fitness function that provides a quantitative value of the solution's utility. The velocity and direction of each particle moving along each dimension of the problem space will be altered with each generation of movement. The movement of particles can be influenced in one of two ways. The first is called is the social behavior in which particle gets attracted to the groups center, i.e. following the group, either updating/foregoing the personal best solution. The second is called the cognitive behavior. In this modeling, the particle follows the cognitive experience via personal best solution foregoing the group solution. A more detailed implementation of the RF System is presented in [7].

The PSO based retrieval engine considers as input the MPEG - 7 feature vector extracted from the complementary resources and based on the visual training model provided by the users, the keyframes from the news archives are indexed and retrieved. The ranked list of images with and without the use of the complementary resources are evaluated in the following Section 5.

5   Experimental Results

Flickr Complementary Resource Analysis

The user query search was performed with the open source JAVA api's and for each query the top 100 results are manually annotated for the evaluation. The evaluation is to provide a measure for the performance of Flickr retrieval engine based on textual queries. As discussed previously the results obtained from the Flickr data base are sensitive to query keywords and therefore carefully chosen query terms are used to retrieve images from the Flickr. In Table 1 performance of the Flickr retrieval engine is presented.

Table 1. Flickr text based retrieval

Query	Retrieval Accuracy (%)
Wind Mill	67
Solar Energy	35
Tidal Energy	44

The results are based on the manual annotation performed on the top 100 documents returned from the Flickr database.

Relevance Feedback Performance measure using Complementary resources

The MPEG - 7 visual descriptors namely Colour Layout Descriptor and Edge Histogram Descriptor are extracted for images from both complementary resources and keyframes. The PSO optimisation implementation includes a combination of cognitive and social behaviour.

In total, 3 different users took part in the user evaluation system and the average performance of the system is presented in Fig. 5 and Fig. 6 with and without the use of complementary resources is presented. Each user interacted with the system As shown in the figures, in both cases PSO based image retrieval provides better results than the SOM and SVM algorithms.

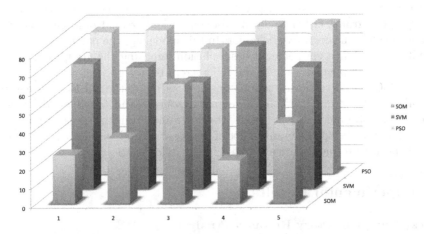

Fig. 5. Average accuracy of relevance feedback results from multiple databases for the queries: Wind Mill, Solar Energy, Tidal Energy without the use of complementary resources

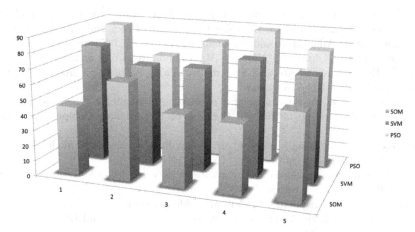

Fig. 6. Average accuracy of relevance feedback results from multiple databases for the queries: Wind Mill, Solar Energy, Tidal Energy with the use of complementary resources

6 Conclusions and Future Work

In this paper, a framework for multimedia indexing is presented using complementary resources. The use of complementary resources are two fold, (i) for building appropriate visual models for the query and (ii) propagating the visual information for indexing keyframes extracted from the video. The evaluation of the relevance feedback algorithm in a multi-user environment is also presented. The experimental results indicate the benefits of creating visual models and propagating the models to index appropriate keyframes. The future work will investigate the possibilities of developing the proposed framework to include P2P network connection. In particular special emphasis will be provided to develop a plugin for Tribler P2P framework. Other possible future work could include the use of SIFT features for image retrieval instead of MPEG - 7 visual features.

Acknowledgement

The research leading to this publication has been partially funded by European Commission under the IST research STREP Papyrus FP7-215874 of the 7th framework programme.

References

1. Goodrum, A.A.: Image information retrieval: An overview of current research. Special Issue of Information Science Research, Information science (2000)
2. Djordjevic, D., Izquierdo, E.: An object- and user- driven system for semantic-based image annotation and retrieval. IEEE Trans. on Circuits and Systems for Video Technology 17(3), 313–323 (2007)

3. Chandramouli, K.: Image classification using self organising feature maps and particle swarm optimisation. In: Doctoral Consortium, Proc. 2nd Int'lWorkshop on Semantic Media Adaptation and Personalization (SMAP 2007), pp. 212–216 (2007)
4. Manjunath, B.S., Ohm, J.R., Vinod, V.V., Yamada, A.: Color and texture descriptors. IEEE Trans. Circuits and Systems for Video Technology, Special Issue on MPEG - 7, 11(6), 703–715 (2001)
5. Eberhart, R., Shi, Y.: Tracking and optimizing dynamic systems with particle swarms. In: Proceedings of the 2001 Congress on Evolutionary Computation, vol. 1 (2001)
6. Reynolds, C.: Flocks, herds and schools: a distributed behavioural model. In: Computer Graphics, pp. 25–34 (1987)
7. Chandramouli, K., Izquierdo, E.: Image Retrieval using Particle Swarm Optimization. CRC Press, Boca Raton (2008) (accepted)
8. Chandramouli, K., Izquierdo, E.: Visual highlight extraction using particle swarm optimisation. In: Latin-American Conference on Networked and Electronic Media (to be published, 2009)

UCMedia 2009

Session 5: Interactive TV

Content Personalization System Based on User Profiling for Mobile Broadcasting Television

Silvia Uribe[1], Iago Fernández-Cedrón[1], Federico Álvarez[1],
José Manuel Menéndez[1], and José Luis Núñez[2]

[1] E.T.S.I. Telecomunicación, U.P.M, Avda Complutense s/n, 28040 Madrid, Spain
[2] Zentym Solutions, Madrid, Spain
{sum,iff,fag,jmm}@gatv.ssr.upm.es,
jln@zentym.com

Abstract. Content personalization is a key element in the media content environment since it contributes to improve the user's experience. In this paper we present a novel intelligent content personalization system for content flow personalization over mobile broadcasting networks and terminals based on user profiling and clustering taking advantage of the consumption data obtained from the user and the information given by user's tastes and behavior.

Keywords: content personalization, mobile television, clustering, profile segmentation, preference, DVB-H, interactivity.

1 Introduction

The digitalization of the media environment in broadcasting networks has opened new possibilities and at the same time, it presents new challenges to the actors involved in the value chain. In fact, consumers expect new incentives in order to contribute to this progress, and for this reason it is needed the development of innovative services that help them to satisfy their needs and to fulfill their expectations.

In this paper we present a novel intelligent content personalization system for content flow personalization over mobile broadcasting networks and terminals based on user profiling and clustering taking advantage of the consumption data obtained from the user and the information given by user's tastes and behavior.

One of the most common ways to improve this experience is by developing content recommendation systems based on user's preferences and behavior. In fact, there are many techniques to implement these recommendations engines: user clustering and profile classification, which are not only used in media environment but also in the web [1] or television applications [2]. But the main difference between these applications and the intelligent system for content personalization presented in this paper is that our system provides a final, clear and transparent solution for the users, by allowing them to receive the content according to their profiles, and not making necessary them to choose the piece of content to be displayed. In this way, since it is developed for mobile broadcasting television, they do not have to choose between several offered options, thus obtaining a more efficient system and real time results.

The rest of the paper is organized as follows. Section 2 outlines the state of the art of recommendation and personalization systems and the techniques used on them.

P. Daras and O. Mayora (Eds.): UCMedia 2009, LNICST 40, pp. 119–126, 2010.

Next section shows the users' clustering design for the solution in this paper. Section 4 presents the general structure of the system and the different modules. The system operation is shown on section 5 and in section 6 we present the conclusions.

2 Overview of the Prior Art in Recommendation and Personalization Systems in Media Environments

One of the most efficient solutions to this problem is the so called recommendation system. In fact, as it was explained in [3], the appearance of new television standards such as DVB-T or DVB-S increased the number of programmes and channels and one effective solution is helping them to find out their favorite contents by giving them several recommendation based on their preferences and tastes. According to this work, these systems are typically composed of four modules:

- User profiling module, which is in charge of generate the profiles of users according to the existing information.
- Program modeling module, which is in charge of extract the content data to classify them.
- Collaborative filtering/content based module: this module creates user's groups by finding out user's neighbors with similar preferences or behaviors or chooses the content to recommend by studying its characteristics and their similarity to user profile. In [4] authors show a wide classification of the existing recommendation techniques.
- Recommendation module: the final module which match content and users.

Within recommendation systems, one of the most important research areas is the content and user activity modeling, since there are many papers focused on the content modeling to allow the recommendation process, mainly based on semantic process of content metadata,, as it is explained in [5] and [6].

But content recommendation systems are also applied not only to television, but also to other environments like web browsing related to media content. In [7], authors present a system that helps users to find out media contents while browsing the Internet. In this case, the content recommended lists are created by analyzing content metadata and users' feedback.

As we have just seen, content recommendation systems are innovative solutions to integrate users' needs and preferences in the media chain. But these solutions require the users to finally make the decision of which content is going to be selected, or in other words, they are non transparent solutions for the users because they need an action by the user. For this reason, the development of content personalization system represents an evolution which contributes to obtain more efficient solutions for real time applications.

Moreover, personalization system can focus on different parts. As it is explained in [8], a structure personalization is needed in order to allow users to access the same content in different devices with different capabilities such as PC, mobile devices and so on. But as well as the structure personalization, content personalization itself represent other way of providing personal flow to the user. In [9], authors focus on the personalization of interactive video content in sports events, but the system presents in this paper does not only focus in a specific kind of events, and it is initially designed to be applied in mobile television, although it could be extended to other environments like other digital television standards and IPTV.

3 Design of a Content Personalization System for Mobile Television: The User Profiling and Content Association

The aim of this system is to improve the user experience by providing targeted content to users according to their preferences, in an automated and real- time way.

User preferences, that is, the center of this system, can be obtained by many ways, as it is explained in [10], where they are predicted by considering user's conformity, user's context and finally user's behavior.

In our system, according to both the application architecture and the available information, the most efficient and effective way to obtain the user' clustering is by the generation of different users' profiles, according to the declared preferences and to the media consumption, and cluster them according to pre-assigned groups.

Given the characteristics of the system, and in order to let the designer of the system to establish threshold values for the belongingness to a cluster of a user we use the fuzzy c-means algorithm. Besides, this solution makes the matching between users and content easier, because advertiser and content provider generate concrete users' classes with specific associated contents.

In this case, the number of groups or clusters (with their centroids) is set a priori.

The initialization of the clustering method is composed of 5 steps, according to the final desired classification:

1. First of all, users have to complete a set of questions about their habits, interests and preferences. Each question is designed to have four different answers, each of them with a different score.
2. Then, we collect the information about media consumption to elaborate a user profile by mixing the result from the declared interests and the result from the measured consumption.
3. We apply the clustering algorithm to obtain the centroids.
4. Once these centroids are established, we keep them fixed and we assign the users to each of the centroids.
5. If a new user joins the system, he/she will be assigned to each of the clusters. When the number of new users reaches the update threshold (UT) then we will perform again the step number 3.

Mathematically, the process from 1 to 5 is explained as follows:

In the first step, 1, we define a set of users $U = \{U_1, U_2, ..., U_n\}$, holding a specific profile U_P. Then we create the function to model the interests and preferences of a user U according to the number of content categories N:

$$U_{Pi} = \sum_{n=1}^{N} L(U_x)C_n \tag{1}$$

Where $L(U)$ is the declared interest in a scale of 0 to 1.

Then we can model the media consumed (M) according to the total time consumed (T) per content category (C_n) and the number of users U. The result is adapted to a scale from 0 to 1, to be compared with the U_{pi}

$$U_{Si} = \sum_{n=1}^{N} M_A(U_i|(T,C_n)) \tag{2}$$

In order to mix both measures, it is needed to establish a weight for each of the components, which can be established by the designer of the system. Then, we compute the fuzzy c-means algorithm, to obtain the centroids and the membership values of the different users to assign them to a cluster. The clustering (in the step 3) is computed through the minimization of the objective function:

$$J_m = \sum_{i=1}^{N} \sum_{j=1}^{C} u_{ij}^m \left\| x_i - c_j \right\|^2 , 1 \leq m \leq \infty \tag{3}$$

Where:

- $m > 1, m \in \mathfrak{R}$;
- u_{ij} is the degree of membership of x_i in the cluster j;
- x_i the *i-th* term of the n-dimensional input data;
- c_j the n-dimension centre of the cluster
- and $\| \cdot \|$ is any norm measuring the similarity between the input data and the centre.

The process is iterative and the optimization of the objective function J is done updating the membership (u_{ij}) and the cluster centres (c_j) as follows:

$$u_{ij} = \frac{1}{\sum_{k=1}^{C} \left(\frac{\left\| x_i - c_j \right\|}{\left\| x_i - c_k \right\|} \right)^{\frac{2}{m-1}}} \tag{4}$$

$$c_j = \frac{\sum_{i=1}^{N} u_{ij}^m x_i}{\sum_{i=1}^{N} u_{ij}^m} \tag{5}$$

The iteration stops when $\max_{ij} \left\{ \left| u_{ij}^{(k+1)} - u_{ij}(k) \right| \right\} < \varepsilon$ where ε is a termination criteria between 0 and 1, with iteration step k.

According to the system structure, content providers use the server module on the server side not only for uploading new content, but also for associating each one of these contents to the different existing users' classes. Thanks to this solution, there is no need of an automatic association or even a content modeling process in order to extract content characteristics, making the system operation simpler.

4   Content Personalization over Mobile Television: System Architecture

A content personalization system over mobile television does not only involve the final users and their preferences, but also the other actors in the media value chain. As it is shown in Fig. 1, in this system there are many components that have to work together in order to obtain the best and most efficient result.

The proposed system is divided into three different parts according to their functionalities: first of all, the broadcaster, which is in charge of providing the content and giving the timestamps of the next change in order to make the swap between common content and personalized one at the right time; second, the personalized content server, which is in charge of doing the assignment and management of the personal content, based on the user clustering and the user-content assignment; and finally, the user side, the user device, which must present the given information and it also allows the bidirectional communication via return channel.

Next subsections explain the functionalities of each architecture module (Fig. 2).

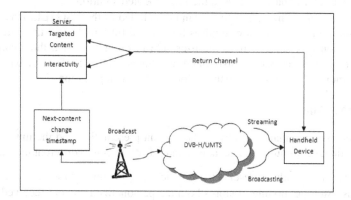

Fig. 1. System architecture

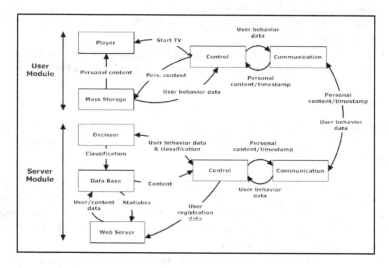

Fig. 2. System modules' interaction

4.1 Broadcaster

Broadcasters are in charge of providing the content flow. In order to replace the common content with the personal one at the right time, a timestamp of commercial breaks is needed. Although there are different algorithms to detect spots in media flows, like the one shown in [12], we consider that these solutions present a high process load for mobile devices, and that is why we propose the broadcaster to include this information, which can be done in two different ways:

- On one side, broadcasters can report to the users the time left to the next break via the return channel. This is the implemented solution.
- On other side, this information can be included in the ESG (Electronic Service Guide), but this solution requires both a real-time update of the guide as well as a constant scan of it in the user side in order to notice the changes. This method is processor intensive and consumes a huge amount of battery, so at the moment is not recommended for a real time application.

4.2 User Module

This module has two main objectives: the content presentation and the implementation of a bidirectional communication channel, and it is divided into different modules:

- Player: this module presents the broadcasted content and the personalized one.
- Mass Storage module: it stores both the personalized content received from the server and the user' behavior data, which can be used to update users' profiles.
- Control module: it is in charge of presenting user's forms for the personalization and registration processes. Once users are logged in, this module starts the TV presentation, asks for the next commercial break and captures users' interactions.
- Communications module: it is in charge of receiving the personal content, and sending back the collected data to the server via an UDP socket.

4.3 Server Module

The main function of this module is making the user clustering and segmentation in order to relate each user's group to the content. This module is divided into the following modules:

- Web server: in this server broadcasters and content providers have the tools to manage the system, by controlling the users' profiles (generation and update) according to the existing information, by creating new commercial campaigns and specifying the target group for them. In this server, broadcasters also find the users' consumptions statistics.
- Media Asset Decision module: based on the method explained before, this module associates each user to the correct group and then selects the tailored content for them. This module has the control of the personalization, and can update the users' profiles.
- Communications module: this module controls the data flow between the server and the user.

- Control module: this module allows the data flow inside the server by giving it the proper format.
- Database: it stores the personal profiles, the data collected from user's behavior, as well as the content given by the content providers.

5 Content Personalization over Mobile Television: System Operation

Based on the structure of the system, its operation (shown of Fig. 3), can be divided into six steps:

1. Clients log in the system, by sending users' connection data (user&password), in order to be identified in the system. Once the server cross-check this information on the database, the user is connected and starts watching TV. If it is the first time of the user in the system and he has completed the first registration form to minimize the so called 'cold start' effect, server receives these data and performs the user's segmentation according to the clustering method designed.
2. Once connected, client device asks the server for both the exact time of the next commercial break, and the personalized content to show.
3. The actions performed by the users are monitored in order to update user's profile. These data are stored on the client side until they are sent to the server side via the return channel.

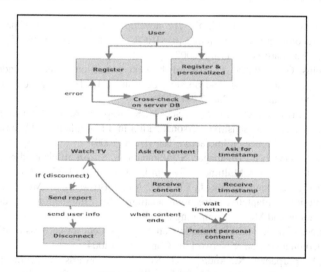

Fig. 3. System operation

4. Once this information is received, it is processed to update the user's personal profile and perform the most efficient personalization.
5. According to the exact time received about the next commercial break, the presentation module on the user side swaps between the common content and

the previous stored personalized one, without any action by the user. Later, when the personalization ends, the system returns to the common media flow.

6. Finally, in order to log out the system, users have to send again their connection information.

6 Conclusions and Future Work

We have presented an efficient system for content personalization. It is based on user clustering according to their preferences and behavior, which means that user's environments are the center of the development. Furthermore, this system represents a useful solution for costumer segmentation and personal advertising, which can be applied in other environments. As it has been explained, user's segmentation includes two main steps: the first one where each user is classified in a group according to his answers to the proposed set of questions, and the second using their content consumption historic data. The system has been extensively tested and is in prototyping by the time of this paper submission.

References

1. Phatak, D., Mulvaney, R.: Clustering for Personalized Mobile Web Usage. In: International Conference on Fuzzy Systems (2002)
2. Virvou, M., Savvopoulos, A.: An intelligent TV shopping Application that provides Recommendation. In: 19th International Conference on Tools with Artificial Intelligence (2007)
3. Xu, J., Zhang, L.-J., Lu, H., Li, Y.: The development and prospect of personalized TV program recommendation system. In: IEEE Fourth International Symposium on Multimedia Software Engineering (2002)
4. Adomavicius, G., Tuzhilin, A.: Toward the next generation of recommender systems: a survey of the state-of-the-art and possible extensions. IEEE Transactions on Knowledge and Data Engineering 17(6), 734–749 (2005)
5. Bellekens, P., Houben, G., Aroyo, L., Schaap, K., Kaptein, A.: User model elicitation and enrichment for context- sensitive personalization in a multiplatform TV environment. In: 7th European Conference on Interactive Television (2009)
6. Velusamy, S., Gopal, L., Bhatnagar, S., Varadarajan, S.: An efficient ad recommendation system for TV programs. Multimedia System Journal 14(2), 73–87 (2008)
7. Liu, Y., Yang, Z., Deng, X., Bu, J., Chen, C.: Media Browsing for Mobile Devices based on Resolution Adaptative Recommendation. In: International Conference on Communcations and Mobile Computing (2009)
8. Yin, X., Lee, W.S., Tan, Z.: Personalization of web content for wireless mobile device. In: Wireless Communication & Networking Conference (2004)
9. Mylonas, P., Karpouzis, K., Andreou, G., Kollias, S.: Towards an integrated personalized interactive video environment. In: IEEE Sixth International Symposium on Multimedia Software Engineering (2004)
10. Boutemedjet, S., Ziou, D.: A graphical Model for Context-Aware Visual Content Recommendation. IEEE Transactions on Multimedia 10(1) (January 2008)
11. Jain, A.K., Murty, M.N.: Data clustering, a review. ACM Computing Surveys (1999)
12. Covell, M., Baluja, S., Fink, M.: Detecting Ads in Video Streams Using Acoustic and Visual Cues. Computer 39(12), 135–137 (2006)

WiMAX TV: Possibilities and Challenges

Omneya Issa, Wei Li, and Hong Liu

Communications Research Centre,
3701 Carling Ave., Box 11490, Station H, Ottawa, ON, Canada
{omneya.issa,wei.li,hong.liu,philippe.bonneau,
simon.perras}@crc.gc.ca

Abstract. WiMAX is an emerging wireless access network offering high data rates and good coverage range. This makes it an appealing last-mile delivery network for IPTV and video applications. However, TV/video delivery must be designed with respect to WiMAX characteristics in different environmental conditions. This paper studies the possibility of providing TV and video over the downlink and uplink of a Fixed WiMAX network without compromising the user perceived video quality. Measurements were taken using professional IPTV and video streaming equipments on commercial IEEE 802.16d equipment. Analysis was done for different video settings and network configurations. A key outcome of this analysis was the feasibility of TV/video delivery over WiMAX with good quality of user experience provided that system limitations are respected.

Keywords: WiMAX, IPTV, High-Definition and Standard-Definition Video Quality Assessment, Quality of Experience (QoE).

1 Introduction

Broadband wireless access has undergone a fundamental change in recent years. WiMAX is a typical example of an emerging wireless access system. At a fraction of the costs of wired access networks, it is currently being deployed across the world; for example, developing countries deploy it as their main network infrastructure while more developed countries exploit it as an alternative to cable and DSL lines in rural and underserved areas.

There are two types of WiMAX systems called Fixed and Mobile. The Fixed WiMAX (IEEE 802.16d) [1] provides point-to-point links to stationary and nomadic (with limited mobility) users. The Mobile WiMAX (IEEE 802.16e) offers full mobile cellular type access. The fixed type represents most of nowadays deployments and is the one studied in this paper.

While WiMAX is being increasingly deployed, the delivery of high resolution video over IP network (i.e. SD/HD IPTV) is becoming a reality thanks to advanced compression technologies. The compression efficiency, in addition to the affordable broadband access and high data rates offered by WiMAX, may make different scenarios of TV applications possible. One featured application is the delivery of IPTV over WiMAX downlinks. Another one might be video live broadcast or video surveillance on the uplink. Figure 1 shows an overview of these scenarios.

P. Daras and O. Mayora (Eds.): UCMedia 2009, LNICST 40, pp. 127–136, 2010.
© Institute for Computer Sciences, Social-Informatics and Telecommunications Engineering 2010

Little or no published material was found on the expected performance and quality of TV and video applications over real WiMAX links. Some studies examined the possibility of delivering H.264 scalable video with small resolution over simulated WiMAX links, such as in [2,3]. Others were limited to analyzing the protocol structure in case of video/IPTV delivery [4-6]. Few papers reported results from trials on real WiMAX links; however, they were limited to throughput and signal strength measurements [7-10].

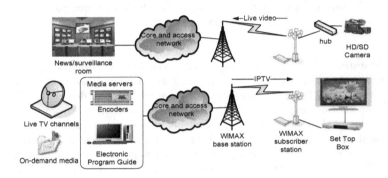

Fig. 1. Scenarios of TV applications over WiMAX

Therefore, there is still work to be done on understanding and analyzing how this broadband wireless access technology can accommodate commercial video and IPTV services, especially when taking advantage of the compression efficiency offered by H.264. The focus of this paper is to study the feasibility of exploiting this technology in different scenarios involving high and standard definition video and TV applications.

The paper is organized as follows. Section 2 describes the testbed design including video and WiMAX network settings. Section 3 covers the characterization of the experimented network in terms of QoS parameters (e.g. packet loss, delay and jitter). Results for downlink and uplink scenarios are analyzed in sections 4 and 5 respectively. Discussion of results and a summary of expected quality and recommended scenarios are presented in section 6. Finally section 7 concludes the paper and gives directions for future work.

2 System Design

A testbed was developed for assessing video quality over WiMAX link in different conditions. The testbed, as shown in Fig. 2, had two main components: an IPTV section and a WiMAX network. The IPTV component consisted of a video server feeding raw video to a professional live encoder. The encoded video is transmitted via the WiMAX network to a professional decoder, emulating a customer set-top box. The decoded video is recorded and stored for post analysis.

The WiMAX network consisted of a base station (BS) and a Customer Premises Equipment (CPE) connected by attenuators, emulating a fixed over-the-air media. The tests were basically done with one subscriber (one CPE) in order to assess single user quality of experience.

WiMAX Network: A real 3.5 GHz Fixed WiMAX network was provided by the WISELAB group [11], within CRC, in order to test the feasibility of TV delivery on WiMAX links. Table 1 gives the important network characteristics. The network provides several FEC code rates. Since video applications require higher bandwidth relative to other multimedia services such as voice and data, we chose a FEC code rate that maximizes the offered throughput within each modulation scheme. ¾-rate FEC setting was used for 64QAM, 16QAM and QPSK. The only code rate available for BPSK is ½.

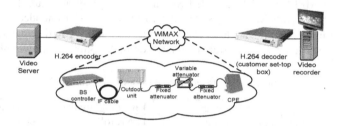

Fig. 2. Testbed overview

Table 1. Network Characteristics

Standard compliance	IEEE 802.16d
Duplex mode	TDD, Full Duplex
Channel size	7 MHz
FFT size / Frame size	256 / 10 ms
Supported modulations	64QAM, 16QAM, QPSK, and BPSK
Transmission powers of BS/CPE	36 dBm / 20 dBm
Total fixed attenuation signal loss	82.6 dB

The WiMAX equipment uses Time-Division Duplex (TDD), which offers the ability to adjust the downlink to uplink bit rate ratio (DL/UL ratio). In the downlink scenarios, the DL/UL ratio was set to 85/15 while a ratio of 25/75 was chosen for testing uplink scenarios.

The maximum throughput on the WiMAX link was measured on both uplink and downlink for different modulations. As shown in Table 2, the throughput depended on the modulation scheme. QAM schemes gave higher bit rates than PSK schemes.

Table 2. WiMAX Link Throughput

Modulation	Throughput (Mbps)			
	DL/UL ratio = 85/15		DL/UL ratio = 25/75	
	DL	UL	DL	UL
64QAM 3/4	19.0	3.2	5.0	17.0
16QAM 3/4	12.7	2.1	3.5	11.5
QPSK 3/4	6.4	1.0	1.6	5.7
BPSK 1/2	2.0	0.3	0.4	1.9

IPTV Settings: The input to the encoder is either a high definition (HD) or standard definition (SD) video. The HD video has 1080i format (1920 x 1080 pixels) at a frame rate of 29.97 fps while the SD video is a 480i (720 x 480 pixels) format at the same frame rate. The video material is a 4-minute sequence. The sequence consisted of 24 10-second clips covering a wide range of picture content and complexity. The encoder is a professional MPEG-4 AVC/H.264 encoder with IP output. Table 3 shows the encoder settings used for HD and SD applications. The GOP length was 32 frames with an IBBBP structure for both definitions. The encoding bit rates were selected to match the maximum throughput we can get on the WiMAX link using different modulation schemes. The lowest bit rates recommended for satisfactory quality of experience of MPEG-4 AVC encoded SD and HD TV services are 1 and 8 Mbps respectively [12]. However, it was decided to lower the HD encoded bit rate to 5 Mbps because the preliminary tests showed that the quality could remain acceptable at that rate. The same test cases were evaluated on both uplink and downlink.

Table 3. Test-cases (Video encoding – modulation)

Definition	H.264 settings			Encoding quality	Corresponding modulation
	Profile	Level	Bit rate (Mbps)	PSNR (dB)	
SD	Main	3	1	33	BPSK 1/2
			4	36	QPSK 3/4
HD	High	4	5	33.62	QPSK 3/4
			8	35.18	16QAM 3/4
			15	37.3	64QAM 3/4

The encoded stream was packetized in 188-byte MPEG2-TS packets before being transmitted in chunks of seven TS packets over RTP/UDP/IP. The protocol overhead is around 3% of the encoded stream bit rate.

The video was decoded by a professional HD/SD MPEG-4 AVC/H.264 decoder. For error concealment, the decoder was configured to replace missing frames (if any occurs due to packet loss) by copying the last decoded frame.

The video quality in each test case was evaluated for different link conditions. The link condition was changed by varying the attenuation. Each test-case was repeated several times (20). We then measured the average quality computed over all decoded samples. The video quality of the received video sequence was measured with the full-reference metric: the PSNR. We also evaluated the quality of encoded sequence (without network transmission) for each test-case, as shown in Table 3, capturing the sole effect of resolution and encoding bit rate.

3 Network Characterization

This section presents a characterization of the network under test in terms of QoS parameters (e.g. packet loss, delay and jitter). The jitter and packet loss were measured for each run. We then computed the average over 20 runs. The same process was done for different channel conditions. The channel condition is reflected in the signal strength, based on which the BS decides to pursue or drop the connection. Thus, the channel condition is represented thereafter by the received signal level (RSL). Note that the accuracy of RSL values is +/-1 dB.

It can be seen in Fig. 3 that the channel had almost no packet loss until the signal strength drops to the threshold of operation, below which packet loss increased dramatically and the signal was rapidly lost. This threshold depended on the modulation scheme.

Thereafter, the signal strength level, below which packet loss rate started to increase, is called signal strength threshold or threshold of operation. Based on results of data loss rate, it was at -71 and -78 dBm for 64QAM and 16QAM schemes respectively on both downlink and uplink. For uplink QPSK and BPSK, the thresholds were -85 and -89 dBm respectively. The downlink test cases of 4 and 1 Mbps, corresponding to QPSK and BPSK schemes respectively, lost the connection at -81 dBm because of the uplink signal loss due to the asymmetric transmission power.

In all the downlink test cases, except the cases where connection was lost abruptly because of uplink signal loss, the packet loss rate on the downlink was in the order of 10^{-4} at 2 dB and of 10^{-3} at 1 dB above the threshold of operation. However, on the uplink, a packet loss rate in the order of 10^{-4} was observed at 4 dB before the signal strength threshold was reached. The next sections present the video quality with respect to these thresholds.

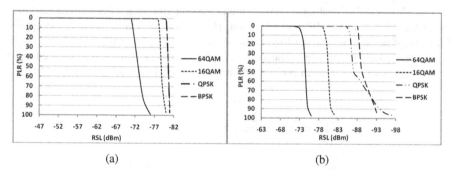

(a)   (b)

Fig. 3. Packet loss rate on downlink (a) and uplink (b) in function of the channel condition for different test cases

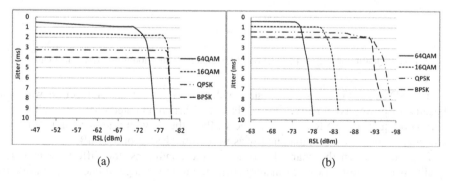

(a)   (b)

Fig. 4. Jitter measurements on downlink (a) and uplink (b) in function of the channel condition for different test cases

Depending on the modulation scheme and, hence, on the bit rate, the jitter varies from 0.5 to 4 ms on downlink and from 0.5 to 2 ms on uplink as illustrated in Fig. 4. This represents an ideal condition for video applications that are sensitive to delay variations.

Concerning the delay (latency) on the WiMAX link between BS and CPE, it was observed to be 35 ms +/-5 ms. In general, reasonable end-to-end delay and jitter values are not problematic due to STB de-jitter buffers, provided the de-jitter buffer size is provisioned to match network and video delay variation.

4 Downlink Scenarios

The video quality for HD test-cases on downlink is shown in Fig.5. The highest HD video quality was observed with 64QAM modulated signal because of the high encoding bit rate of 15 Mbps. However, the PSNR of 37 dB at high RSL fell to 27 dB at a RSL of -70 dBm (1 dB above the threshold). The 8Mbps test-case on 16QAM channel resulted in slightly lower quality than that of 64QAM, with a PSNR of 35 dB. This quality was observed until the RSL was 3 dB above the threshold.

The 5Mbps QPSK test-case showed the lowest quality with a PSNR of 33 dB until the connection is lost at RSL of -81 dBm. However, this quality remains acceptable for most of scene types, according to the subjective study in [13]. The 5 Mbps might not be sufficient to encode some complex HD scenes. Fortunately, these complex scenes represent only a small portion of the types generally broadcast for TV [14].

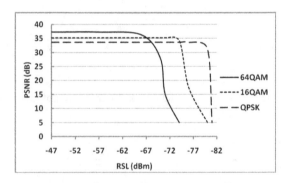

Fig. 5. Quality of HDTV in function of the channel condition for different test cases

The SDTV test-cases gave constant video quality until the connection was lost because of power asymmetry. The 4 Mbps on QPSK modulated channel resulted in a very good quality (PSNR of 36 dB). The 1 Mbps, the only bit rate that can fit a BPSK channel, showed an acceptable quality as well with a PSNR of 33.1 dB.

In general, when the channel had acceptable conditions, the difference in video quality between test-cases was mainly attributed to the change of encoding bit rate to suit the modulation scheme.

It is worth noting that, above signal strength thresholds, the perceived video quality in the entire test cases was almost the same as the encoded video (coded at the same bit rate and then decoded without transmission on network). This means that the WiMAX channel might have minimal effect on video quality if channel limits are respected and suitable modulation scheme is selected accordingly.

When QPSK and BPSK modulation schemes were selected on downlink, the video quality remained stable until the connection was lost. However, the video quality started decreasing before the signal strength threshold was reached when the 64QAM and 16QAM schemes were used. Thus, it is advised to operate the network 2-3 dB above the threshold of operation to guarantee a good video quality when these QAM modulation schemes are selected.

5 Uplink Scenarios

The video quality on uplink was measured for scenarios such as Electronic News Gathering, up to news room, etc. The same HD and SD video test-cases were evaluated for different conditions. Recall that the uplink was not affected by the transmission power asymmetry.

The HD quality is shown in Fig. 6(a) for the three HD test-cases. The highest HD video quality with PSNR of 37 dB, observed for 64QAM modulation, only lasted over a narrow range of high RSL. Although an uplink 64QAM channel can be maintained around a RSL of -71 dBm, the 15 Mbps video quality starts to drop at -67 dBm.

The same observation can be reported for the 8 and 5 Mbps HD test-cases, except that the video quality showed more steep water fall region. The best video quality achievable for each test-case was observed when the RSL was 3 dBm above the signal strength threshold of each corresponding modulation. Even if an uplink signal continued to be transmitted near the threshold the video was not watchable.

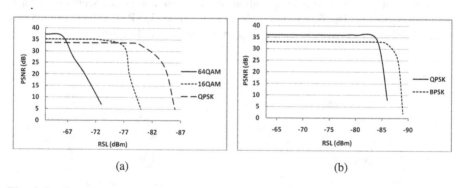

(a)   (b)

Fig. 6. Quality of HDTV (a) and SDTV (b) in function of the channel condition for different test cases

Fig. 6(b) shows the video quality results for SD test-cases. The 4 Mbps test-case can maintain a very good SD video quality (PSNR of 36 dB) up to an RSL of -84 dBm, which is only 1 dBm apart from the threshold. This is very close to the signal threshold above which QPSK modulation is supported. However, the 5 Mbps HD test-case, that used the same QPSK modulation, required larger distance from the threshold (3 dBm and higher) to preserve the same quality. This means that SD video was more tolerant than HD video to higher packet loss rates, experienced just above the signal strength threshold.

The 1 Mbps SD test-case showed a constant acceptable quality (PSNR of 33 dB) that started to degrade when RSL was 2 dB above the threshold (-89 dBm).

In general, in best conditions, when the RSL was above enough the threshold in each test-case, video quality was the same on both uplink and downlink. However, on uplink, if HD video service is to be provided, it is recommended to design the link budget in such a way to maintain the RSL at 3-4 dB above the thresholds of operation. The recommendation can be relaxed to 1-2 dB in case of SD video service on uplink.

6 Discussions

After we evaluated the expected video quality perceived by individual users at different link conditions and network settings, we discuss in this section factors that can influence the business case of TV/video over WiMAX.

In fact, video resolution and quality that can be offered to customers mainly depend on the bit rate allocated to the TV/video service. The bit rate per WiMAX user is controlled by several factors: the modulation scheme, the DL/UL ratio and the BS load of CPEs. As seen before, the supported modulation depends on the link condition, which is reflected in RSL and is resulting from signal strength gains (transmission power, antenna gain) and attenuation (i.e. path loss, interference).

Based on previous analysis, we can estimate the aggregated video bit rate available for all CPEs per BS with respect to RSL. Recall also that the value of RSL above which a good video quality can be sustained, depended on the video resolution and bit rate. Fig. 7 summarizes the aggregated throughput (excluding protocol overhead) that can be allocated to IPTV/video applications in downlink and uplink scenarios in each modulation scheme. The figure shows also the recommended RSL values for high and standard definition services in each case on both uplink and downlink.

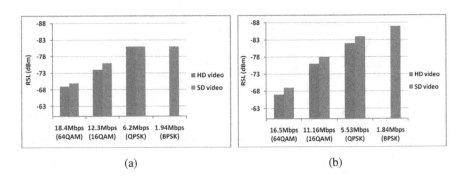

Fig. 7. Total video throughput with respect to RSL on (a) downlink and (b) uplink

The signal strength depends on the BS location. Basically, RSL of a line-of-sight (LOS) channel is above -70 dBm and most of the time below -70 dBm in non line-of-sight (NLOS) cases [9, 10]. So according to the received signal strength measured in the area of deployment, engineers can know a priori what video service may be supported and the perceived quality that can be expected.

When all users agreed on the same service (i.e. same rate setting of the same QoS class), throughput is equally divided on them. Another plausible scenario would be users having different service agreements. In this case, throughput will be attributed using different bit rate settings of QoS classes and, hence, some users may have SD service while others may get HDTV with the same BS.

Nevertheless, despite the capacity to deliver a good TV/video quality, WiMAX link is quite different from a classic broadcasting cable (ATSC or DVB). The WiMAX link can only support a limited number of TV channels at once. This means that fewer video streams (or one stream in some cases) would be sent from the IPTV local office to each subscriber due to the limited bandwidth. When a user changes the TV channel on his STB, it does not tune a channel like a cable system, but it switches to another stream. This way, only channels that are currently being watched are actually sent from the local office to users and so, the WiMAX available bandwidth may be enough for delivering IPTV services.

Even so, a distinguishing capability from traditional broadcast networks is that WiMAX equipments offer an adaptive modulation feature. This feature enables the transmitter and receiver to negotiate the highest mutually sustainable modulation (data rate), then dynamically changes the modulation scheme to adapt to RF conditions.

The adaptive modulation feature can help in sustaining TV delivery in most of link conditions. That is, when the radio channel is good, the 64QAM, 16QAM or QPSK provides a very good TV quality. However, when BPSK is the only supported modulation due to channel conditions, clients can still get a standard definition TV or video service with acceptable perceived quality.

The adaptive modulation feature must be coupled with H.264 encoders capable of dynamically changing the encoding bit rate to match the rate change of the channel. Such encoders already exist on the market; they are able to vary the encoding bit rate transparently while keeping the same definition. Thus, the adaptive modulation backed by these featured H.264 encoders can guarantee a continuous good service delivery. Scalable video coding may also be a good solution in case of software encoders. Video base layer can be transmitted in low bit rate cases while both base and enhanced layers can be sent when higher bit rate channel can be afforded.

7 Conclusions

This paper presented a study on possibilities and limitations to provide high and standard definition TV/video services on WiMAX technology in different scenarios. The expected video quality was analyzed with respect to channel condition, traffic direction, encoding bit rate and resolution of video service. Results have shown a potential of providing a good video quality, satisfying IPTV/video QoE requirements, when signal strength limits are respected. These limits were identified with respect to modulation scheme, scenario and service type. Service adaptation and factors affecting user perceived quality were also discussed in order to give better options of service provisioning.

Acknowledgments

The authors would like to thank the CRC WISELAB, in particular, Simon Perras and Philippe-André Bonneau, for providing the WiMAX system used for the testing, as well as the technical involvement and support in implementing the WiMAX testbed. Also, the authors gratefully acknowledge the contributions of Ron Renaud from CRC for providing video sequence and helping in setting the video testbed.

References

1. Air Interface for Fixed Broadband Wireless Access Systems. IEEE STD 802.16 (2004)
2. Hillested, O., Perkis, A., Genc, V., Murphy, S., Murphy, J.: Adaptive H.264/MPEG-4 SVC Video over IEEE 802.16 Broadband Wireless Networks. In: Proc. Packet Video, pp. 26–35 (2007)
3. Huang, C., Hwang, J., Chang, D.: Congestion and Error Control for Layered Scalable Video Multicast over WiMAX. In: EEE Mobile WiMAX Symposium, pp. 114–119 (2007)
4. Retnasothie, F., Ozdemir, M., Yucek, T., Celebi, H., Zhang, J., Muthaiah, R.: Wireless IPTV over WiMAX: Challenges and Applications. In: Proc. IEEE Wamicon, Clearwater, FL (2006)
5. Uilecan, I., Zhou, C., Atkin, G.: Framework for Delivering IPTV Services over WiMAX Wireless Networks. In: IEEE EIT, pp. 470–475 (2007)
6. Tsitserov, D., Markarian, G., Manuylov, I.: Real-Time Video Distribution over WiMAX Networks. In: Proc. Annual Postgraduate Symposium, Liverpool, UK (2008)
7. Filis, K., Theodoropoulou, E., Lyberopoulos, G.: The Effect of a Rapidly Changing Urban Environment on Nomadic WiMAX Performance. In: Proc. IST Mobile and Wireless Communications Summit, pp. 1–5 (2007)
8. Grondalen, O., Gronsund, P., Breivik, T., Engelstad, P.: Fixed WiMAX Field Trial Measurements and Analyses. White paper, Unik - University Graduate Center (2007)
9. Buschmann, J., Vallone, F., Dettorre, P., Valle, A., Allegrezza, G.: WiMAX Experimentation and Verification in Field in Italy. Journal of Systemics, Cybernetics and Informatics 5(5), 16–20
10. Lipfert, H., Zistler, A., Vogl, A., Keltsch, M.: Performance Testing in a WiMAX-Pilot at Intitut fur Rundfunktechnik. In: Proc. IEEE BMSB, Spain, pp. 1–6 (2009)
11. WISELAB, WiMAX Activity in Canada. CRC Canada (2007)
12. ITU FG-IPTV-DOC-0814: Quality of Experience Requirements for IPTV Services (2007)
13. Speranza, F., Vincent, A., Renaud, R.: Bit-Rate Efficiency of H.264 Encoders measured with Subjective Assessment Techniques (submitted for publication, 2009)
14. Nishida, Y., Nakasu, E., Aoki, K., Kanda, K., Mizuno, O.: Statistical Analysis of Picture Quality for The digital Television Broadcasting. In: IEE International Broadcasting Convention, pp. 337–342 (1996)

Personalized Semantic News: Combining Semantics and Television

Roberto Borgotallo[1], Roberto Del Pero[1], Alberto Messina[1], Fulvio Negro[1],
Luca Vignaroli[1], Lora Aroyo[2], Chris van Aart[2], and Alex Conconi[3]

[1] Rai Radiotelevisione Italiana, Italy
[2] VU University Amsterdam, The Netherlands
[3] TXT Polymedia, Italy

Abstract. The integration of semantic technologies and television services is an important innovation in traditional broadcasting in order to improve services delivered to end users in an extended home environment: new methods emerge for getting TV content via the Web and interacting with TV content on end users devices. This paper gives a short description of a Personalized Semantic News scenario in the context of the NoTube project illustrating the use of semantics for personalized filtering and access of news items.

Keywords: NoTube, television services, user-centric, personalized content, device adaptability, semantic annotation, content enrichment, home ambient.

1 Introduction

The TV industry landscape is developing into a highly-interactive distributed environment in which people interact with multiple devices portable devices and home equipment, as well as with multiple applications. People more than ever become early adopters of technology. The Web and these 'new technologies' are steadily transforming this state of the TV industry. New methods emerge for getting TV content via the Web and interacting with TV content on set top boxes. Companies are already attempting to bundle Electronic Program Guides (EPGs) into their software, along with personal recommendation services based on users viewing habits. However, most of those services are still bound to one platform only, e.g. either set top box or Web and stay rather TV-centric. Additionally, users are also increasingly involved in multiple virtual environments (e.g. MySpace, Flickr, YouTube, Amazon, entertainment sites) in each of them with a different identity (e.g. login information, preferences). There is very limited integration and reuse of these user data, or if there exists integration it is not always under the control of the user and there is a lack of transparency in the use of personal data between different applications.

The NoTube[1][1] project aims to overcome these deficiencies and to cover these new requirements using Semantic Web languages and technologies. The ultimate goal

---

[1] NoTube (*Networks and Ontologies for the Transformation and Unification of Broadcasting and the Internet*) is an EU FP7 Integrated Project (2009-2012).

P. Daras and O. Mayora (Eds.): UCMedia 2009, LNICST 40, pp. 137–140, 2010.
© Institute for Computer Sciences, Social-Informatics and Telecommunications Engineering 2010

of the project is to develop flexible/adaptive end-to-end architecture, based on semantic technologies, for personalized creation, distribution and consumption of TV content. We take a user-centric approach to investigate fundamental aspects of consumers' content-customization needs, interaction requirements and entertainment wishes, which will shape the future of the "TV" in all its new forms. The project explores three different scenarios: (1) personalized semantic news, (2) narrowcasting advertisement and (3) community based content selection.

In this paper, we focus on the first scenario, i.e. personalized semantic news, provided by RAI Research Center in Torino. We present an overall sketch of the data, services and users involved. We address issues related to the use of semantic tools for the context-aware management of multimedia archive content and its exploitation beyond the creation of traditional TV products. With the advancement of Web technologies and with the convergence of various platforms for the access of multimedia content, new added value services are explored to enable the exploitation of the so-called long tail phenomenon.

2 Personalized Semantic News Scenario

The Personalized Semantic News scenario focus on the creation, distribution and usage of personalized news services that will be able to (1) acquire news items from generic broadcast streams, (2) understand the meaning of video news items, (3) understand the physical context in which news items are shown and (4) apply criteria for matching the user profile with the available news items (see Figure 1 for data flow).

The creation of personalized news services is performed at the *Service Provider Environment*, considering service provider editorial requirements and generic and privacy non-sensitive user profile information; as well as at the *Home Ambient Environment*, considering local context and user information and data enrichment services. The Home Ambient Environment, an extended home environment, consists of two parts:

- *Physical Home Ambient:* a portion of the environment where the user has access to personalized services (context is determined by various sensors input) distributed through the home LAN and enjoyed through different devices.
- *Logic Home Ambient:* virtual space, in which the users, content, metadata and services are identified, filtered and stored. In this space also the content semantics and operational rules are defined.

Three main services are provided at the Home Ambient Environment, e.g. "My News Agency", which automatically generates a personalized local news multimedia channel; "News Alerts", which issues alerts for an incoming News Items relevant to the user's interests; and "News Search" which provides searching capabilities based on semantic filtering of the available News content. Only News Items that match the Home Ambient rules are stored in the home ambient (service level semantic filtering). Each News Item stored in the home ambient has a period of expiration defined either at the provider side or at the user side and it should be locally enriched with metadata and resources automatically retrieved from local repository or from predefined (following Home Ambient rules) area of the Web (Home Ambient enrichment). Home Ambient Services can be automatically created by grouping News Items (My News Agency, News Search) or based on in-coming events (News Alerts). Following User, Device and Environment rules, dynamic device adaptation is performed on different user devices, e.g. Sofa Television and Hand Television.

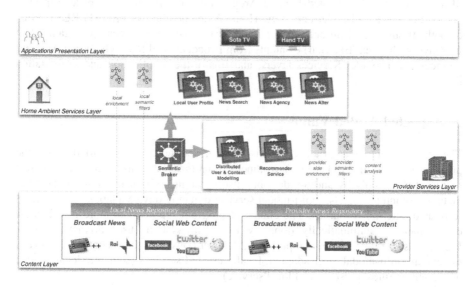

Fig. 1. Basic data flow of the Personalized News scenario

In the *Service Provider Environment* all audio-video segments detected as "News Item" are extracted from the available broadcast content (by means of the Automatic Newscast Transcription System ANTS[2][3]). Each extracted News Item is further enriched with information extracted from internal and external Web resources, as well as with metadata from different related domain vocabularies. Service Provider rules are defined for this enrichment step. The Service Provider rules take in consideration single and group user behavior (in a privacy preserving manner) and apply semantic filtering methods according to the users' (or stereotypes) interests and preferences.

At the Service Provider side we can identify a set components supporting the provider in the preparation of the TV contents, e.g. User and Context services (generic user/context profile categories), Content Annotation and Enrichment services (meta-tagging of TV contents), Metadata Management services (aimed at adapting metadata to professionals/non-professionals) and Model and Semantic services (common background for metadata annotation and enrichment).

The *Semantic Broker* is a core part of the architecture. It discovers the basic internal services organizing them through a specific ontology; it discovers external sources that are semantically related to the application scenario; it exposes meta-services to the upper level (the Application Logic layer) providing an individual entry point for the system, regardless of the physical location of the underlying services; and it composes services in order to perform more complex tasks.

At the Home Ambient side another instance of the Semantic Broker manages internal services, provides semantically related content and services from external sources (typically the Web) and combines locally stored data, e.g. news, advertising, web feeds, etc. with the user and context profiles with the final goal to produce a personalized experience to the end user.

The NoTube infrastructure described in this use case is service-oriented, follows SESA (Semantically Enabled Service Architecture) paradigm and exposes services, both at the Service Provider and at the Home Ambient side, with SOAP, REST APIs, or both. It supports different type of middleware. All the services are organized in four layers: (Internal/External) Contents from Broadcast and the Web, Home Ambient Services and Provider Services layers, and finally the Application and Presentation layer.

3 Expected Added Value

The main added value granted by this kind of service is to give to the user the possibility to get in his home environment programs segments of his own interest represented not only by audio-video, but also by audio-only or text-only or by other metadata (content objects, multi-modal service, dynamic device adaptation) and links to related external resources. The choice of program segments to store in the home environment is automatically done by the local system (user privacy granted) following some user semantic input rules; the user consumes these contents by means of several multi-modal, user adapted, locally created services (semantic local service user/ambient rules, personalized local service).

References

1. The NoTube project web-site, http://www.notube.tv
2. Messina, A., Borgotallo, R., Dimino, G., Boch, L., Airola Gnota, D.: An Automatic Indexing System for Television Newscasts. In: Proc. of IEEE ICME 2008 (2008)
3. Messina, A., Montagnuolo, M.: A Generalised Cross-Modal Clustering Method Applied to Multimedia News Semantic Indexing and Retrieval. In: Proc. of WWW 2009 (2009)

Service and Content Metadata Systems in Interactive TV Domains

Mark Guelbahar[1] and Josip Zoric[2]

[1] Institut für Rundfunktechnik GmbH, Platforms for Broadcast Services,
Munich, Germany
guelbahar@irt.de
[2] Telenor Research and Innovation and Norwegian University of Science and Technology,
Trondheim, Norway
josip.zoric@telenor.com

Abstract. In this paper we discuss information interoperability, focusing on design and implementation of service and content metadata solutions for cross-domain IPTV services. We approach the problem by discussing *experience sharing* in interactive TV, and relate it to necessary service platform facilitators, such as: *interactive TV sessions*, *cross-domain service framework* and *metadata integration* (service, content and session metadata). We detail the design and an implementation of such *a metadata infrastructure*, and communicate our practical experiences.

Keywords: Networked electronic media, Cross Domain, Metadata, Service Discovery.

1 Introduction

We discuss the metadata functionality required for providing interactive TV services with enhanced cross-domain service support, where *cross-domain* can be described as "providing TV, Internet, Home-domain and Telco services as an integrated part of the user TV session". Cross-domain services cover user centric media discovery, delivery and consumption (with emphasis on social communities). Enabling interactive TV service maturity in an access- and network transparent manner is not a trivial task, and we will discuss why. The iNEM4U [1] vision of a networked electronic media framework enables individuals and groups of users to *share their interactive media experiences* in an intuitive and a seamless manner, regardless of their choice of service domains, networks, and devices. However, today's landscape of networked electronic media consists of a number of *non-interoperable technology islands* that were designed for different types of users, services, content, and devices. Examples are the consumer electronics devices in the home, and different kinds of network and service environments for mobile, IPTV, and broadcast usage.

The degree of connectedness within and outside of homes is improving significantly - however, available networks are still confronted with interoperability issues, regarding content and metadata formats, encodings and presentation (devices simply contain several network interfaces, but these as such don't make the contents interoperable and

P. Daras and O. Mayora (Eds.): UCMedia 2009, LNICST 40, pp. 141–150, 2010.
© Institute for Computer Sciences, Social-Informatics and Telecommunications Engineering 2010

manageable via single points of access). The reason for this is quite obvious – within various domains, different content and metadata formats, distribution systems and standards are being used. Not only formats differ, but even similarity of metadata for the same content item offered by two different parties is not guaranteed.

Especially, when the Internet and the Web2.0 world is considered to play a new content provisioning role, things get even more complicated – the Web is in most cases totally unmanaged and un-standardised, especially regarding content and metadata formats. On top of that, the quality of content and metadata is not controlled, which makes good viewing experiences hard to find. Recommendations are one way out of this, where other projects try to manage these problems by using complex algorithms to allow separation of high and low quality offerings (although still in an early stage of development).

The iNEM4U [1] vision is accomplished by providing means to combine multiple service domains to offer services or content to end-users, allowing seamless integration of professional and user-generated content and services, and making the result accessible across terminals, locations and networks. Personalised interaction with services and content is supported, as well as synchronous community-based sharing of content and experiences. Interoperability is a major obstacle for such systems, spanning several dimensions, as network, and service interoperability, network and service roaming, as well as information interoperability, to which we dedicate this paper.

Metadata for services and enablers plays a crucial role in various service platform (SP) mechanisms, supporting the service delivery, e.g. service discovery (SD), composition, brokering and mediation - as already mentioned, various metadata formats and standards exist. DVB has adopted a profile of metadata defined by the TV-Anytime Forum (ETSI TS102323). TV-Anytime [2] is an XML-based solution, with the additional functionality to offer personalised recommendations on what to watch. It is a subset of MPEG-7 [3] and also an ETSI standard, which was included in several DVB standards, like DVB-IP, DVB-S/C/T and DVB-H.

Several projects dealt with the cross domain metadata compatibility and provided the metadata entities, which might be partly reused in the interactive TV service platforms. Worth mentioning here are Daidalos [4], SPICE [5], Mobilife [6], NoTube [7] and SAVANT [8]. They all worked on metadata solutions, so they provide useful input for domain-specific solutions, e.g. Mobile service platforms, Web service platforms etc. But as none of them offers an approach that suits the cross-domain interactive service platform solutions, we only can take over fragments with metadata descriptions of service context, user profiles, service and enabler descriptions, and reuse them in our cross-domain framework.

This paper is organized in the following way: after having introduced the problem, its practical relevance and related work, we introduce the concepts of cross domain services and experience sharing in section 2. In section 3 we explain requirements and design issues related to cross-domain service architectures enabling interactive TV services, focusing on functionality for Metadata (MD) retrieval and management. Section 4 presents the iNEM4U service and content MD infrastructure, while section 5 communicates our implementation approach and practical experiences. Section 6 concludes this work.

2 Experience Sharing, Cross-Domain Services and Interactive Sessions

We dedicate this section to the iNEM4U approach towards cross-domain experience sharing. We start by *(1)* structuring the concept of experience sharing, considering simultaneously the user and the business aspects, *(2)* continue by relating the shared experience to the system entities which realize it, i.e. services and enablers, and finish by *(3)* detailing their composition and orchestration by introducing the concept of an interactive session.

2.1 Experience Sharing

The aim of the iNEM4U project is to enable individuals and groups of users to share interactive media experiences in an intuitive and seamless manner across domains and within cross-domain communities. *Sharing a live event experience*, such as a concert or a soccer match, with a group of people that have gathered in an ad-hoc way around that life event, contains the following business processes:

- *Sending recommendations* to *watch the concert* from various users' locations.
- Creating an iNEM4U sessions in order to *watch the event together*, in a synchronized way.
- *Opening up a synchronized video, voice or text chat session*, as an overlay video on top of the event broadcast.
- *Merging several user sessions* with the actual iNEM4U session.
- *Sharing notices and invitations across domains and devices.*
- *Watching* the event and sharing the experience with others.
- *Storing and replaying the sessions* (e.g. which consists of the broadcast video of the event and associated user generated content (UGC), comments, annotations and various service interactions) at any place and time, and making it available through information portals, community websites and other media channels.

2.2 Cross-Domain Services and Architectures

Above-mentioned business processes have to be mapped to their system counterparts, which specify the involved system platform functionality. System processes are realized by a composition and orchestration of service bundles, as discussed below.

iNEM4U enables both end-users and professional content providers to make use of interactive media services across domains, without the need to replicate the services that are currently provided by the existing infrastructures in these domains. Rather, iNEM4U uses these services to provide new "bundles" of services and applications, and adds new features such as cross-domain recommendations and experience sharing. These new services build upon the iNEM4U cross-domain service architecture, which forms a *convergence layer* on top of domain-specific ("native") service infrastructures. Service platforms, which offer interactive TV sessions, combine services, enabling technology, content and users in well-structured and synchronized service sessions. Such "iSessions" provide *experience sharing*, which facilitates both, on-line participation and off-line retrieval, and the possibility to replay previous sessions. One key enabler for this (besides domain-independent synchronization functions and content formats) is the presence of a *well- defined and lightweight metadata system*. We will firstly define the above-mentioned service and platform entities, and secondly discuss their metadata descriptors and their retrieval and management techniques.

The key concept to deliver cross-domain services independent of the user's access network is based on the underlying assumption that all iNEM4U services are delivered using Internet Protocol (IP). This assumption implies they can be delivered using a variety of different networks and channels, as long as they support IP connectivity.

For example, services and applications that are residing:

- in the home (e.g. WiFi and UPnP devices),
- on the "open internet" (e.g. Web2.0 Web Sites/Applications)
- as IPTV services on set-top-boxes, IP-enabled TV's or PC hardware and
- within Mobile services (e.g. IMS-based multimedia services).

Furthermore, it is also possible for services, content, sessions and applications to be provisioned via managed networks, for example, an end-to-end network managed by an operator using technologies such as Next Generation Networks or, more specifically, the IP Multimedia Subsystem (IMS [9]). Users of the iNEM4U system are not bound to specific delivery channels and are free to utilise applications that span the four main delivery channels outlined above and depicted in Figure 1. Services and enablers also hold important metadata, which is used by various service

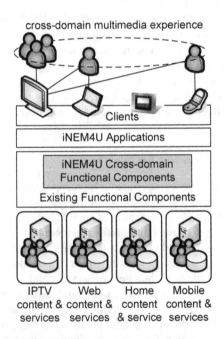

Fig. 1. iNEM4U High Level Architecture

platform mechanisms (service discovery, composition and brokering), and usually includes the information about the service and enabler capabilities, the type of the information / service they can deliver, the requirements / quality of service, the targeted terminals, where the services will be played etc.

2.3 Interactive TV Sessions

Let us now discuss interactive TV sessions (iSessions) as a way of bundling and synchronizing (spatially and temporarily) cross-domain services. iSessions can be used in nearly all situations where information has to be interchanged and a certain level of history should be tracked during the session lifetime (or even afterwards). We focus on sessions used to deliver multimedia content and related communication services in a synchronized manner to the end user. Metadata about sessions should include the metadata subtypes about content, services and enablers, applications, and users participating in sessions.

The iNEM4U iSession management, illustrated in Figure 2 is designed to be independent of the underlying domain technology. iSessions can be created either by users or by Service Providers, giving the possibility to create an iSession and to share it with others, i.e. via a community website or the users' buddy list. Users can discover sessions through invitation from other users, by subscribing to iSessions of specific types or genres, by selecting an iSession from an Electronic Program Guide (EPG) containing iSession information, or by downloading it from a web server. An overview of the basic iSession management architecture is given in [9]. In Figure 2, the session management architecture is illustrated in a simplified way as a single module in the server / backend side (the upper part of the architectural model). The *lifecycle actions* of an iSession include the following states: creation, initialization, running, pausing, resuming, stopping, storing. *The modification actions* on an iSession include: adding / removing content, adding / removing users, enabling / disabling content preview, modifying layout, modifying playlist, modifying session timing.

During the creation phase, the creator can add Content Sources [9] and users to a session and is able to define the layout and timing of the content within the session in one or more individual ways. During the initialization phase, the iSession description document is downloaded to one iNEM4U client. Once all Content Sources have been discovered, the iSession server notifies all of the other invited users and all iNEM4U clients can start to render the iSession content.

As stated, sessions can be paused, stored and *replayed at any later point in time*. One approach for storing an iSession is to store only the iSession description i.e. in a session repository, and there, Content Sources are just stored by reference - if they disappear (e.g. their URI changes or the source is going offline), the session cannot be played back any more, which is especially problematic whenever live content is part of the session. The extreme opposite of this approach is a complete recording of the session and all its involved Content Sources to ensure replay-ability of the complete session later. This involves the storage of all content items that form the session (including live broadcast content, video chats and text messages), as well as timing-related modifications during runtime (e.g. a content item has been added at a given time after the start of the session). This infers availability of huge data stores, but also could also enable new business models (i.e. the role of a "premium session provider").

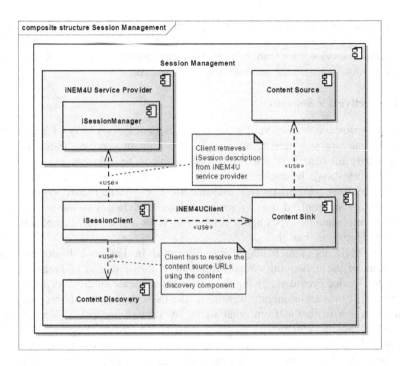

Fig. 2. Overview of Session Management

3   iNEM4U Cross-Domain Metadata Solution

As mentioned above, experience sharing is obtained by composing interactive TV sessions and orchestrating their cross-domain services. However, service and content interoperability require sharing of various metadata entities. We can identify the *need for a bridge* between those domain-specific implementations. In this section, we present both, design and implementation of the iNEM4U metadata solution.

The *iNEM4U metadata solution* bridges these differences in a way transparent to its clients and can handle proprietary formats as well as formats used for internal and external representation of multimedia content. As an internally used data format, we chose TV-Anytime Phase II [2], because it is an existing standard, and de facto the only reliable and most flexible one – it is even *by design* capable of describing applications and "iSessions". Since it is based on XML, TV-Anytime is also an extendable format, which is an important requirement to be able to support future metadata languages. Additionally and as already stated, TV-Anytime already *is* part of existing domains, and therefore, there's no need to create bridging modules to and in between those.

The interactive TV session infrastructure illustrated in Figure 2 requires support from several service platform mechanisms. We will outline just the ones related to retrieval and management of MD:

- SP functionality dedicated to metadata mapping,
- SP mechanisms for discovery of services and content,
- Notification systems for changes in content, services and sessions.

A precondition for efficient service composition, orchestration, and user interactivity, is the existence of simple and well structured metadata descriptions, as described in the following subsection.

Metadata Mapping

An important facilitator of knowledge/information exchange among domain-specific solutions is mapping the various MD formats/standards. Service domains live their own life, just slightly influence each other (e.g. broadcast and telecom service domain), so it is difficult to expect that they might comply with the same standard. Instead we can expect the coexistence of various MD standards.

The only solution to enable cross domain functionality thus is to build an extensible framework that allows for the translation of existing and future metadata models. If an application would want to provide i.e. a listing of available multimedia content across all those domains, it could – in the iNEM4U case, as depicted in Figure 3 - send a generic, TV-Anytime compliant query to the iNEM4U service discovery & metadata component. This component would then pass this query to the domain-specific implementation modules (i.e. a module managing YouTube MD, another module managing DVB-IP MD, etc), which translate it into domain-specific queries and execute it. The query results are then returned via the framework to the requesting application, combining the resulting content and iSession information from several domains in one smart overview.

Additionally, the domain-handlers are triggered to subscribe to metadata updates, which results in the user automatically being informed whenever new services (with MD similar to queried one) are available and interesting content or iSession information has changed. Technically, this is realised by the use of XMPP [10] or JMS [11] (depending on the clients' capabilities) in an asynchronous way, and messages containing the changed information will be sent to the clients.

Cross-Domain Discovery and Notification

During interactions in iSessions we have to discover changes in content, users, services but also in iSessions. The discovery system has to be cross-domain - aware, which implies a federation of queries and subscription/notification mechanisms, as discussed in this subsection. In order to realize the system scenario presented in Figure 3, we have used the identical way to access all domains, namely through several interfaces:

- ***Cross-CRUD interface*** (create, retrieve, update, delete) and corresponding domain-specific CRUD interfaces, allowing the mapping of cross-domain retrieval and management functions to their domain-specific counterparts (Hierarchical multilayer-multistep search, i.e.: (i) a search for relevant sessions (a bundle of services and content), (ii) a search for services, which can provide a relevant content, and (iii) a retrieval of the content.
- ***Cross-SUBSCRIBE / NOTIFY interface*** and corresponding domain-specific SUBSCRIBE / NOTIFY interfaces, which are used for notification-based

interactions (notifications about new content, services, user who want to join or new sessions).

- *Generic cross-domain metadata API*, which is mapped to domain-specific metadata standards.

Discovery and notification mechanisms can be used either during composition and initialization of sessions, or during their orchestration (when events and notifications give a chance to react on new content, users or services, as well as other changes in the service context). We are evaluating our design approach by implementation and usage tests, as discussed below.

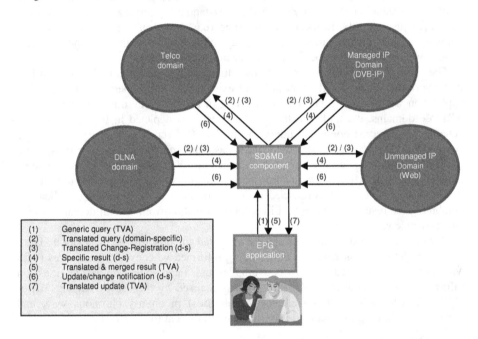

Fig. 3. Cross domain EPG scenario

4 Practical Experiences

The iNEM4U metadata *solution* has been implemented as a modular web service that can run in any technology domain, for instance within a home environment (e.g., on a DLNA gateway), or at a third-party service provider in the Web. Its implementation is Java-based, and uses JMS and XMPP for update notifications. Implementation of cross-domain service discovery and notification had a *cross-domain part* and a *domain-specific counterpart*, as well as a mapping between them. For querying we have used the web service technology, while for the notifications we have provided a XMPP - compatible solution in order to "attach" the SD/Notification to the technology used by the session management functionality. In such a way these service platform mechanisms can be used by any functional entity which subscribes to the information.

The Cross-Domain Personal EPG *application* is a lightweight, browser-based Java-Servlet application that can be used on any device providing (CE-) HTML support and IP connectivity. There, a profile- and context- based cross-domain EPG is realised and presented to the user, showing a personalised guide to content from broadcast, web, and mobile domains. It typically provides functions allowing a viewer to discover, navigate, and select services, content and iSessions, filtered by time, title, channel, genre etc. The users can control the application via their remote control, a keyboard, or other input devices such as a phone keypad. The implementation of the EPG has shown that the iNEM4U Metadata API allows application developers to create a cross domain application in a quick and easy manner, as in example it took the developers less than 15 lines of code to get a list of available EPG items, covering 3 different domains. Presenting this EPG in a graphical way requires much more work, especially in the Java Servlet case, but retrieval of cross domain metadata is very much simplified.

User tests and open demos at exhibitions (i.e. IBC 2009) have shown that the broad mass welcomes the domain-transparent way of content representation, and especially younger people tend to demand exactly for this network and source abstraction, as they grow up with a high degree of connectedness to several networks and content sources, and they don't care *where* the content they're actually interested in comes from, as long as they can get it.

5 Conclusion

Interactive cross-domain TV sessions require robust and efficient metadata support. We believe in the coexistence of domain-specific solutions, integrated by cross-domain service platform mechanisms, while the metadata interoperability gets ensured by metadata mapping techniques. A combination of several functional entities is required for the information interoperability in interactive TV, at least: (a) interactive session entities, (b) cross-domain session / service / content search and retrieval mechanisms, and (c) metadata mappings. We are continuing design improvements and implementation tests, the results of which we will communicate in the close future.

Acknowledgements

This work was partially funded by the EU FP7 grant (FP7/2007-2013), agreement no. 216647. We thank our colleagues within iNEM4U project for their support and insights. Authors are grateful to task workforce for all the input and suggestions they received, especially from D.Goergen, J.O'Connell and O.Friedrich.

References

1. iNEM4U, EU FP7 project - Interactive new experiences in multimedia for you,
   http://www.inem4u.eu
2. TVAnytime Forum, http://www.tv-anytime.org

3. MPEG, MPEG-7 Overview (2004),
   http://www.chiariglione.org/mpeg/standards/mpeg-7/mpeg-7.htm
4. Daidalos I and II, EU IST – FP6 project, http://www.ist-daidalos.org/
5. SPICE, European IST-FP6 project - Service Platform for Innovative Communication Environment, http://www.ist-spice.org/
6. Mobilife, EU IST – FP6 project, Services and applications from the user and terminal perspective, http://www.ist-mobilife.org/
7. NoTube, EU FP7 project, http://www.notube.tv/
8. SAVANT, EU FP5 project,
   http://www.ist-world.org/
   ProjectDetails.aspx?ProjectId=cf85b5506fde4e5697cdbaa5184364fb
9. Friedrich, et al.: Deliverable: D1.3, Overall Architecture of the iNEM4U Platform, iNEM4U, EU STREP project, FP7-ICT-2007-1-216647 (2008),
   https://doc.telin.nl/dsweb/Get/Document-95057
10. XMPP: Extensible Messaging and Presence Protocol (XMPP), http://xmpp.org
11. JMS: Java Message Service, http://java.sun.com/products/jms

UCMedia 2009

Session 6: Content Delivery

Efficient Scalable Video Streaming over P2P Network

Stefano Asioli, Naeem Ramzan, and Ebroul Izquierdo

School of Electronic Engineering and Computer Science,
Queen Mary University of London,
London, United Kingdom
{Stefano.asioli,naeem.ramzan,ebroul.izquierdo}@elec.qmul.ac.uk

Abstract. In this paper, we exploit the characteristics of scalable video and Peer-to-peer (P2P) network in order to propose an efficient streaming mechanism for scalable video. The scalable video is divided into chunks and prioritized with respect to its significance in the sliding window by an efficient proposed piece picking policy. Furthermore the neighbour selective policy is also proposed to receive the most important chunks from the good peers in the neighbourhood to maintain smooth content delivery of certain Quality of Service for the received video. Experimental evaluation of the proposed system clearly demonstrates the superiority of the proposed approach.

Keywords: Scalable video coding, Peer to peer, Bittorrent.

1 Introduction

Multimedia applications over the Internet are becoming popular due to the widespread deployment of broadband access. However, the conventional client-server architecture [1] severely limits the number of simultaneous users, especially for bandwidth intensive applications such as video streaming. On the other hand, Peer-To-Peer (P2P) networking architectures [2] receive a lot of interest, as they facilitate a range of new applications that can take benefit of the distributed storage and increased computing resources offered by such networks. In addition, P2P systems also represent a scalable and cost effective alternative to classic media delivery services. Their advantage resides in their ability for self organization, bandwidth scalability, and network path redundancy, which are all very attractive features for effective delivery of media streams over networks.

In conventional client server applications, the video server requires video contents of different fidelities, such as high quality material for storage and future editing and lower bit-rate content for distribution. In traditional video communications over heterogeneous channels, the video is usually processed offline. Compression and storage are tailored to the targeted application according to the available bandwidth and potential end-user receiver or display characteristics. However, this process requires either transcoding of compressed content or storage of several different versions of the encoded video.

Scalable Video Coding (SVC) [3] promises to partially solve this problem by "encoding once and decoding many". SVC enables content organization in a hierarchical

P. Daras and O. Mayora (Eds.): UCMedia 2009, LNICST 40, pp. 153–160, 2010.

manner to allow decoding and interactivity at several granularity levels. That is, scalable coded bit-streams can efficiently adapt to the application requirements. The SVC encoded bit-stream can be truncated at different points and decoded. The truncated bit-stream can be further truncated to some lower resolution, frame rate or quality as well. Thus, it is important to tackle the problems inherent to the diversity of bandwidth in heterogeneous networks and in order to provide improved quality of services.

In this paper, we exploit the features of scalable video coding and P2P network to perform an efficient video communication over distributed network. The layered structure of the scalable video bit-stream allowed us to achieve efficient on-the-fly rate adaptation. Additionally, SVC gives us the foundation for efficient use of network bandwidth on a P2P network [4] by enabling intermediate high capacity nodes in the overlay to dynamically extract layers from the scalable bit-stream to serve less capable peers. We also proposed an efficient piece picking policy and neighbourhood selection in P2P network for efficient scalable video streaming.

In this paper, Section 2 explains the proposed framework. The main parts of the proposed frame work and proposed optimization are also explained in Section 2. Section 3 provides the experimental evaluation of the proposed technique. Finally, Section 4 concludes this paper.

2   Proposed Framework for Scalable Video over P2P Network

The proposed system is based on two main modules: scalable video coding and the P2P architecture. In this system, we assume that each peer contains the scalable video coder and the proposed policy of receiving chunk is to make sure that each peer at least receives the base layer of the scalable bit-stream for each group of picture (GOP). Under these circumstances, peers could download different layers from different users, as shown in Figure 1. In this section, first we will explain the used scalable video coding and P2P architecture and then the proposed modification to adapt scalable video in P2P network.

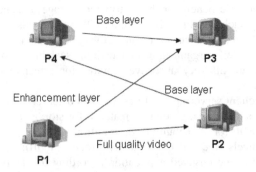

Fig. 1. An example of the proposed system for scalable video coding in P2P network

2.1 Scalable Video Coding

A wavelet-based scalable video aceSVC [5-6] is employed in this research work. Architecture of aceSVC features spatial, temporal, quality and combined scalability. Temporal scalability is achieved through repeated steps of motion compensated temporal filtering [5]. To achieve spatial scalability, each frame is decomposed using a 2D wavelet transform. Coefficients obtained through spatio-temporal decomposition are coded through the process of bit-plane coding [6] which provides basis for quality scalability. The main features of the used codec are: hierarchical variable size block matching motion estimation; flexible selection of wavelet filters for both spatial and temporal wavelet transforms on each level of decomposition, including the 2D adaptive wavelet transform in lifting implementation; and efficient bit-plane coder.

aceSVC bit-stream organisation: The input video is initially encoded with the maximum required quality. The compressed bit-stream features a highly scalable yet simple structure. The smallest entity in the compressed bit-stream is called a layer, or Atom, which can be added or removed from the bit-stream. The bit-stream is divided into GOPs. Each GOP is composed of a GOP header, the layers and the allocation table of all layers. Each layer contains the layer header, layer data that can be motion vectors data (some layers do not contain motion vector data) and texture data of a certain sub-band. The bit-stream structure is shown in Figure 2.

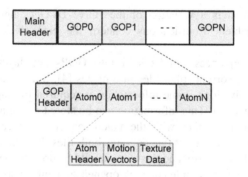

Fig. 2. Detailed description of used scalable bit-stream

In the main header of the aceSVC bit-stream, the organization of the bit-stream is defined so that the truncation is performed at different user points with low complexity.

These atoms are adapted with respect to the user requirements, bandwidth allocation in P2P network for the proposed system.

2.2 P2P Network

BitTorrent [4] is a widely used peer-to-peer protocol developed by Bram Cohen in 2003. The main idea behind it is that users' download rates should be proportional to their upload rates, in order to provide a fair mechanism and motivate users to share more. Free-riding is occasionally accepted, but only if there is enough spare capacity in the system [2]. In the original version of the protocol, this is achieved using a tit-for-tat mechanism, in which peers mainly upload data to peers they are downloading from. Moreover, peers occasionally behave altruistically, in order to discover potentially good neighbours.

In this research work, Tribler [2] is used as a BitTorrent client. It is based on ABC [7] (Yet Another BitTorrent Client), which is itself based on BitTornado [8]. New features include exploitation of social relationships among peers and content discovery through exploration of the network, instead of browsing a torrent repository. Finally, Tribler supports video on demand for non-scalable sequences. In this case a give-to-get algorithm is used, instead of tit-for-tat, which means that peers will upload to peers that have proven to be 'generous' towards third parties, instead of those peers that have a high upload rate.

2.3 Proposed Modification for Efficient Scalable Video Streaming

In this section, we formulate how the scalable layers are prioritized in our proposed system. First we explain how the video segments or chunks are arranged and prioritized in our proposed system and then efficient selection policy of good neighbour for the most important chunks is elucidated.

a. Piece Picking Policy

The proposed solution is a variation of the "Give-To-Get" algorithm [2], already implemented in Tribler. Modifications concern the piece picking and neighbour selection policies.

Scalable video sequences can be split into GOPs and layers as explained in section 2.2, while BitTorrent splits files into pieces [4]. Since there is no correlation between these two divisions, some information is required to map GOPs and layers into pieces and vice versa. This information can be stored inside an index file, which should be transmitted together with the video sequence. Therefore, the first step consists of creating a new torrent that contains both files. It is clear that the index file should have the highest priority and therefore should be downloaded first.

Once the index file is completed, it is opened and information about offsets of different GOPs and layers in the video sequence is extracted. At this point, it is possible to define a sliding window, made of W GOPs and the pre-buffering phase starts. Pieces can only be picked among those inside the window, unless all of them have already been downloaded. In the latter case, the piece picking policy will be the same as the original BitTorrent, which is rarest [piece] first.

Inside the window, pieces have different priorities. First of all, a peer will try to download the base layer, then the first enhancement layer and so on. Pieces from the base layer are downloaded in a sequential order, while all the other pieces are downloaded rarest-first (within the same layer).

The window shifts every $t_{(GOP)}$ seconds, where $t_{(GOP)}$ represents the duration of a GOP. The only exception is given by the first shift, which is performed after the pre-buffering, which lasts $W * t_{(GOP)}$ seconds.

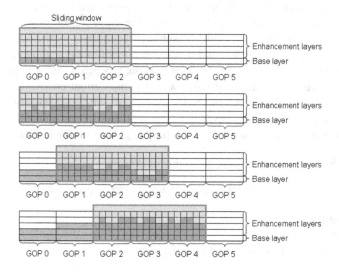

Fig. 3. Sliding window for scalable video bit-stream a) Pre-buffering phase start, b) Pre-buffering phase end, c) The window shifted after GOP0, d) The window shifted after GOP1

Every time the window shifts, two operations are made. First, downloaded pieces are checked, in order to evaluate which layers have been completely downloaded. Second, all pending requests that concern pieces belonging to a GOP that lies before the window are dropped. An important remark is that the window only shifts if at least the base layer has been received, otherwise the system will auto-pause. Figure 3 shows the behaviour of the system with $W = 3$. An early stage of the pre-buffering phase is showed in Figure 3a. The peer is downloading pieces from the base layer in a sequential way, while in Figure 3b the first two layers have been downloaded and pieces are being picked from the enhancement layer 2 according to a rarest-first policy. These pieces can belong to any of the GOPs in the window. In Figure 3c, the window has shifted as not all the pieces of enhancement layer 2 of GOP 0 have been received; this layer and higher layers are discarded. Inside the window, before downloading any other pieces from GOP 1 or GOP 2, the system will pick pieces from GOP 3 until the quality of the completed layers is the same. In other words, before picking any pieces that belongs to enhancement layer 2, pieces belonging to base layer of GOP 3 and enhancement layer 1 of GOP 3 have to be picked. In Figure 3d all the GOPs have the same number of complete layers and pieces are picked from enhancement layer 3. Another issue is the wise choice of the neighbours.

b. Neighbour Selection Policy

It is extremely important that at least the base layer of each GOP is received before the window shifts. Therefore, pieces belonging to the base layer should be requested from good neighbours. Good neighbours are those peers that own the piece with the highest transfer rates, which alone could provide the current peer with a transfer rate that is above a certain threshold. This threshold is determined by the current number of base layer pieces in the window and it is the minimum rate that allows this layer to be received on time. During the pre-buffering phase, any piece can be requested from any peer. However, every time the window shifts, the current download rates of all the neighbours are evaluated and the peers are ranked. After this operation, the base layer is only requested from the peers that have a download rate above this threshold value. The following algorithm explains the best neighbour policy in detail.

Algorithm: Sort neighbours according to download rate

Threshold = (# of pieces in layer 0 inside the window) * (piece length)/ (duration of the window)
for i in sorted list :
  if i owns the piece
    total download rate += download rate [i]
  end if
  if total download rate > Threshold:
    all the pieces with a download rate lower than i are bad peers
  end if
end for

3   Experimental Results

The performance of the proposed framework has been extensively evaluated to transmit wavelet-based SVC encoded video over P2P network. The network used for the experiments consists of three peers: two seeders and one leecher. No restriction was applied to download bandwidth of the leecher, while, as far as the two seeders are concerned, upload bandwidth was limited to 50 kbytes/s and 25 kbytes/s respectively. The leecher downloaded three five-minutes video sequences (Crew; City; Soccer) at CIF resolution and 30fps. The results are shown in "Download Rate" of the receiver and the "Received Video Bitrate" against time. Some selected results are shown in Figure 4-6.

All three graphs show similar results. In none of the cases the download rate is high enough to allow the transfer of all the layers and we have quality degradation. Moreover, the quality is not constant and follows the behaviour of the download speed. Finally, in all three cases we have a higher quality immediately after the prebuffering, because these GOPs are in the window for a slightly longer period, and at the end of the sequence, when we cannot shift the window anymore and we need to shrink it.

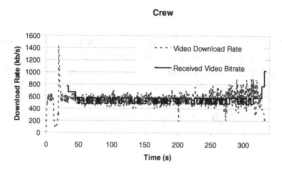

Fig. 4. Received download rate and received video bitrate for Crew CIF sequence

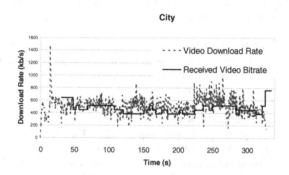

Fig. 5. Received download rate and video bitrate for City CIF sequence

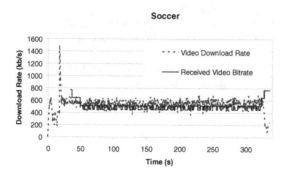

Fig. 6. Received download rate and video bitrate for Soccer CIF sequence

These are the preliminary results of the proposed technique when we have enough download rates to download the video. However we are working on scenarios when we don't have enough download speed to always download the base layer of scalable video. One solution will be to autopause the video until we get the full quality or we can downgrade the video to its lower resolution or frame rate. As these are the open questions in the community, so we are also working in this direction.

4 Conclusion

In this paper, we have presented a novel technique for scalable video streaming over P2P network. We proposed the new piece picking policy and neighbour selection policy in P2P network for efficient streaming of P2P network. Some selected results were presented, which show the efficiency of the proposed system. The future work will concentrate on how to efficiently receive layers of scalable video when the network doesn't allow us enough download speed to download the base layer of the scalable video.

References

1. Ramzan, N., Zgaljic, T., Izquierdo, E.: An Efficient Optimisation Scheme for Scalable Surveillance Centric Video Communications. Signal Processing: Image Communication 24, 510–523 (2009)
2. Pouwelse, J.A., Garbacki, P., Wang J., Yang, J., Iosup, A., Epema, D., Reinders, M., van Steen, M.R., Sips, H.J.: Tribler: A social-based based peer to peer system. In: 5th Int'l Workshop on Peer-to-Peer Systems (IPTPS) (February 2006),
   http://citeseerx.ist.psu.edu/viewdoc/
   summary?doi=10.1.1.60.8696
3. Mrak, M., Sprljan, N., Zgaljic, T., Ramzan, N., Wan, S., Izquierdo, E.: Performance Evidence of Software Proposal for Wavelet Video Coding Exploration Group. Technical Report, ISO/IEC JTC1/SC29/WG11/MPEG2006/M13146 (2006)
4. Cohen, B.: Incentives build robustness in BitTorrent. In: Proc. of First Workshop on Economics of Peer-to-Peer Systems, Berkeley, CA (June 2003)
5. Mrak, M., Izquierdo, E.: Spatially Adaptive Wavelet Transform for Video Coding with Multi-Scale Motion Compensation. In: IEEE International Conference on Image Processing, September 2007, vol. 2, pp. 317–320 (2007)
6. Zgaljic, T., Sprljan, N., Izquierdo, E.: Bit-Stream Allocation Methods for Scalable Video Coding Supporting Wireless Communications. Signal Processing: Image Communications 22, 298–316 (2007)
7. Tucker, T., Pate, D., Rattanapoka, C.: Yet Another Bittorrent Client,
   http://pingpongabc.sourceforge.net/
8. http://www.bittornado.com

Rate Adaptation Techniques for WebTV

Hugo Santos[1], Rui Santos Cruz[1], and Mário Serafim Nunes[2]

[1] Instituto Superior Técnico, Lisboa, Portugal
hugo.santos@tagus.ist.utl.pt, rui.cruz@ieee.org
[2] IST/INESC-ID/INOV, Lisboa, Portugal
mario.nunes@ieee.org

Abstract. This paper presents a novel hybrid **RTSP**, **SIP** and **HTTP** solution for providing converged WebTV services with automatic Rate Adaptation of multimedia content. The prototypical solution, being developed under the scope of the *My-eDirector 2012* european project, was deployed in a high-bandwidth LAN and some preliminary tests were performed.

Keywords: IPTV, WebTV, DCCP, Multimedia Streaming, Content Adaptation, Rate Adaptation.

1 Introduction

Telecommunications evolved from System centric to Network centric, and nowadays to User centric (allowing users to access services in a customized way and with a single authentication, always in a unique user session) [4].

An increasing large number of people is already having access to all kinds of multimedia content from all kinds of networks using a large variety of terminal devices, requiring different types of content formats with different quality levels.

Some adaptation systems have been created, enabling users to watch multimedia content on their terminal devices with the maximum quality possible, but in a very strict way, largely dependent on the devices/networks of each Service Provider. The biggest challenge is still the choice of the adequate adaptation technique to use for each multimedia content, in each of the Access Networks of the Provider and for each of the terminal devices connected.

While IPTV is a technology with capabilities to provide live video streaming and television services, over managed IP access networks, with mechanisms to ensure interactivity and appropriate Quality of Service (QoS) and of Experience (QoE) [9,11,18], WebTV addresses similar objectives but over the global Internet. While IPTV is usually based on Multicast only requiring IGMP, WebTV requires a more complex signaling protocol in order to initiate, modify and terminate sessions, a control protocol for multimedia contents and transport protocols, responsible for transporting the multimedia streams to the end device.

The Session Initiation Protocol (SIP) [15] and the associated Session Description Protocol (SDP) [7] are the protocols in the ETSI's TISPAN [3] architecture for signaling and session control of IP multimedia applications in NGN [12]. Coupled with Real Time Streaming Protocol (RTSP) [16] methods, for multimedia

P. Daras and O. Mayora (Eds.): UCMedia 2009, LNICST 40, pp. 161–168, 2010.
© Institute for Computer Sciences, Social-Informatics and Telecommunications Engineering 2010

control, they become a very flexible solution for managing multimedia sessions over IP and for E2E QoS signaling, related to the quality being experienced by the user, to achieve continuous QoS guarantees throughout the session.

This paper presents a novel hybrid **RTSP**, **SIP** and **HTTP** solution for providing converged WebTV services with automatic Rate Adaptation of multimedia content, and shares the experience of its development and evaluation, realized within the scope of the european project *My eDirector 2012* [2]. Several preliminary tests were carried out using high-bandwidth LAN connections.

In this paper, Section 2 describes the State of the Art and Section 3 presents the architecture and functionalities of the WebTV solution prototype. Section 4 presents some preliminary performance evaluation of the prototype. Section 5 concludes the paper.

2   State of the Art

There are several adaptation techniques that can be applied to the multimedia contents that flow over the network. But the efficiency of the adaptation techniques, rely on information that needs to be known before the adaptation technique takes in [13]:

- Information about the **characteristics of the client device**, like the size of the display, the colors that the device support and the size of the buffer.
- Information about the **content**, like the size of the buffer, the minimum streaming bitrate and compression formats.
- Information about the **network to which the client device is connected**, like bandwidth, jitter, packet loss, delays and all the variations of the characteristics that happen in the channel.

From the perspective of an Adaptation System there are three components to consider: the adaptation of the Content, the adaptation to Network conditions and the adaptation for an adequate Quality of Service and of Experience for the End user.

2.1   Video Content Adaptation

The adaptation techniques for multimedia contents can be typically included in three classes [17]:

Format Conversion: This type of technique simply transcodes the original content into another format (e.g., MPEG-4 to MPEG-2).

Selection/Reduction: This type of technique aims to reduce the number of frames of a video or to reduce the resolution of a stream. The most common are the **Resolution Reduction, Frame Dropping, DCT coefficients dropping** or a combination of the last two techniques.

Substitution: This type of technique replaces the original content by certain elements of it, being the most common the **Video-to-Text**, the **Video-to-Audio** and the **Video-to-Image**.

The points where the adaptation techniques can be used may be located directly in a Service Platform (Server) at the Provider Core, in the End user device (client application) or in a Proxy server (between the Service Platform and the End user device).

2.2   Adaptation to Network Conditions

The Datagram Congestion Control Protocol (DCCP) [10] is a standardized protocol that fills the gap between *TCP* and *UDP* protocols. Unlike *TCP*, it does not support reliable data delivery and unlike *UDP*, it provides a *TCP-friendly* congestion control mechanism in order to behave in a fair manner with other *TCP* flows. *DCCP* includes multiple congestion control algorithms that can be selected, depending on user *QoS* requirements. *DCCP* identifies a congestion control algorithm through its *Congestion Control ID* (CCID).

The most appropriate congestion control algorithm for video streaming is the CCID3 [5, 6], allowing a simple and fast computation of the most adequate transmission rate, and can be expressed by the following equation:

$$THR = \frac{s}{RTT.\sqrt{(\frac{p.2}{3})} + RTO.\sqrt{(\frac{p.27}{8})}.p.(1 + 32.p^2)} \tag{1}$$

where: THR is the transmission rate in bytes/second, s is the packet size in bytes, RTT the round trip time in seconds, p is the loss event rate (between 0 and 1.0) and RTO is the TCP retransmission timeout value in seconds.

2.3   Adaptation for End User Quality Perception

End user perception is crucial for the successful deployment of WebTV services. A practical way to measure End user quality is to use a parametric objective opinion model to estimate the subjectively perceived quality of the video, taking as quality factors the coding quality and potential problems due to transport on the network.

ITU-T has standardized a parametric computational model, as Recommendation G.1070 [8], for evaluating QoE of video-telephony. The model estimates the QoE of video-telephony services based on quality design/management parameters in terminals and networks. This model can be extended to live streaming WebTV environments to compute the best video quality V_q (MOS) at any specific moment in time, by the equation:

$$V_q = 1 + I_{coding} \exp\left\{-\frac{P_{plv}}{D_{P_{plv}}}\right\} \tag{2}$$

where I_{coding} is the basic video quality affected by the coding distortion, $D_{P_{plv}}$ is the packet loss robustness factor, expressing the degree of video quality robustness due to packet loss, and $P_{plv}[\%]$ is the packet-loss rate.

3   Outline of the Proposed Solution

The architecture of the streaming server for the WebTV solution (named In-
Stream Server) is being developed in the scope of the *My-eDirector 2012* european project. A final decision is not yet being taken, but a prototype being
developed by the authors considers a hybrid **RTSP**, **SIP** and **HTTP** Server
with an **Adaptation System** implemented on it. The basic functionality of the
architecture is the following:

- The *Encoders* send live streaming contents to the *InStream Server* but not
  necessarily all of them at the same time. Each live streamed content is encoded in MPEG-4 and/or H.264 and transmitted with several resolutions.
- A large variety of terminal devices, with a *SIP* module installed, can connect to the *InStream Server*, that starts a *SIP* session with the client after
  receiving the *RTSP* live streaming content request.
- The quality/bitrate of the streamed content is selected by the Adaptation
  System implemented in the *InStream Server* based on the periodic reports
  sent by the client during the session.
- If the user wants to switch to a different content channel, a *HTTP* request
  is sent to the *InStream Server*, requesting the new stream (channel).

3.1   Adaptation System Model

The main scope of the Adaptation System is to know the exact moment when to
switch a stream to a client. For the Adaptation System to be cost-effective and
lightweight, not requiring much computational resources during operation, and
provide good prediction performance within its scope, the best choice falls on a
parametric model. The model that was designed for the Adaptation System uses
information from the Network conditions by computing THR (equation 1) from
Congestion Control ID 3 (CCID3) algorithm of DCCP [10] and the estimate of
the subjective MOS for Video Quality, V_q (equation 2), adapted from ITU-T
G.1070 standard [8], with parameters related to the base quality for a given
codec, the bitrate and packet loss rate.

V_q will be computed at the same time of THR. With the results, the Adaptation System is able to identify the bitrate/video quality that the Client must
receive for the transmission channel conditions. The Stream-Switch Decision algorithm of the Adaptation System at the InStream Server is the following:

- A: The Server collects network information sent by clients.
- B: With that information, THR is computed.
- C: The THR value is then compared to the standard bitrate streams
     on the Server (from 128 kbps to 1024 kbps) to find the closest upper
     and lower standard bitrates.
- D: The two values are then used (one at the time) for V_q computations.
- E: With the results, a specific stream bitrate, with the closest higher
     video quality is then selected.

3.2   The WebTV Streaming Server Prototype

The *InStream Server* prototype has four main modules (Figure 1): the *RTSP* module, the *SIP* module, the *Link* module and *Adaptation System* module.

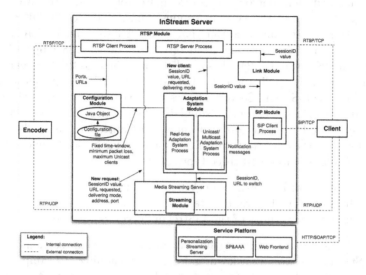

Fig. 1. The InStream Server prototype architecture modules

- **The RTSP Module:** This module is responsible for the content negotiations between the *InStream Server* and the *Encoders*, and also for the negotiation and the streaming of each multimedia content to the clients.
- **The SIP Module:** This module is used to exchange (send/receive) signaling and control messages with clients.
- **The Link Module:** This module provides the liaison between the *RTSP* and the *SIP* modules.
- **The Adaptation System Module:** This module implements the adaptation algorithms for the selection of the most appropriate stream for each client at any moment in time.

The SIP messages: The specific SIP messages used during each session are the **Subscribe Message** and the **Notify Message** [14].

- **Subscribe Message:** This message is created by the *InStream Server* and sent to the client device, to initiate a *SIP* session. The header contains the Unique Identifier created in the *RTSP* session and the body is empty.
- **Notify Message:** This message is sent by the client, periodically (every second), and contains the identifier of the *SIP* session (previously received in the Subscribe message) and, in the body, the information about the transmission channel conditions.

4 Performance Evaluation

The goals while testing the WebTV Streaming Server/Adaptation System prototype were to evaluate the functionalities it provides. A basic Test Architecture was implemented, in a controlled network environment, and a Test Methodology was designed to collect the system response times. The test architecture consists of a Content Streaming Server that stores five multimedia streams of the same content, but with different fixed bitrate/qualities (128 Kbps, 256 Kbps, 512 Kbps, 768 Kbps and 1024 Kbps) and streams them to the InStream Server. The End user client application is a hybrid RTSP+SIP player able to request streams from the InStream Server. The tests were carried out on a private High-Speed 100 Mbps LAN network. To simulate different network speeds and degradation (and introduce some loss) a Bandwidth Controller was implemented at the client system (Bandwidth Controller Standard Edition for Windows [1]). The Test Methodology considered three types of tests:

1. Test the effect of Packet-Loss on Video Quality
2. Test the Video Quality vs. bitrate
3. Test the Response times of the InStream Server vs. bitrates

The type 3 Test consisted on a suite of ten different client sessions where the Session Startup, Session Teardown and Stream Switching (10 samples per session) times were collected. A session starts when a stream (request) is ordered to the InStream Server and ends with a Teardown request. For each initial request the InStream Server starts sending the lowest quality stream to the client. The adaptation algorithm, upon receiving feedback information, in a certain time-frame, from the client (player) takes decision about the most appropriate quality class (bitrate/quality) stream to be served to the client.

Test Results: The effect of Packet-Loss for different bitrates can be observed in Figure 2. As expected, the increased packet loss reduces the perceived quality

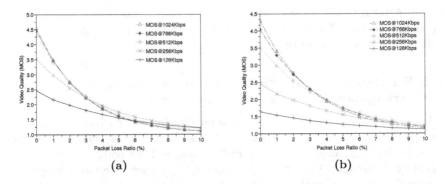

Fig. 2. Video Quality vs. Packet-Loss rate (different bit-rates): (a) H.264; (b) MPEG-4

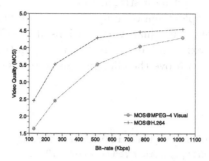

Fig. 3. Video Quality vs. bit-rate

Table 1. Response time of InStream Server

	Stream Switching time (ms)		Session Startup time (ms)	Session Teardown time (ms)
	MPEG-4	H.264		
Average	522.2	40.0	1055.4	1.0
Minimum	14.0	11.0	335.0	1.0
Maximum	2519.0	74.0	1390.0	1.0
Standard Deviation	374.7	17.1	357.2	0.0

in a non-linear way. In addition, the quality ranking is in agreement with the bit rate ranking. For the same packet loss rate, the 1024 Kbps sequence exhibits the best quality, whereas the 128 Kbps shows the worst quality. Figure 3 shows the video quality with no losses, where it can be observed that the Video Quality increases with the bitrate, as expected. Table 1 summarizes the results from Response times of the InStream Server. On average, the stream switching time at the InStream Server is around half a second for MPEG-4 and less than 0.1 second for H.264. For the Session Startup time an average of around one second is a very interesting figure.

All the measurements were collected at the InStream Server, reflecting just the time that the application (InStream Server/Adaptation System) takes to react to a certain command or request. No measurements were taken for the network transit time (from client to server, or server to client).

5   Conclusion and Future Work

This project is currently under development. Almost all components are implemented, but not fully tested. One of the key issues in the developed environment is related to stream synchronization. All the streams that come from the *Encoders* should bear the same start time reference (or close, just separated by a few milliseconds). If this doesn't happen, when the stream switching is made, there is a possibility of selecting a stream that is delayed. Future work on this project will include the deployment of components not yet implemented (for scalability of the solution, objective quality estimations at the Client, etc.) the fix of eventual bugs that may exist and an exhaustive test plan (with both Functional and Performance tests), to prove the effectiveness of this solution.

References

1. Bandwidth Controller, http://bandwidthcontroller.com/index.html
2. My eDirector (2012), http://www.myedirector2012.eu
3. ETSI: TISPAN - Telecoms and Internet converged Services and Protocols for Advanced Networks, http://www.etsi.org/tispan/
4. ETSI: Specialist Task Force 360: QoS management at the Network Interfaces to optimize quality of the service delivered to the User (2009), http://portal.etsi.org/STFs/STF_HomePages/STF360/STF360.asp
5. Floyd, S., Handley, M., Padhye, J., Widmer, J.: RFC 5348-TCP Friendly Rate Control (TFRC): Protocol Specification. IETF (2008)
6. Floyd, S., Kohler, E., Padhye, J.: RFC 4342-Profile for Datagram Congestion Control Protocol (DCCP) CCID 3: TCP-Friendly Rate Control (TFRC). IETF (2006)
7. Handley, M., Jacobson, V.: RFC 2327-SDP: Session Description Protocol. IETF (1998)
8. ITU-T: Rec. G.1070 Opinion model for video-telephony applications (2007)
9. Kerpez, K., Waring, D., Lapiotis, G., Lyles, J., Vaidyanathan, R.: IPTV service assurance. IEEE Communications Magazine 44(9), 166–172 (2006)
10. Kohler, E., Handley, M., Floyd, S.: RFC 4340-Datagram Congestion Control Protocol (DCCP). IETF (2006)
11. Lee, C.S.: IPTV over Next Generation Networks in ITU-T. Broadband Convergence Networks, 1–18 (2007)
12. Mikoczy, E.: Next generation of multimedia services - NGN based IPTV architecture. In: IWSSIP 2008, pp. 523–526 (2008)
13. Mohan, R., Smith, J.R., sheng Li, C.: Adapting Multimedia Internet Content for Universal Access. IEEE Transactions on Multimedia 1, 104–114 (1999)
14. Roach, A.B.: Session Initiation Protocol (SIP)-Specific Event Notification. RFC 3265, Updated by RFC 5367 (2002), http://www.ietf.org/rfc/rfc3265.txt
15. Rosenburg, J., Shulzrinne, M., Camarillo, G., Johnston, A., Peterson, J., Sparks, R., Handley, M., Schooler, E.: RFC 3261-SIP: Session Initiation Protocol. IETF (2002)
16. Schulzrinne, H., Rao, A., Lanphier, R.: RFC 2326-Real Time Streaming Protocol. IETF (1998)
17. Shanableh, T., Ghanbari, M.: Heterogeneous video transcoding to lower spatio-temporal resolutions and different encoding formats. IEEE Trans. Multimedia 2, 101–110 (2000)
18. Volk, M., Guna, J., Kos, A., Bester, J.: IPTV Systems, Standards and Architectures: Part II - Quality-Assured Provisioning of IPTV Services within the NGN Environment. IEEE Communications Magazine 46(5), 118–126 (2008)

An Adaptive Control System for Interactive Virtual Environment Content Delivery to Handheld Devices

Gianluca Paravati, Andrea Sanna, Fabrizio Lamberti, and Luigi Ciminiera

Politecnico di Torino, Dipartimento di Automatica e Informatica,
C.so Duca degli Abruzzi 24, I-10129, Torino, Italy
gianluca.paravati@polito.it,
andrea.sanna@polito.it,
fabrizio.lamberti@polito.it,
luigi.ciminiera@polito.it

Abstract. Wireless communication advances enable emerging video streaming applications to mobile handheld devices. For example, it is possible to show and interact with a complex 3D virtual environment on a "thin" mobile device through remote rendering techniques, where a rendering server is in charge to render 3D data and stream the corresponding image flow to the demanding client. However, due to bandwidth fluctuating characteristics and limited mobile device CPU capabilities, it is extremely challenging to design effective systems for interactive streaming multimedia over wireless networks. This paper presents a novel approach based on a controller able to automatically adjust streaming parameters basing on feedback measures coming back from the client device. Experimental results prove the effectiveness of the proposed solution to cope with bandwidth changes, thus providing a high Quality of Service (QoS) in remote visualizations.

Keywords: Video Streaming to Mobile Devices, Closed-Loop Controller, QoS, Remote Visualization.

1   Introduction

The combination of advances in wireless communication and the rapid evolution and growing popularity of mobile handheld devices can converge into appealing applications within the field of 3D graphics and virtual reality. For instance, virtual guiding malls, multiplayer games and collaborative virtual environments are emerging applications now available also on "thin" devices. The increased capability of wireless connectivity allows to display graphically attractive environments also on devices with limited HW capabilities through remote rendering techniques, where a server is in charge to render 3D data and stream the corresponding image flow to the mobile client [1,2]; the user on the client side can interact with the 3D scene by sending navigation commands to the server.

P. Daras and O. Mayora (Eds.): UCMedia 2009, LNICST 40, pp. 169–178, 2010.
© Institute for Computer Sciences, Social-Informatics and Telecommunications Engineering 2010

This work presents the design of a novel feedback-based controller in the context of image streaming systems with interactivity constraints; the controller is designed to automatically adapt streaming parameters to both bandwidth fluctuations and device characteristics such as maximum supported resolution and processing capabilities. The proposed technique works at the application level and it can be applied to any kind of client device; however, in this work, we specifically focus on common handheld devices, such as PDAs and smart phones, because of their limited computational capabilities. Preliminary results indicate the effectiveness of the proposed approach, although it is only a starting point for future work investigating control-based remote rendering.

The main features of the proposed system include: increase in visualization quality through smoothly changes in streaming parameter values, adaptation of streaming parameters to bandwidth fluctuations, and adjustment to end device capabilities. The proposed control technique is general and it can be used in any image compression-based streaming scenario.

The remainder of the paper is organized as follows. Section 2 reviews previous works related to the management of streaming parameters and control-theoretic approaches. Section 3 presents the details of the proposed control framework. Experimental results are discussed in Section 4. Finally, conclusions are drawn in Section 5.

2   Related Work

Real-time and interactive streaming to handheld devices over variable bandwidth channels has to deal with different Quality of Service (QoS) issues ranging from network features (such as bandwidth fluctuations and channel latency) to mobile device capabilities (such as receiver decoding performances). The concept of QoS addressed in this paper refers to the issues pertaining the achievement of interactive frame rates and low latencies for interactive remote visualization applications. These issues are investigated in depth in [3]; to deploy effective remote visualization applications, a few QoS requirements need to be met. If these requirements cannot be met, the user will be probably unable to get the expected results. The QoS requirements that are most directly related to remote visualization are: low delay, high throughput/bandwidth, low latency. Motion-JPEG (M-JPEG) has proven to be an effective means for obtaining very low latencies and low processor overhead, although at the expense of an increased bandwidth [4,5]. In fact, M-JPEG parameters like resolution, frame rate and image quality (that determine the bandwidth occupation of the streaming flow) can be combined and tuned to suit network characteristics. The relationship among these parameters can be modeled as in [6]:

$$f = \frac{B}{w \cdot h \cdot C_d \cdot \frac{1}{C_r}}; \tag{1}$$

where f is the achievable frame rate, w and h denote image resolution, B is the currently available bandwidth, C_d is the color depth in terms of bits per pixel, and C_r is the compression ratio of an image with respect to the same uncompressed picture (i.e., C_r is strictly related to the image quality).

Control-theoretic approaches to performance management for computer systems such as Internet web servers, databases and storage systems, have been successfully applied in the past. In [7], a study on the admission control for an Internet web server is presented; a linear-parameter-varying (LPV) approximation for the modeling of the dynamic relationship from the request rejection ratio to the response time for the admitted requests is used. The main characteristic of an LPV controller is the possibility to control state variable non-linear dynamics in different working conditions, due to external agents influencing the system. Although this approach initially seemed to be suitable to control the modification of the above parameters basing on the available bandwidth, the main drawback that prevents from using it in a M-JPEG streaming scenario is due to the image quality dynamics; indeed, its impact on the image size to be delivered (and hence on the bandwidth occupation) is unpredictable.

Since working conditions in interactive streaming to mobile handheld devices can differ with available channel bandwidth, the attention was focused on adaptive control techniques [8,9], in order to build a controller able to modify its control parameters according to state variations. Adaptive control techniques estimate the behavior of the system through linear regression algorithms, which functionality is based on the refinement at each step to asymptotically reach a set of parameters able to represent the unknown initial system. In video streaming context, these methods are not suitable because of the large fluctuations of the parameters; an estimator based on these techniques cannot identify an asymptotically stable system.

After establishing that the application of complicated control techniques does not favor the solution of the problem, it has been preferred to design a PID-based controller; its robustness and reliability are particularly useful to control a system characterized by unpredictable fluctuations and to correctly exploit the feedback channel of the system. In order to achieve performance specifications, gain-scheduling techniques were used. Gain scheduling [10] is based on the idea of using different (a-priori performed) calibrations in different circumstances, thus realizing a parameter calibration system able to adapt to the state of the system.

3   The Control Algorithm

The design of the proposed controller aims at taking advantage of automatic control techniques to concurrently tune all the parameters involved in a remote rendering scenario (i.e. resolution, image quality, frame rate) without any a-priori knowledge about the precise effects caused by altering these parameters. The main requirements identified in the design phase of the controller are: sensitivity to feedback measures, robustness to non linearity, independence to the quality of the underlying network, and optimal usage of available resources.

The control system can be sketched as in Figure 1. The controller is fully implemented on the server side, meaning that the client has only to compute its

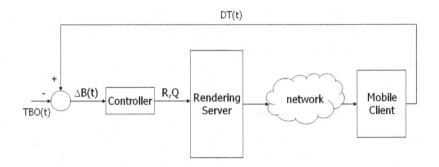

Fig. 1. The logical control system

own device throughput and to periodically feed this information to the streaming server. It is worth remarking that, in this paper, the term robustness is related to the degree to which the controller can function correctly in the presence of uncertainties affecting our model of the system.

The reference of the control system is the theoretical bandwidth occupation $TBO(t)$ that depends on the current encoding parameters and can be computed reversing eq. 1. The measured output of the system (on the client side) is the device throughput $DT(t)$. The general structure of the controller system is characterized by a single input (the bandwidth error $\Delta B(t) = DT(t) - TBO(t)$) and two separate outputs: a resolution (R) and a quality (Q) factor. These outputs are used by the rendering server to create, compress, and stream through the network an image representation of the 3D scene to the mobile device.

The controller deals with the bandwidth control trying to balance the system around a target frame rate (default or supplied by the user); the value of the target frame rate influences $TBO(t)$. Basically, the actual frame rate of the client (computed using eq. 1) is continuously compared to the target frame rate; the controller tries to minimize the error/difference between these two values; the system variables are controlled so as to undergo changes that are proportional to the error measured as input. In more details, the system changes the parameters of the encoded flow of images in such a way to reach the desired frame rate exploiting the minimum between the current available bandwidth (that depends on network status) and the current device throughput (that depends on device capabilities).

The reciprocal interaction between the controlled variables (R and Q) causes non linear and unpredictable fluctuations in the required bandwidth to be used for the transmission of the data stream to the client. This issue can be tackled through the isolation of the relationship that binds each variable to the image size, so that we can establish the degree of variation needed to increase or decrease the image size (and so the bandwidth occupation) to a determined quantity.

The basic idea is to subdivide the available bandwidth in different portions to be assigned to each system variable; thus, the input bandwidth error ΔB

can be splitted into two separated quantities: ΔB_R and ΔB_Q. The knowledge of mathematical relations able to govern the growth or the reduction of a system variable, as the bandwidth changes, allows to individually set each variable by keeping fixed all the other parameters. A mathematical relationship that links a bandwidth quantity to be filled (or released) and a resolution increment (or decrement) has been defined:

$$\Delta R = \frac{\Delta B}{B_R^{up} - B_R};$$

(2)

where ΔR is the amount of the increment or decrement in resolution, ΔB is the bandwidth error measured by the controller, B_R and B_R^{up} are respectively the bandwidth to be used with the current resolution index and the bandwidth to be used with the immediately upper resolution index. The proportionality between resolution and bandwidth is clearly visible from eq. 2; on the other hand, eq. 1 makes explicit the relationship between frame rate and bandwidth, showing how a resolution change can be reflected on a frame rate change through a proportionality coefficient k_R:

$$\Delta R = k_R \cdot \Delta f.$$

(3)

The controller deals with another system variable represented by the compression quality. The identification process of a control relationship for this variable has been more complicated, mainly because it is difficult to evaluate, even approximately, a mathematical relationship between image quality and compression factor, which changes reflect to the compressed image size through an inverse proportionality relationship. Thereby, it has been decided to exploit the controller feedback measure, slowly varying the quality level at each loop and then correcting the excessive bandwidth repercussions if necessary. The idea is to subsequently control a variable that cannot be controlled a priori because of its unpredictability. Since a precise bandwidth variation implies a precise frame rate variation, a direct relationship between image quality and frame rate is established. Actually, these variables are not linearly dependent because the incidence on the frame rate is determined by the compression factor C_r, that has a non-deterministic dependence on image quality. Such a relationship can be modelled as:

$$\Delta Q = k_Q \cdot \Delta f.$$

(4)

The instability both of the available bandwidth and of the device throughput makes unstable the input reference error ΔB of the controller; thus, it becomes vacillating and it does not asymptotically approach to a stable equilibrium value. Every instant can be characterized by different working conditions. In order to manage this instability an adaptive control approach can be used. It allows to find some specific coefficients for the control equations; these coefficients are relative to external system variables and the current state of the system. The k_R coefficient can be expressed as:

$$k_R = \frac{k_p}{k_B} = \frac{k_p}{B_R^{up} - B_R};$$

(5)

where k_p is a proportional constant of a PID-based system, k_B is an adaptive parameter that depends on the current resolution. In this way, k_R is constituted by a part used to compute the bandwidth occupied by a change in resolution, represented by k_p, and a part used by the system to tune the bandwidth quantity to be assigned to the resolution. As a result, the input bandwidth is subdivided into a portion that is assigned to the resolution variable. As the coefficient k_p increases a greater portion of bandwidth is reserved to allow a change in resolution, enhancing the sensitivity of the system to bandwidth fluctuations and the quick response to reference tracking as a matter of fact. The unpredictable nature of bandwidth fluctuations, particularly in wireless connections, and the precision that is possible to reach estimating the bandwidth as a function of the resolution index, lead the integral-derivative part of a PID-based controller useless at the moment.

The subdivision of the bandwidth between the system variables is a key aspect for the controller (calibration phase). Indeed, the two variables (resolution and image quality) cannot be precisely tuned at the same time because of their reciprocal interactions; on the other hand, it is possible to act singularly on each of them through the relationships between these variables and the reference parameter, i.e. the target frame rate. Thus, the control phase has been split in two stages. In every stage, it is possible to act on a single variable by proceeding in cascade, having a different bandwidth as an input at each stage and keeping fixed all the parameters to modify only the considered variable. At each stage the input bandwidth is used to modify the variable of interest and it progressively decreases till exhausting, thus minimizing the input error as a matter of fact. In this way the system can be characterized as a SISO model. A first stage receives as input the bandwidth error ΔB and uses part of it to regulate resolution (according to eq. 2); the new resolution value causes a change in theoretical bandwidth occupation and the achievable frame rate computed with the current parameters, thus it is used as input of a second stage to recompute the bandwidth error. The second stage receives as input the new bandwidth error that is used to regulate quality according to eq. 4.

4   Tests and Results

The proposed controller has been implemented and tested in a remote rendering scenario. The rendering server runs on a Dual-Core AMD Opteron CPU 2.60 GHz workstation equipped with 3.50 GB of RAM and with an NVIDIA Quadro FX 3500 graphics card; it has been developed using the C++ language and it is based on the OpenSG library. The client program runs on a HTC TyTN II smart phone, connected to the rendering server through a 802.11g wireless access point. The client program has been developed using J2ME (Java Micro Edition), as it provides a flexible environment for applications running on mobile and other embedded devices. Figure 2 shows a remote rendering session; the server (shown in the background) is in charge of rendering the 3D scene and of streaming a M-JPEG flow to the "thin" client (shown in the foreground).

Fig. 2. A remote 3D rendering session on a TyTN II smart phone

Figure 3 shows the behavior of the controller under varying system conditions; the monitored parameters are (from top to bottom): resolution, image quality, actual frame rate reached by the mobile device, and measured throughput (the feedback feature). Figure 3 (a) shows the ability of the controller to react to changes in network conditions. The system is configured to maintain a target frame rate ($FR_{target} = 15$ fps) during the whole test. Indeed, the client application allows the user to express his/her preference in terms of frame rate; a higher frame rate enhances motion smoothness and impacts on perceived interactivity. During the first part of the simulation (period between $t = 0$ and $t = t_1$) the system worked in stationary conditions exploiting the maximum throughput of the device ($DT \approx 60$ KB/s); a trade-off between resolution and quality was reached allowing to maintain the target frame rate. At time $t = t_1$ a maximum bandwidth limitation was imposed on the server side ($BW_{limit} = 40$ KB/s) to simulate a network bottleneck; the frame rate on the client side rapidly dropped down (from $FR \approx 15$ fps to $FR \approx 9$ fps) and the controller reacted by gradually reducing the parameter values. While changing parameters, the controller continuously compared the target frame rate with the effective frame rate. After the $t = t_2$, the controller leaded the system to a different steady state reaching again the target frame rate, thus mitigating the effects of the bandwidth bottleneck previously introduced. If the target frame rate was higher, parameters would have continued to be reduced until the target frame rate was reached again. Figure 3 (b) proves the ability of the controller to follow a specified target

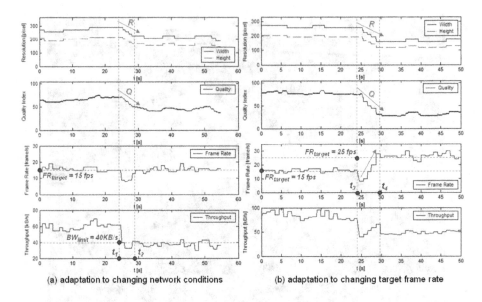

(a) adaptation to changing network conditions   (b) adaptation to changing target frame rate

Fig. 3. Performance on a TyTN II smart phone: (a) adaptation to changing network conditions and (b) adaptation to changing target frame rate

frame rate. Initially, the system maintained a target frame rate equal to 15 fps (period between $t = 0$ and $t = t_3$); the system was steady around high values of resolution ($R = 288 \times 216$ pixels) and quality ($Q \approx 80$) parameters. At time $t = t_3$ the target frame rate was set to 25 fps through the client GUI. During the period between $t = t_3$ and $t = t_4$ the controller reacted by gradually reducing resolution ($R = 176 \times 132$ pixels) and quality ($Q \approx 28$). After the transitory, the system reached a different steady state around the new frame rate value; although the growth of the number of frames per second received, the device throughput is lower than before because images are more compressed, thus indicating that performances are limited by the decoding capabilities of the mobile device.

In this work, a set of experimental tests of user experience were carried out in order to evaluate its relationship with QoS parameters. A group of 56 subjects was asked to carry out two sets of tests on the designed system and later answer a questionnaire aimed at collecting a feedback on system performances. Each user was individually trained in order to know how to use features of the remote visualization application running on an HTC TyTN II mobile device. The trainer performed two sessions (with and without the proposed controller) showing the aspect and behavior of the user interface; in particular, the possibility of changing the target frame rate was emphasized. After the training phase, each user was allowed to use the device alone. Each user was asked to perform the two tests by navigating the 3D scene; during the first test the controller was disabled, while during the second test it was enabled. During the first test, the image stream

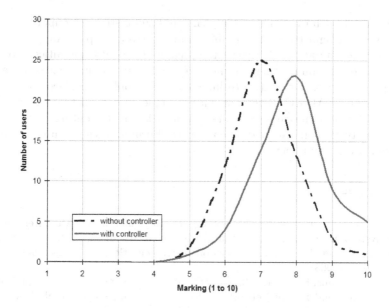

Fig. 4. User feedbacks from user experience tests

characteristics were fixed; in particular, encoding parameters were configured to match the maximum device resolution with the best image quality. This choice led to a poor frame rate. During the second test, the controller was enabled and users were allowed to change the target frame rate. In the questionnaires, users were then asked to assign marks (in a range 1-10) to the perceived performance of the visualization tool both with and without the controller. Figure 4 shows the results of this subjective evaluation. The x-axis represents the possible mark values, while the y-axis indicates the number of users assigning the given mark. The broken line is related to the score distribution obtained by experimenting the visualization system without the proposed controller. On the other hand, the unbroken line refers to the score distribution achieved by enabling the control system. Figure 4 clearly shows that application of the control system presented in this paper noticeably improved the user experience, that is the perceived performance of the visualization. The average mark obtained using the described methodology is equal to 7.89, compared to 7.1 when it is not used. This can constitute an effective proof of the relationship between the user experience and QoS parameters.

5   Conclusions

Due to bandwidth-demanding characteristics and limited mobile device capabilities, it is extremely challenging to design effective systems for real-time and interactive streaming multimedia over wireless networks for handheld devices. This paper presents a controller able to automatically adjust M-JPEG streaming parameters; the solution has been implemented and tested in a remote rendering scenario, where smart phones are able to interactively display complex 3D virtual environments although their limited computational capabilities. However, the proposed solution is general and can be applied to any interactive streaming scenario; moreover, this control technique can be applied to video compression schemes such as MPEG. The controller is implemented on the server side; it exploits a feedback measure from clients thus it is able to continuously adapt to changing network conditions, different device throughput capabilities and different interaction requirements.

References

1. Pazzi, R.W.N., Boukerche, A., Huang, T.: Implementation, Measurement, and Analysis of an Image-Based Virtual Environment Streaming Protocol for Wireless Mobile Devices. IEEE Transactions on Instrumentation and Measurement 57(9), 1894–1907 (2008)
2. Quax, P., Geuns, B., Jehaes, T., Lamotte, W., Vansichem, G.: On the Applicability of Remote Rendering of Networked Virtual Environments on Mobile Devices. In: International Conference on Systems and Networks Communications, p. 16 (2006)
3. Stegmaier, S., Diepstraten, J., Weiler, M., Ertl, T.: Widening the Remote Visualization Bottleneck. In: Proc. Third Int'l Symp. Image and Signal Processing and Analysis, pp. 174–179 (2003)
4. Endoh, K., Yoshida, K., Yakoh, T.: Low delay live video streaming system for interactive use. In: IEEE International Conference on Industrial Informatics, pp. 1481–1486 (2008)
5. Nishantha, D., Hayashida, Y., Hayashi, T.: Application level rate adaptive motion-JPEG transmission for medical collaboration systems. In: The 24th International Conference on Distributed Computing Systems Workshops, pp. 64–69 (2004)
6. Paravati, G., Sanna, A., Lamberti, F., Ciminiera, L.: A novel approach to support quality of experience in remote visualization on mobile devices. In: Eurographics 2008 short paper Proceedings, pp. 223–226 (2008)
7. Qin, W., Wang, Q.: An LPV approximation for admission control of an internet web server: Identification and control. Control Engineering Practice 15, 1457–1467 (2007)
8. Lu, Y., Abdelzaher, T., Lu, C., Tao, G.: An adaptive control framework for QoS guarantees and its application to differentiated caching services. In: The 10th IEEE International Workshop on Quality of Service, pp. 23–32 (2002)
9. Wu, K., Liljia, D.J., Bai, H.: The applicability of adaptive control theory to QoS design: limitations and solutions. In: The 19th IEEE International Parallel and Distributed Processing Symposium (2005)
10. Slotine, J.J., Li, W.: Applied Nonlinear Control. Prentice-Hall, Englewood Cliffs (1991)

UCMedia 2009

Session 7: Security, Surveillance and Legal Aspects

VirtualLife: Secure Identity Management in Peer-to-Peer Systems

Dan Bogdanov[1,2] and Ilja Livenson[1,3]

[1] University of Tartu, Liivi 2, 50409 Tartu, Estonia
[2] AS Cybernetica, Akadeemia tee 21, 12618 Tallinn, Estonia
[3] NICPB, Akadeemia tee 23, 12618 Tallinn, Estonia
db@ut.ee, ilja@kbfi.ee

Abstract. The popularity of virtual worlds and their increasing economic impact has created a situation where the value of trusted identification has risen substantially. We propose an identity management solution that provides the user with secure credentials and allows to decrease the required trust that the user must have towards the server running the virtual world. Additionally, the identity management system allows the virtual world to incorporate reputation information. This allows the "wisdom of the crowd" to provide more input to users about the reliability of a certain identity. We describe how to use these identities to provide secure services in the virtual world. These include secure communications, digital signatures and secure bindings to external services.

Keywords: identity management, virtual worlds, security, trust and reputation.

1 Introduction

Online virtual worlds are popular among users and organizations. Virtual environments like Second Life and Active Worlds are actively used by companies and organizations to promote their products and services[1]. Establishing a visible presence in such a world has become a marketing strategy. The users are interested in virtual worlds for the social interaction and entertainment possibilities. Building a virtual world to attract both users and service providers requires a strong technical framework and a well-defined focus.

In our work we address the issue of identity verification and trusted service provision. Most of the online worlds currently in active use put little effort on the identification of participants. This is a problem for anyone who has to trust the presented identity of their communication partner. One motivating example is a business transaction, where parties need to identify each other to enter an agreement. Another is a system that verifies the users' age to restrict access to age-specific content or provides age information to communication partners. The last example can be extremely motivating for parents whose children engage in online chats. Also, if a user conducts a criminal act inside the virtual world, then it can be claimed that the responsibility lies on the virtual world provider, because it did not fully identify the user.

Our contribution. We present a holistic solution to identity management and its applications in an online virtual world. We propose a way to handle the assignment

P. Daras and O. Mayora (Eds.): UCMedia 2009, LNICST 40, pp. 181–188, 2010.

and storage of identity information, how to prove identities to other participants and how to build services that use this information. We also describe techniques to make the system more intuitive for users by providing visual indicators of the strength of identity and trust information. The solution has been developed in conjunction with the VirtualLife virtual world [2] and it has been implemented within that world.

The usage of the proposed identity management system relies on the following assumptions: the capability to use X.509 security infrastructure and the capability to establish network connection to any node in the system. Although the solution is generic and can be used in any multi-user system, it was developed and adjusted for use in 3D worlds. It relies on a custom peer-to-peer messaging layer, that is complicated to implement in browser-based virtual worlds.

In this paper we introduce VirtualLife and its identity management system. We discuss how a variety of services can be built using this system and how they benefit from its properties. This is the underlying work for further research that may be conducted once the VirtualLife system is online and actual user experiences can be taken into account.

2   The Architecture of VirtualLife

The design of VirtualLife [2] is based on the idea of a connected network of peers. Every peer can act both as a provider and a user to the services in the virtual world. However, in practice it makes sense to distinguish some more powerful peers that provide additional services. Also, a minimal amount of transactions should rely on a trusted third party, e.g. a server. For these reasons VirtualLife has been designed to use a hybrid peer-to-peer network topology.

Each node runs a selection of services. Based on the services running in a particular node we separate the nodes into *clients*, *zones* and *nations*. The client node is run by a user and it provides the means for connecting to the virtual world and letting the user interact with it. The zone node is responsible for running a part of the virtual world. It maintains the world state, performs the necessary simulations and relays information between nodes. The nation node manages a group of zones and provides them with rules that make all zones in a nation more consistent. The deployment of the VirtualLife network is illustrated on Figure 1.

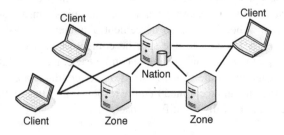

Fig. 1. An example deployment of the VirtualLife network

The security layer of VirtualLife provides a wide range of services that are used by other VirtualLife platform components. These services include data structures and operations for cryptographic primitives, certificate management and authorization. The system is built on top of the X.509 public key infrastructure standard[3], which is widely used for securing access to the sensitive services, for example e-mail accounts and online banks. It defines a hierarchy of trusted third parties called Certification Authorities that issue temporary digital documents binding together a public key and identity information. These digital documents are called *certificates*.

The networking layer of VirtualLife is designed to support the peer-to-peer topology and the identity system. All VirtualLife connections may contain multiple logical streams for different services. For example, chat, world data and user avatar coordinates can use different streams in the same connection. A stream can also be transparently encrypted and authenticated to provide secure transport between peers.

3 VirtualLife Identities

3.1 Identity Information

Every node in the VirtualLife system has an identity that is used for identification in services. The identity is a collection of profile information, security credentials, trust and reputation. VirtualLife uses X.509 key pairs consisting of a public and a private key. The identity is created and managed by every node itself. The key pairs may be added to the identity as needed. In the VirtualLife network, identities are registered at the nation node to simplify lookups during verification procedures. The structure of a VirtualLife identity is shown on Figure 2.

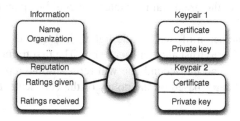

Fig. 2. An example of a VirtualLife identity with two key pairs

The X.509 infrastructure is commonly used for identification and secure transport in existing applications. For example, companies like VeriSign and Thawte are issuing X.509 certificates that can be used for securing websites, e-mail connections and other services. Certification authorities have *certification policies* that describe the procedures taken when issuing a certificate. Some authorities will verify the personal information on the certificate by requiring the user to appear in person or send a copy of a document. Others may issue certificates more freely.

VirtualLife is making use of these established trust relations in VirtualLife to give hints to the users. If it is known that the certification policy of an authority requires certificate holders to prove their identities with a document, VirtualLife can tell its users that if someone who cryptographically proves the ownership of a verified certificate than that someone is more probably who he or she claims to be. This of course holds only to the information in the certificate. All other information provided by the user that has not been verified by the certification authority cannot still be trusted.

We note that the identities contain key pairs that contain private keys. In order for the proposed security measures to work, the private keys must be stored in the client software and not uploaded to the any other node. Any node with access to a private key of an identity can effectively claim to own that identity. If the user wants to use the identity from multiple computers, the key pairs must be transported using a portable storage device.

3.2 Intuitive Identity Verification

There is always a trade-off between the security of the system and its usability. If a system has a strict security policy then the users must perform additional tasks to ensure their security. In the context of identity verification such a task occurs when the user is trying to find out the reliability of a communication partner. In most systems, the only proof of the identity of the other party is the user name provided by the server relaying information. This approach is convenient, but allows an attacker to easily forge an identity as the user gets no proof of identity and has to rely on the information provided by the server.

VirtualLife ensures that every authenticated connection between two parties also provides a cryptographic proofs of the parties' identities. However, even when an identity is established, the users still have to decide whether the provided information is reliable. This can be complex due to the amount of associated technical details. To overcome this obstacle, the graphical user interface could display the summary of the identity information.

We propose the inclusion of two "traffic lights" in locations where the user must make a decision whether to trust another user or not. The first traffic light will represent the strength of identity and the second one will represent the reputation. The identity traffic light has three states based on the policy of the certificate authority that issued the user certificate:

1. red (entrusted)—guest or temporary user;
2. yellow (weakly trusted)—certified by an authority that does not verify people's identities using a document and/or physical appearance;
3. green (trusted)—certified by an authority that verifies user identities.

The reputation traffic light has three states as well, depending on whether the user has negative, neutral or good reputation. An example of this design is illustrated in Figure 3.

Clicking on each of the "traffic lights" opens a pop-up with a detailed information. The identity pop-up will include certificate details such as the name of the certification authority, expiration date and certified user information. The reputation pop-up will use visualization to illustrate the reputation status of the user. The identity information must originate from a direct connection with the other user. This is the case in possibly security-critical scenarios such as text or voice chat, file exchange and contract signing.

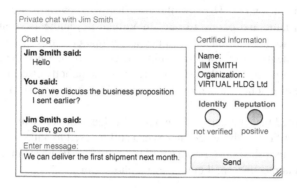

Fig. 3. Identity verification in a chat. The identity light is yellow and reputation is green.

3.3   Trust and Reputation Management

A strong identification method can be used to improve trust and reputation management. In social networks as well as virtual worlds users can rate each other. While the aspect being rated may differ from system to system, the gathered information can be used by other users to make decisions. Trust and reputation are more important in systems with a commercial component such as auction or sales environments. If many other users consider a seller to be trusted, more users have a reason to trust that seller. On the other hand, a seller with a negative reputation will be avoided by new customers.

In a typical case a seller with a bad reputation will abandon the account and make a new one. This will be impossible if the identity is based on an external source of trust. VirtualLife can restrict the user from creating another identity with the same certificate. We point out that the proposed enhancements are achieved, if the use of outside trust is actively encouraged or strictly required. If the user has a certificate from an authority that issues certificates freely, then any user can create a new identity with a new certificate that has no connection to the previous one. In the latter case the new identity will have no ratings and a default reputation.

The exact algorithms that will be used in VirtualLife for trust and reputation calculations are still being developed.

4   Using Identities in Services

We will now give an overview of services and interactions enabled by the suggested identity and reputation system. We concentrate on the basic secure operations in virtual worlds—authentication, authorization and secure communications. We also describe the use of digitally signed contracts in virtual worlds and linking outside databases to improve in-world services.

4.1   Authenticating Users

Authentication is the process of determining the identity of a user. Virtual worlds authenticate their users to introduce them to other users and look up stored information like profiles and inventory. Usually, the user has to provide a security token, for example, a username and password pair. In VirtualLife, authentication is performed by opening a secure connection between the parties that provides each endpoint with a proof about the identity used for establishing the connection. Since the client machine is the only one with access to the relevant private keys, nobody else can claim to have this identity because nobody can provide the necessary cryptographic proof without the private key.

It must be noted that if a certification authority does not verify the user's identities while issuing certificates, these certificates can still be used to authenticate users if the other party has established the correctness of the certificate using an alternative channel such as a personal meeting, mail message or a phone call.

4.2   Authorization

Authorization is used to verify the user's permission to use a service provided by a peer in the system. For example, nations and zones have to authorize clients before they can join. In a client-server virtual world authorization is essentially a server-side check. This approach does not translate well into a peer-to-peer world, where every node might want to establish its own authorization policy.

In VirtualLife, each node can define its own authorization policy. These policies can be synchronized between nodes that want to behave similarly. For example, in VirtualLife, zones belonging to a nation can ask the nation for its access control list and enforce it also in the zone. VirtualLife has a built-in support for using *whitelists* and *blacklists*. If a whitelist is used, only the users in the list are authorized to use the service and everybody else is denied access. If a blacklist is used, everybody except for the identities in the list are authorized. If we add an identity to a blacklist, we add all the associated certificates too. This way the user cannot circumvent the ban by creating a new identity without losing the chance to use the trusted certification.

4.3   Communication between Users

Interaction between the users of a virtual world must be secure if the world is expected to support business transactions. Textual chat, voice chat, file exchange and other collaborations must be authenticated. When a user is chatting or exchanging files with another user, he or she may not want to disclose private information without verifying the other party's identity. The user interface for all communication services will contain visual indicators for determining the strength of identity as presented in Section 3.

In VirtualLife, private communication channels are implemented using secure streams between the two users. Also, the secure stream is automatically encrypted and verified to prevent eavesdropping or active tampering.

4.4   Electronic Documents

Service providers and clients have to be able to sign contracts to engage in business transactions. The availability of X.509 key pairs allows VirtualLife to provide a digital signature service that conforms to the XML-DSIG[4] and XAdES[5] standards. These standards specify digital documents that can have a number of attached signatures.

A single person may sign a document containing a manifest, a guarantee or an invoice. Two people may sign a contract where one promises to perform some work in the virtual world and the other promises to pay for that work. Any number of people could also sign a declaration and present it to relevant parties. An integrated electronic document system can allow the user to digitally sign the contents of a chat session.

Any contract must state the identity used by each party for signing the contract. This way it is possible to verify that the people who had to sign the contract have indeed done so. A digital contract can be verified as long as at least one copy of the document with signatures exists.

4.5   Binding External Services

It might be the case that the outside certification authority has a database of additional information about the certificate holders. If a service provider in a virtual world has access to that database, it can be used to look up verified user data. For example, a certification authority can store the birth date of the certificate holder. If the certificate holder authenticates to a node with access to this database, this node can use the verified identity to verify the age of the user. Figure 4 illustrates this concept.

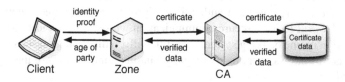

Fig. 4. Using an external certificate authority database for determining age

5   Implementation

In this chapter we give some details of how the proposed identity infrastructure is implemented within the VirtualLife world. VirtualLife has three modules that handle security protocols—**vlsec**, **vlnet** and **vlprotocol**. **vlsec** contains security-related data structures and services. **vlnet** provides the required networking services like stream management and secure communications. **vlprotocol** contains all the application-specific protocols of VirtualLife. Such a distinction is made to have a separation of duties and also allow the re-use of the security and network modules in other software systems.

Figure 5 shows the role of the *Identity* information in the class model. Several instances of *Certificate* and *Keypair* may be bound to a single *Identity*. The *Identity* structure is then used in services to distinguish between the users. In authorization, the *Whitelist* and *Blacklist* contain references to *Identity* instances. Signatures on the *ElectronicDocument* class are identifiable through the *Identity* class. Personal key pairs are encrypted and password-protected. The certificates of other users are cached locally with their trust information to minimize certificate lookup queries to other nodes.

At the time of writing this paper there is no publicly available version of VirtualLife as the system is still in development.

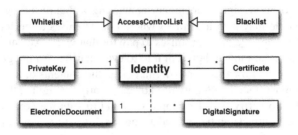

Fig. 5. A selection of identity-related classes

6 Conclusion

In this paper we present an identity management technique for peer-to-peer virtual worlds. The system is based on public key cryptography. Every peer proves its identity when accessing services provided by another peer. The proposed system has been implemented in the VirtualLife virtual world platform. VirtualLife is capable of using already established trust relations in the form of X.509 certificates for notifying its users about the trustworthiness of other users and service providers.

Peers in VirtualLife can have authenticated private channels that allow the users of the virtual world to have secure communication with similar security levels as the one provided in online banking systems. Additionally, built-in support for standardized digital signatures allows users to sign legally binding contracts where the signers can be identified.

References

1. Second Life Work, showcase of the industry usage of the Second Life platform, http://work.secondlife.com (Last checked: 28.09.2009)
2. Secure, Trusted and Legally Ruled Collaboration Environment in Virtual Life, EU FP7 project, http://www.ict-virtuallife.eu (Last checked: 13.07.2009)
3. Housley, R., Polk, W., Ford, W., Solo, D.: Internet X. 509 public key infrastructure certificate and certificate revocation list (CRL) profile. RFC 3280 (2002)
4. Bartel, M., Boyer, J., Fox, B., Lamacchia, B., Simon, E.: XML-Signature Syntax and Processing, IETF/W3C XML Signature Working Group, http://www.w3.org/TR/xmldsig-core/ (retrieved 13.07.2009)
5. Cruellas, J.C., Karlinger, G., Pinkas, D., Ross, J.: XML Advanced Electronic Signatures (XAdES), In: World Wide Web Consortium, Note NOTE-XAdES-20030220 (2003)

Overview of the Legal Issues in Virtual Worlds

Gerald Spindler, Katharina Anton, and Jan Wehage

University of Göttingen,
Platz der Göttinger Sieben 6, 37073 Göttingen, Germany

Abstract. The following paper contains a summary and brief assessment of the legal problems that may arise in virtual worlds. The results presented herein were gained during the legal research in order to establish a virtual world platform named VirtualLife. Virtual Life is an FP7 ICT research project conveyed by the EU. The project aims to provide an immersive and secure environment, combining a high quality 3D virtual experience with the trustiness of a secure communication infrastructure based on a peer-to-peer architecture. The paper cannot provide a comprehensive solution to all problems, but has to be seen as a basic overview of the legal aspects of virtual worlds. The ultimate answer to any legal problem will hinge on the exact scenario and therefore be subject to the law of the individually competent legal system and the deciding court.

Keywords: Virtual worlds' legal issues, virtual world property, user evaluation systems, data protection, copyright, trademarks, advertising, liability, online dispute resolution (ODR).

1 Introduction

Although the user takes part in a Virtual World by controlling an image of himself, the *Avatar*, which therefore is the only way the user gets in touch with other users, the only legal subject in a virtual world environment itself is the user who registers to this world, while the Avatar would simply be regarded as the means the user acts through.

The applicable law in virtual worlds differs depending on which relationship between the involved parties (i.e. user, provider of the software, service provider) is affected. All these parties first of all fall under the jurisdiction of a certain state whose law applies to the respective relationship. This state is defined by national and international rules of conflict of laws. Further customization of the virtual platform's legal framework is possible by means of contractual agreements [1].

2 Virtual World Property

The issue of virtual world users reclaiming virtual world property also in the real world may be decided on the contractual agreements between the software company and the user, real world criminal [2] or intellectual property law. This legal framework does not offer complete protection, e.g. protection by copyright requires a certain level of creativity or originality. Having property in virtual items would give

P. Daras and O. Mayora (Eds.): UCMedia 2009, LNICST 40, pp. 189–198, 2010.
© Institute for Computer Sciences, Social-Informatics and Telecommunications Engineering 2010

users an absolute rather than a mere contractual right, which would be effective against everyone not only the contractual partner and which does not hinge on the requirements of the aforementioned framework. However, the grant of protection will eventually hinge on national law and on the type of virtual item in question, as the different national legal systems have different requirements regarding definition of property and regarding the scope of protected objects. There are sufficient justifications to recognize property rights in virtual items: Virtual world property already is a real life commodity. Non-recognition would lead to an artificial restraint on the transference of wealth between real and virtual worlds and may prevent the production of further virtual property since people want to be rewarded for the labour and effort they have invested irrespective of existing intellectual property rights in the object.

Up to now, courts, users and software companies still have to rely on privately enforceable contractual agreements between software companies and consumers, law of contracts and intellectual property laws.

3 User Evaluation Systems

Social interaction depends on the development of ongoing relationships of trust which is formed through iterative interaction that gives rise to shared values and norms. Specifically reputation is a marker of this trust [3]. Hence evaluation systems are an essential feature of a platform with the exclusive use of electronic means such as a virtual world as they are the only possibility to gain trust without personal interaction.

From a legal point of view the feedback on a transaction with another user gives rise to the issues of the freedom of expression as opposed to the right to the protection of one's personality. The granting of a court order for a feedback removal will primarily depend on the result of the balancing of these legal guarantees in the individual case. Only if the plaintiff's personal rights outweigh the defendant's right to express his opinion, the court will order the removal from the feedback profile.

To be protected by the user's right of freedom of expression usually the feedback must not be false or reach a degree of abusive criticism which qualifies it as defamation. The individual decision may differ from jurisdiction to jurisdiction. Not only does it depend on the individual wording of the comments and the context of the case, but also on the weighting of those rights by the constitution and the case law in the competent jurisdiction.

4 Data Protection

To achieve trust in one's contractual partner not only the aforementioned user evaluation systems but also authentication methods have to be implemented.

These methods will to some extent involve the collection of information concerning the individual, who at the same time has the right of "informational self-determination", i.e. the right to have a say in how data relating to oneself are processed by others. Data protection is regarded as a fundamental human right in Europe [4]. The legal requirements for data processing hinge on the concrete scenarios as well as on the respective law in force. Legal requirements will be stated by the applicable national law of a certain state as well as by international regulations such as the Data Protection Directive of the European Community [5] respectively its transformation into national law.

5 Copyright Issues in Virtual Worlds

Content provided by users in virtual worlds can be divided into two categories. First, there are items created for the virtual world itself, such as the user's Avatar and virtual items like buildings or clothing. Secondly there may be content that exists independent of the software but is communicated via the means of the virtual world, e.g. videos, photos, music or narratives posted on discussion boards.

5.1 Legal Framework

Copyright law has developed nationally and is founded on very different concepts and justifications for protection.

Protection Criteria. To qualify for protection, a work has to meet several requirements as to form, content and status. Following from the idea/expression concept the mere idea behind a work is not protected. Laws require that the work must have a certain form, i.e. a perceptible expression [6], in the case of literary works by an expression in words. The criterion of content comprises two aspects: First, it concerns the particular subject matter. National laws may have exhaustive (usually the common law countries) or non-exhaustive (the civil law countries) lists of protected subject matter. Secondly, it is generally accepted that a work must be classified as original or creative. Protection further depends on the status of the author, i.e. his nationality or place of residence, or the status of the work, i.e. its place of publication.

Beneficiaries of Protection. Beneficiaries of protection are those persons the laws on copyright indicate as the person in favour of whom the rights are granted, or by whom rights may be exercised.

Objects of Protection. The software, the virtual world's (game) concept, and its multimedia representation in the form of graphics and sound may merit copyright protection. So, the developers of a platform will only be afforded protection of the platform software, not of the user-generated content.

In detail, object of protection can be the audiovisual representation that may be protected as computer program or as film. Protection as computer program requires that the interfaces lie within the scope of the applicable law. Regarding the protection as film the design of a virtual world platform will meet the defining factor of creating the impression of a moving image. Traditionally however, films constituted a closed system where scenes, actions and dramatic development could not be altered. Therefore protection as film must not be limited only to unalterable sequences of images as a virtual world does not predefine certain alternative actions. It rather is the users who make distinct contributions by individual decisions that influence the scenes and development. There is also little to no creative input on the part of the software developer regarding the dramatic development. Nonetheless, a scene that takes place in a virtual world once recorded will be protected under copyright law as moving image, if the applicable law offers protection for this kind of work at all.

As copyright does not protect ideas but only the expression of an idea it likely does not protect the concept of virtual worlds as a platform for users to interact.

5.2 Eligibility for Protection of In-World Created Items

Avatar. The visual depiction as seen by other Avatars interacting with the user and the unique person of an Avatar may fulfil the requirements of copyright protection.

It may generally fulfil the criterion of originality or creativity as it is the aspiration of the user to design his Avatar as individual as possible to be recognizable. This will only be achieved by investing at least a certain amount of labour, skill and judgment and creativity respectively. In jurisdictions with a closed list of protected subject matter the visual depiction has to meet this list. If the Avatar's unique persona, i.e. the Avatar's features of appearance, speech or moral attitude has become exceptionally distinctive, it may enjoy protection – though only in some jurisdictions.

Protection in common law countries is likely to fail since there is no category of work that would embrace an Avatar's visual depiction or his persona [7]. If protection is available, in light of the creation process and the aforementioned justifications for copyright the beneficiaries of protection are the users.

Virtual Items. Furthermore virtual items created by users of virtual worlds may fulfil the requirements for copyright protection. First, they should qualify as "artistic" works, even if artistic is defined rigidly requiring a distinctive element of aesthetic creativity not only investing labour and capital. Secondly, they may show the required level of creativity/originality which would be questionable regarding the representation of a real-world object. Thirdly, a certain fixation, if required by the law applicable, is given the moment text, pictures etc. are electronically incorporated in the random access memory or hard disk of a computer or server [8]. As it is the user that created the Avatar or virtual item he should be the initial beneficiary of protection.

5.3 Copyright Infringements and Exploitation of Rights

Copyright infringements will in particular surface by unauthorized use of copyrighted material, e.g. virtual items may be copied without authorization and traditional media content such as audio and video clips and photographs may be made available without the rightowner's authorisation. To minimize litigation from the outset EULAs [9] of providers of a virtual world may contain provisions addressing the question of use of copyright works in the virtual world. Additionally the graphical user interface could provide an easy to understand licensing system, maybe based on the CC [10] licenses, that enables users to share their creations with other users.

6   Trademark Issues in Virtual Worlds

As companies like Toyota, Dell and Reebok have opened virtual outlets in the virtual world Second Life [11], trademark issues will come to virtual worlds as well as purely "in-world" trademarks will be created and may receive substantial profits. Furthermore, content creators in virtual worlds may use real-world trademarks for their virtual products, with or without the real-world trademark owner's authorization. In this context, all of the attendant concerns of brand reputation and disparagement will be present in this new medium just as they are in the real world.

6.1 Protection of In-World Trademarks

There are three basic justifications for offering legal protection of trademarks, which relate to the different economic functions of a trademark [12]. First is to guarantee origin, i.e. it operates as an indicator of the trade source from which goods or services come thus enabling customers to easily distinguish between different goods and products. Secondly, trademarks perform an investment or advertising function since they are symbols around which investment in the promotion of a product is built. Thirdly, it is said that trademarks should be protected as manifestations of creativity.

In-world trademarks in principle fulfil the requirements for protection as well as they are not excluded from protection from the outset. So even an Avatar's name and likeness as such are registrable, just as their real-world counterparts are.

6.2 Protection against Virtual Knock Offs

Besides the classical type of infringement where a user uses a real-world trademark on virtual items users could apply the trademark to other virtual items the trademark owner usually is not associated with. In general infringements for example require using a trademark for which registration is sought identical or similar to an earlier registered trademark.

These infringements in the virtual world can only be qualified as infringements also in the real world or another virtual world if one does not have to distinguish between these worlds. Since users may participate in different virtual worlds the border between these worlds will as well be blurred as the border to the real world as the virtual world has developed to a platform for transactions and interactions which also affects the real world. Therefore it seems indispensible to consider other virtual worlds as well as the real world when examining a trademark infringement [13].

As concerned the identity or similarity of signs, the same considerations as in the context of real-world trademarks have to be made. Thus there are no particularities regarding virtual worlds. Determining the identity or similarity of goods or services may be more of a contentious issue. With regards to goods, in most cases it will be very difficult to talk about identity in virtual worlds. Avatar clothes are definitely not identical to real clothes and a virtual vehicle is not identical to a real car [14]. It also seems difficult to establish similarity of goods unless the judge is willing to accept the argument that driving a car in the virtual world is the same as driving the car in the real world. The products do not compete with each other. Rather, the sale of a t-shirt effectively is the provision of a service consisting of the creation of code that displays the t-shirt on the Avatar. Claims for trademark infringement may thus fall short of identity or similarity of goods. By way of contrast, services may as well be identical.

Assuming that the goods or services were similar and the marks were similar, it is questionable whether there is a likelihood of confusion, as the average user would have to be likely to assume a material trade link between the goods or services offered in virtual worlds and the real-world trademark owner. On the one hand, it is not common yet for trademark owners to establish some presence in virtual worlds. Hence there will be no confusion. On the other, if a trademark owner had opened a business in the virtual world, it seems that the assessment would have to be different. The use of an identical or similar sign will further infringe an earlier trademark where the use of the sign would take unfair advantage of or would be detrimental to the distinctive character or the repute of an earlier registered trademark. Anyway, all of the infringement cases require the use in the course of trade.

Apart from trademark law, a real-world trademark owner could further try to invoke national unfair competition law to protect his mark against virtual knockoffs.

7 Advertising in Virtual Worlds

To reach customers offerors need advertisement to make products known. Similarly to the rising market of in-game advertising it can be expected that advertising in virtual worlds will be a market of the same scale, as a virtual world combines the possibilities of a realistic "game-graphic" with the possibilities of a social network – the options of digital and real life.

7.1 Possible Types of Advertising

Comparing the advertising opportunities in virtual worlds to in-game advertising, advertisements are generally divided into three general types, namely static advertisement, dynamic advertisement and product placement [15]. Additionally, as a fourth type the so-called multi dynamic advertisement is particular to virtual worlds.

Static advertisement is firmly incorporated in the source code. So each user will see the same advert. In dynamic advertising the content of advertising space changes dynamically. It could be individualised depending on the Avatar "looking at" the advert. Additionally, product placement is likely to be heavily used in virtual worlds as an Avatar might be able to dress himself in clothes of a specific brand. Multi dynamic advertising offers the possibility to create advertisement that will not only change by time or user, but also by place or a user's movements in the virtual world.

7.2 Legal Assessment

There is no particular law regulating advertising in virtual worlds but restrictions extend from other fields of law into them. These particularly are restrictions on unfair commercial practices, restrictions resulting from data protection law or the laws on the protection of minors, restrictions based on the regulation of audiovisual media services, restrictions relating to specific products such as intoxicants, and limitations following from third parties rights such as trademarks.

If the applicable law does not contain any constraints regarding its scope, virtual world advertisements thus have to meet the same obligations as in the real world taking into account the particularities of experiencing a virtual world.

8 Provider Liability for User-Generated Content

Virtual worlds are used to distribute user-generated content like videos, photos, music, book reviews or personal narratives. This, however, may infringe other content providers' rights or community sensibilities in many ways. For instance, it could infringe copyright or trademarks, it could be an act of unfair competition or it could amount to defamation. Generally, the user will be liable for his infringements. Because of the anonymous structure of the Internet it may be difficult to determine the user who had uploaded the material and, even then, any effort to claim substantial damages may be failed by their lack of sufficient assets. More importantly, an action against the originator is unlikely to prevent further dissemination of the content as another user may post it again. So, the question of liability of third parties providing the infrastructure the users interact in, i.e. intermediary service providers, arises.

To translate this into a virtual world, it is not only the provider of the software but also other service providers, including the user's individual Internet access provider, the host providers and other service providers that provide virtual environments in which Avatars meet without actually storing data for their user. These third parties could be held liable to remove user-generated content, to cease and desist the distribution of such and to pay damages.

As access providers offer mere conduit services their responsibility for unlawful material is rather limited. Anyhow, parties have generally tried to obtain injunctions to block or remove illicit content.

The provider of software does not qualify as a provider of mere conduit service since he only provides the tool with the assistance of which users can initiate a transmission within an already existing network [16]. Therefore, he will only be held liable for illicit content, if it is sufficient by the law applicable to provide the software in the absence of which an infringement would not be possible at all. Thus providing software may qualify as inducement, contribution or may lead to secondary liability for interferences. Though it has to be taken into account that a virtual world software is generally intended to be used for lawful purposes.

Host providers offer the creation and maintenance of server space as well as the organisation of content. Their immunity from liability for third party content therefore hinges on whether they have constructive knowledge of the infringing act and, if so, react expeditiously to remove it.

Summing up, the question whether a virtual world service provider is liable for unlawful acts by his users depends on the service in question and on the courts' construction of the national implementation of immunities of the E-Commerce Directive.

9   Dispute Resolution in Virtual Worlds

As a virtual world platform offers the possibility to let important transactions happen it also provides a platform to solve disputes that may arise from these transactions or do not even have such a connection at all. As it will be assessed a virtual world platform combines all the advantages of Alternative Dispute Resolution (ADR) and combines it furthermore with the advantages of modern communication technologies. Therefore it is almost predestinated being a means to solve legal disputes.

9.1   Online Dispute Resolution

Introduction and Terminology. Alternative Dispute Resolution (ADR) refers to processes other than judicial determination in which an impartial person assists those in a dispute in order to resolve the issues between them. Its characteristic feature can thus be described as providing a platform for the settlement of legal disputes outside the traditional bodies of the State's judicial system. ADR avoids the problem of the increasing caseload of traditional courts, imposes fewer costs than traditional litigation before state courts and meets the preference for confidentiality and the desire of having greater control over the selection of the person who will decide their dispute.

Online ADR refers to ADR processes assisted by information technology, particularly the Internet. Therefore it is often seen as being the online equivalent of ADR for which ODR (Online Dispute Resolution) has emerged as the preferred term.

Methods and Advantages of ODR. ODR combines the benefits from ADR and the new information technologies. It significantly reduces the transaction costs of dispute resolution as there is no need for the parties to convene in a particular location and present evidence in written form. Lengthy documents can be sent as e-mail attachments, thereby avoiding copying fees as well as facsimile and/or postal charges [17]. There are no additional costs from the payment of human thirds [18]. ODR is accessible twenty-four hours a day, seven days a week around the world. This helps to eliminate scheduling conflicts and potential problems with time zone discrepancies. As a result, the dispute will be resolved much faster. Using e-mail, discussion groups and Web sites, agreements can be written, posted and responded to when convenient.

Risks of ODR. In essence, three risks have been identified that will particularly arise from the use of information technologies for the purpose of settling conflicts: Confidentiality, transparency and authenticity.

The principle of confidentiality requires the parties and the arbitrator not to reveal information and documents related to the case, and not to disclose the final agreement. The parties involved in the dispute need to be sure that the virtual arbitration body they are dealing with is trustworthy and completely impartial.

As the evidence is submitted electronically by the parties, ODR largely relies on a further development towards more advanced electronic authentication methods to proof the authenticity of a document.

9.2 Legal Aspects

Legal issues in the context of ADR/ODR primarily concern the validity of ADR clauses, the suspension of limitation periods and the enforceability of arbitral results.

As regards the enforceability of the result it is crucial to keep in mind it is nothing but a contract between the parties at dispute. Thus, non-compliance of the settlement agreement constitutes a breach of contract, which would have to be pursued before a traditional court as well as the judgment would have to be enforced [19]. Additionally the validity of the ADR clause itself may be questionable, i.e. if the parties can reach a valid agreement as to use ADR to solve their disputes outside traditional courts.

10 Conclusions

As every lawsuit differs in its facts from other lawsuits, the legal assessment does so as well. Therefore, regarding the legal issues of virtual worlds, it is only possible to determine which problems may occur and which law may provide the solution to these problems. The final assessment is up to the national courts. While trying to anticipate these decisions, it has to be taken into account how the environment of a virtual world affects the legal assessment of a lawsuit. Most of the legal problems that may arise in virtual worlds are known from the real world. Examining possible copyright protection of virtual items, for example, does not differ from examining a real world item as both have to fulfil the same criteria of protection. Besides most of these differences just result from the new form of interaction a virtual world offers, which therefore has to be subsumed under common scenarios the current law addresses.

Besides, virtual worlds offer new possibilities of interaction whose risks have to be taken into account.

References

1. See for an overview of the jurisdiction in the USA: Kennedy, C.H.: Making Enforceable Online Contracts. CRi 38 (2009)
2. See, for instance, Ernst, S.: Recht kurios im Internet - Virtuell gestohlene Phönixschuhe, Cyber-Mobbing und noch viel mehr. NJW 1320 (2009); District Court of Leeuwarden (NL) – Criminal Section. CRi 59 (2009)
3. Noveck, B.S.: Trademark law and the social construction of trust: Creating the legal framework for online identity. Washington University Law Quarterly 83, 1733, 1751 (2005) speaks of 'reputation as the key to identity in the cyberspace'
4. See, for instance, Article 8 of the Charter of Fundamental Rights of the European Union
5. Directive (EC) 95/46 of the European Parliament and of the Council of 24 October 1995 on the protection of individuals with regard to the processing of personal data and on the free movement of such data (1995) OJ L281/31
6. Sterling, J.A.L.: World Copyright Law, 2nd edn. Sweet & Maxwell, London (2003); 7.03
7. Protected subject matter in the UK is original literary, dramatic, musical and artistic works; films; sound recordings, broadcasts and typographical arrangements and databases; see s 1 (1) Copyright, Designs and Patents Act (CDPA), c. 48 (1988)
8. Sterling, J.A.L.: World Copyright Law, 2nd edn. Sweet & Maxwell, London (2003); 7.03
9. End User License Agreement
10. Creative Commons; creative commons has developed several types of copyright licenses, which allow creators to easily communicate which rights they reserve and which rights they waive for the benefit of other creators, http://creativecommons.org/about/license
11. WIPO: IP and Business: Second Life – Brand Promotion and Unauthorized Trademark Use in Virtual Worlds. WIPO Magazine 6/2007, http://www.wipo.int/wipo_magazine/en/2007/06/article_0004.html
12. Cornish, W., Llewelyn, D.: Intellectual Property: Patents, Copyright, Trade Marks and Allied Rights, 5th edn. Sweet & Maxwell, London (2003); 15.22
13. These borders is already referred to as "porous membranes": Kotelnikov, A.: Trade Marks and Visual Replicas of Branded Merchandise in Virtual Worlds. I.P.Q. 110, 121 (2008)
14. See also Varas, C.: Virtual Protection: Applying Trade Mark Law within Virtual Worlds such as Second Life. Ent LR 5, 9 (2008)
15. Lober, A.: Spielend werben: Rechtliche Rahmenbedingungen des Ingame-Advertising. MMR 643, 643 et seq (2006); Göttlich, P.: Online Games from the Standpoint of Media and Copyright Law. IRISplus 10.2, 8 (2007)

16. Spindler, G., Leistner, M.: Secondary Copyright Infringement: New Perspectives in Germany and Europe. IIC 788, 794 (2006); McAleese, D., Cahir, J.: A European Perspective on the Peer-to-Peer Model post-Grokster. CRi 38, 42 (2006)
17. Alford, R.P.: The Virtual World and the Arbitration World. Journal of International Arbitration 449, 457 (2001)
18. Lodder, A., Bol, S.: Essays on legal and technical aspects of Online Dispute Resolution. Papers from the ICAIL 2003 ODR Workshop, June 28, 9 (2003)
19. Hörnle, J.: Alternative Dispute Resolution and the European Union, Essays on legal and technical aspects of Online Dispute Resolution. Papers from the ICAIL 2003 ODR Workshop, June 28, 1, 4 (2003)

Selective Motion Estimation for Surveillance Videos

Muhammad Akram, Naeem Ramzan, and Ebroul Izquierdo

Electronic Engineering Department,
Queen Mary University of London,
United Kingdom
{muhammad.akram,naeem.ramzan,ebroul.izquierdo}@elec.qmul.ac.uk

Abstract. In this paper, we propose a novel approach to perform efficient motion estimation specific to surveillance videos. A real-time background subtractor is used to detect the presence of any motion activity in the sequence. Two approaches for selective motion estimation, GOP-by-GOP and Frame-by-Frame, are implemented. In the former, motion estimation is performed for the whole group of pictures (GOP) only when moving object is detected for any frame of the GOP. While for the latter approach; each frame is tested for the motion activity and consequently for selective motion estimation. Experimental evaluation shows that significant reduction in computational complexity can be achieved by applying the proposed strategy.

Keywords: Fast motion estimation, surveillance video, background subtraction, block matching algorithm.

1 Introduction

In surveillance applications, video captured by CCTV is usually encoded using conventional techniques, such as H.264/AVC. These techniques have been developed in view of conventional videos. With growing number of surveillance system deployments, there is a need to introduce surveillance centric coding techniques. Goal of this paper is to propose an efficient motion estimation approach specific to surveillance videos.

Motion is main source of temporal variations in videos. High compression efficiency is achieved through special treatment for motion based temporal variations. Motion estimation (ME) is a process that estimates spatial displacements of same pixels in neighboring frames. This spatial displacement is described through the concept of motion vector (MV). Almost all video coding standards deploy motion estimation modules to aid in removal of temporal redundancies. The process of ME divides frames into group of pixels known as block. Block matching algorithms (BMAs) are used to find out the best matched block from the reference frame within a fixed-sized search window. The location of the best matched blocked is described by MV. So, instead of encoding the texture of the block, only MV of the block is coded. While decoding video, motion vectors are used to replace the original blocks with its best matched block searched through motion estimation. Encoding complexity is

P. Daras and O. Mayora (Eds.): UCMedia 2009, LNICST 40, pp. 199–206, 2010.

dominated by the ME if full search is used as BMA. FS matches all possible displaced candidate blocks within search window to find a block with minimum block distortion measure (BDM).

Several fast BMAs have been introduced to beat FS in terms of computational complexity. These include new three step search (N3SS) [1], four-step search (4SS) [2], diamond search (DS) [3], kite-cross diamond search (KCDS) [4], and modified DS (MODS) [5], etc. In this paper, we propose a novel approach to reduce computational complexity for encoding surveillance videos. Proposed approach utilizes a real-time background subtractor (BGS) [6] to detect the presence of the motion activity in the sequence. In typical surveillance videos, scene remains static for a long period of time. Performing motion vector search for these frames is wastage of computing resources. Motion vector (MV) search is performed only for frames which have some motion activity identified by BGS.

In this paper, Section 2 introduces the implemented approach to perform efficient MV search for surveillance videos. Workflow of the proposed system is presented. Section 3 describes experimental results and presents a comparison of the proposed search with conventional search approach. Finally, Section 4 concludes this paper.

2 Selective Motion Estimation

The generic architecture of the implemented system is shown in Fig. 1. Surveillance video is presented to background subtraction and video encoding modules of the system. The real-time background subtractor detects motion activity present in the sequence. This information is passed onto the motion estimation module of the encoder. Motion estimation module utilizes the motion detection information to perform selective motion estimation. After motion compensated temporal filtering (MCTF) step, spatial transformation is performed to remove the spatial redundancies. Finally, entropy coding techniques are used to improve compression efficiency.

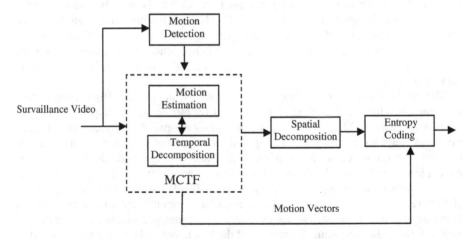

Fig. 1. Architecture of implemented system

2.1 Real-Time Background Subtraction

Motion detection module must be efficient in terms of processing power otherwise; the complexity of the motion estimation module shall be reduced at the cost of increased complexity for motion detection. A real-time video background subtraction module based on Gaussian mixture model [6] is used to detect motion activity present in the video. This method is able to deal robustly with light changes, bimodal background like swaying trees and introduction or removal of objects from the scene. Value of each pixel is matched against weighted Gaussians of mixture. Those pixels are declared as foreground whose value is not within 2.5 standard deviation of the background Gaussians. Foreground pixels are grouped into segmentation regions and bounded by rectangular boxes throughout the sequence. Output of BGS module for hall video is illustrated in Fig. 2. Pixels which are static for a number of frames are modeled as background; therefore they do not fall within the boundary of boxes. The presence of bounding box is an indication of motion activity present in the frame. This indication is used to perform selective motion estimation.

Fig. 2. BGS result for frame 122 of hall video

2.2 Selection Policy

As aforementioned, video captured from the CCTV camera is processed through a real-time background subtraction (BGS) module to detect the presence of motion activity in the sequence. Presence of motion activity for each frame of the sequence is marked and recorded. This information is utilized by the motion estimation module of the encoder to perform selective motion estimation. Two different selective motion estimation approaches, GOP-by-GOP and Frame-by-Frame, are implemented to improve the efficiency of motion estimation process in terms of saving processing power and processing time. In GOP-by-GOP approach, BGS information is analyzed for all the frames of every GOP. So, a single decision of performing selective motion estimation is made for each GOP. If BGS detects any moving object in any frame of the GOP then motion estimation is performed for the GOP otherwise motion vectors for all the frames of the GOP are set to zero. Workflow of the proposed system is shown in Fig. 3. GOP-by-GOP selective motion estimation performs better when there is no motion activity for a large number of frames in the sequence. Its efficiency is lower when there is some pattern of activity present in the sequence not allowing to bypass the ME module. Also, this approach is dependent on GOP size set for encoding the sequence where smaller GOP size has better efficiency.

In Frame-by-Frame approach, BGS information for each frame of the sequence is analyzed. Decision of performing selective motion estimation is made for each frame. If any motion activity is present in a particular frame then motion estimation is performed otherwise motion vectors for the frame are set to zero. This approach improves processing efficiency by performing ME only for the frames where it requires and bypassing ME module for static frames. Thus, based on BGS analysis, no compromise is made for the frames which are important from surveillance standpoint and still complexity can be reduced by applying the proposed approach.

```
for (frame=1 to end of sequence)
   if (motion activity found)
      frameMotion [frame] = 1
   otherwise
      frameMotion [frame] = 0
end for

switch (motion estimation mode)
case GopByGop:
   for (each GOP of the sequence)
      for (first frame of GOP to GOP size)
         if ( frameMotion [frame] is 1)
            Perform Motion estimation for this GOP
            ME_performed = 1
            Break
      end for
      if ( ME_performed is not equal to 1)
         All the motion vectors are set to zero
   end for

case FrameByFrame:
   for (each frame of the sequence)
      if ( frameMotion [frame] is 1)
         Perform Motion estimation for this frame
      otherwise
         All the motion vectors are set to zero
   end for
```

Fig. 3. Algorithm for selective motion estimation

2.3 Video Coding

For actual encoding of the surveillance video, a wavelet-based scalable video codec – aceSVC [7] is employed. The scalable video coding (SVC) framework helps to improve utilization efficiency of available resources such as transmission bandwidth and storage, etc. Furthermore, SVC has potential for surveillance videos as in [8],[9]. Architecture of aceSVC features spatial, temporal, quality and combined scalability. Temporal scalability is achieved through repeated steps of motion compensated temporal filtering [10]. To achieve spatial scalability, each frame is decomposed using a 2D wavelet transform. Coefficients obtained through spatio-temporal decomposition are coded through the process of bit-plane coding [11] which provides basis for quality scalability.

3 Experimental Results

Performance evaluation of the proposed approach is carried out on three different surveillance sequences: "Dance", "Hall" and "Street" with 500, 300 and 750 frames, respectively. All of these sequences have CIF (352x288) spatial resolution and frame rate of 30 Hz. Background in all the sequences is static throughout the length of the sequences. "Dance" sequence contains an animated person dancing with fast leg and arm motion. In "Hall", two persons are walking in opposite directions in a corridor. In "Street", with an outdoor street background, different animated objects move through the street.

While performing the experiment, sum of absolute difference (SAD) is used as block distortion measure (BDM). Block size is 16x16 while the search window size is 15 (+- 15 pel displacement is possible in vertical and horizontal directions). All the videos are compressed for 256 kbps bit-rate. Quarter pixel search approach is used for all the sequences. The true processing time is used to evaluate the performance of the proposed approach, while Y-PSNR is calculated to assess the image quality. As the surveillance video quality must be good enough for visual perception of the end user/observer; therefore, a subjective quality assessment test is also performed for the reconstructed sequences. The evaluation is performed using different GOP sizes. Each GOP contains at least one intra-coded frame. Thus, increasing the GOP size for the same sequence reduces the intra-coded frames in the whole compressed bit-stream. Consequently, higher GOP size has higher processing time.

All the tests are performed on machine with Intel Core(TM) 2CPU 6600@2.40GHz (2 CPU) processor and 2 GB RAM. First of all, BGS module has to be real-time to improve the efficiency of the proposed system. For this, Table 1 shows that the motion detection process is real-time where processing time for each surveillance sequence is given in seconds. BGS processes almost 30 frames in each second on the above described machine. Although BGS performance is real-time, still time consumed by BGS is included in overall encoding time for the evaluation of proposed selective motion estimation approach. In all the tables, PSNR results are in dB's and time is measured in seconds.

Table 1. Real-time motion detection

Sequences	Total Frames	Time	Frames/Sec
Dance	500	17	29.41
Hall	300	10	30.00
Street	750	25	30.00

Experimental results for full search based motion estimation are summarized in Table 2. These results are used as reference to compare the proposed approach. Table 3 shows the results for GOP-by-GOP motion estimation. Different GOP sizes are selected to perform the experiment. With each GOP, MCTF is performed in such a way to produce maximum number of estimated frames. The processing time saving, compared to full motion estimation, achieved for GOP-by-GOP approach is shown in Table 4. Results show that the nature of the sequence has great influence on the efficiency of the proposed approach. One drawback with GOP-by-GOP motion estimation is that motion estimation is performed for all the frames of the GOP even if only one frame has the foreground object. Thus to refine and improve the performance, Frame-by-Frame selective motion estimation is implemented. Motion estimation is performed only for frames for which foreground object is detected. Table 5 and Table 6 show the experimental results for Frame-by-Frame approach. Results show significant improvement over GOP-by-GOP approach.

Table 2. Full motion estimation

Seq	Gop Size=8		Gop Size=16		Gop Size=32		Gop Size=64	
	Time	PSNR	Time	PSNR	Time	PSNR	Time	PSNR
Dance	565	44.65	719	46.3	872	47.07	1027	47.37
Hall	438	35.53	556	37.62	678	38.59	819	39.19
Street	834	29.03	1085	32.13	1369	34.59	1692	36.19

Table 3. GOP-by-GOP selective motion estimation

Seq	Gop Size=8		Gop Size=16		Gop Size=32		Gop Size=64	
	Time	PSNR	Time	PSNR	Time	PSNR	Time	PSNR
Dance	409	44.65	520	46.27	677	47.08	765	47.38
Hall	430	35.53	544	37.61	677	38.59	816	39.19
Street	810	29.04	1051	32.14	1334	34.59	1635	36.19

Table 4. Processing time saving for GOP-by-GOP selective motion estimation

Seq	Gop Size=8	Gop Size=16	Gop Size=32	Gop Size=64
Dance	27.61	27.68	22.36	25.51
Hall	1.83	2.16	0.15	0.37
Street	2.88	3.13	2.56	3.37

Table 5. Frame-by-Frame selective motion estimation

Seq	Gop Size=8		Gop Size=16		Gop Size=32		Gop Size=64	
	Time	PSNR	Time	PSNR	Time	PSNR	Time	PSNR
Dance	320	44.61	398	46.24	481	47.04	600	47.35
Hall	430	35.53	549	37.61	665	38.58	804	39.18
Street	662	29.05	834	32.14	1053	34.61	1351	36.22

Table 6. Processing time saving for Frame-by-Frame selective motion estimation

Seq	Gop Size=8	Gop Size=16	Gop Size=32	Gop Size=64
Dance	43.36	44.64	44.84	41.58
Hall	1.83	1.26	1.92	1.83
Street	20.62	23.13	23.08	20.15

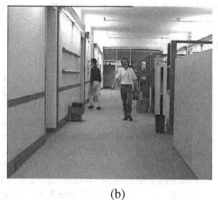

(a)   (b)

Fig. 4. Visual comparison Hall frame 225 (a) Full ME (b) Frame-by-Frame ME

For assessing user perception based on visual quality, subjective quality evaluation based on the double stimulus impairment scale [12] method is performed as in Fig. 4. Different users participated in this test. Videos from full motion estimation, GOP-by-GOP motion estimation and Frame-by-Frame motion estimation were organized in random. User had to assign any number from 1 to 5 after watching the videos.

Table 7. Subjective quality result

Seq	Full ME	GOP-by-GOP	Frame-by-Frame
Dance	2.57	2.71	2.71
Hall	4.39	4.25	4.25
Street	2.82	2.68	2.53

Table 7 shows the results for visual evaluation of the sequences. These are average numbers where 5 is the maximum number representing the best quality. Results show that applying the proposed approach has no much effect on the visual perception of the video which is important from the surveillance standpoint. This shows that the processing efficiency for the proposed approach is improved without compromising on visual quality of the surveillance videos.

4 Conclusion

In this paper, we have presented a novel technique to perform fast motion estimation specific to surveillance applications using the information of a real-time video background subtraction. Selective motion estimation is performed for GOP-by-GOP and Frame-by-Frame approaches. Performance of the implemented selective motion estimation approach is compared against motion estimation performed for all the frames. A high relative saving in processing time is achieved by using the proposed technique. Results obtained through experimental evaluation show that processing speed can be improved significantly by using the proposed approach while maintaining the surveillance sensitive information.

References

1. Li, R., Zeng, B., Liou, M.L.: A New Three-Step Search Algorithm for Block Motion Estimation. IEEE Trans. Circuit Syst. Video Technol. 4, 438–442 (1994)
2. Po, L.M., Ma, W.C.: A Novel Four Step Search Algorithm for Fast Block Motion Estimation. IEEE Trans. Circuit Syst. Video Technol. 6, 313–317 (1996)
3. Zhu, S., Ma, K.K.: A New Diamond Search Algorithm for Fast Block-Matching Motion Estimation. IEEE Trans. Image Processing. 9, 287–290 (2000)
4. Lam, C.W., Po, L.M., Cheung, C.H.: A Novel Kite-Cross-Diamond Search Algorithm for Fast Block Matching Motion Estimation. In: IEEE ISCAS, vol. 3, pp. 729–732 (2004)
5. Yi, X., Ling, N.: Rapid Block-Matching Motion Estimation Using Modified Diamond Search. In: IEEE ISCAS, vol. 6, May 2005, pp. 5489–5492 (2005)
6. Stauffer, C., Grimson, W.E.L.: Learning Patterns of Activity Using Real Time Tracking. IEEE Transaction on Pattern Analysis and Machine Intelligence 22, 747–757 (2000)
7. Mrak, M., Sprljan, N., Zgaljic, T., Ramzan, N., Wan, S., Izquierdo, E.: Performance Evidence of Software Proposal for Wavelet Video Coding Exploration Group. Technical Report, ISO/IEC JTC1/SC29/WG11/MPEG2006/ 13146 (2006)
8. Zgaljic, T., Ramzan, N., Akram, M., Izquierdo, E., Caballero, R., Finn, A., Wang, H., Xiong, Z.: Surveillance Centric Coding. In: 5th International Conference on Visual Information Engineering (VIE 2008), July 2008, pp. 835–839 (2008)
9. Akram, M., Ramzan, N., Izquierdo, E.: Event Based Video Coding Architecture. In: 5th International Conference on Visual Information Engineering (VIE 2008), July 2008, pp. 807–812 (2008)
10. Mrak, M., Izquierdo, E.: Spatially Adaptive Wavelet Transform for Video Coding with Multi-Scale Motion Compensation. In: IEEE International Conference on Image Processing, September 2007, vol. 2, pp. 317–3320 (2007)
11. Zgaljic, T., Sprljan, N., Izquierdo, E.: Bit-Stream Allocation Methods for Scalable Video Coding Supporting Wireless Communications. Signal Processing: Image Communications 22, 298–316 (2007)
12. Recommendation ITU-T BT 500.10: Methodology for the Subjective Assessment of the Quality of Televisions Pictures (2000)

PerMeD 2009

Session 1

Personalization of Media Delivery in Seamless Content Delivery Networks

Marta Alvargonzález[1], Laura Arnaiz[1], Lara García[1], Faustino Sanchez[1],
Theodore Zahariadis[2], and Federico Álvarez[1]

[1] Universidad Politécnica de Madrid (GATV)
Avenida Complutense 30, Madrid, Spain
{mad,lav,lgv,fsg,fag}@gatv.ssr.upm.es
[2] Synelixis Solutions Ltd
10 Farmakidou Av, Chalkida, GR34100, Greece
zahariad@synelixis.com

Abstract. In this paper, we propose an innovative system that aims to adapt to the user needs and preferences the media content transmissions within IP and P2P environments. To personalize the manner the content is displayed to the final user, this proposed network allows the transmission of multiple views and different layers for each media content piece. In addition, we suggest an approach on how to deal with the problem of contents transmission over P2P networks while preserving the author's rights. In this document, the system architecture is presented, especially the structure concerning the different streams sent over it and the security involved. This research path is being investigated within "SEAmless Content delivery" (SEA) project [1].

Keywords: MVC, SVC, layered video coding.

1 Introduction

Our objective is to present an innovative and interoperable solution to share and transfer personalized media content over the different identified networks, and receive the data in the possible different devices where a user may like to view the media content. Additionally, we are aiming to preserve the creator's rights while sharing their contents. Taking into account the benefits of peer-to-peer networks, our proposed solution leans on these network structures.

Figure 1 illustrates the different networks, and end-user devices with which we have developed our system [1]. As the end-user terminal characteristics can be very different and there are several kind of networks, in order to display the media content correctly some adaptations may be required to be done. An important element of the network is the sHMG, the gate to the home environment. Its functionality is to interconnect various access networks (e.g. DVB-T, DVBS/ S2, xDSL, WiMAX) with indoor networks (e.g. WiFi, Ethernet, Powerline, ZigBee), as Figure 1 shows [2].

P. Daras and O. Mayora (Eds.): UCMedia 2009, LNICST 40, pp. 209–213, 2010.
© Institute for Computer Sciences, Social-Informatics and Telecommunications Engineering 2010

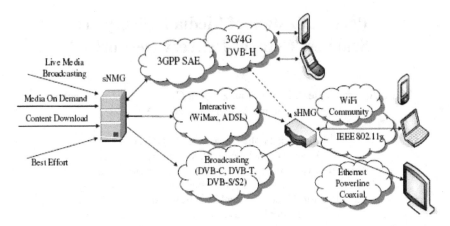

Fig. 1. Network elements and interfaces (figure taken from [2])

This paper is organized in three parts. Following this introduction we analyze the different approaches suggested to achieve the desirable personalization in the media delivery system of the proposed network, and we illustrate the method we use to securely transfer media and to protect the contents author rights. Finally, conclusions are drawn.

2   Content Delivery and User Personalization

As we have explained in the introduction, we are aiming the interoperability of different access networks and devices whilst adapting the contents displayed depending on different imposed or decided causes.

There are mainly two coding solutions that we are investigating and implementing to allow the desired personalization in our network: Scalable Video Coding (SVC) and Multiview Video Coding (MVC). The first one does layered temporal/spatial/ quality scalability, and the later allows the transmission of different views embedded in a single video stream. Considering the need and desire of a high security system, each different layer and view stream is encrypted separately.

We propose to carry out the personalization and adaptation under three points of view: end-user preferences, end-device capabilities and network congestion.

2.1   End-User Preferences

Our first target is to take into account the end-user preferences. The user can choose the layers and views he wants to receive. The content shared in our network will have at least one associated license, and will have more than one if various possibilities on how this content can be viewed are given. We have analyzed different approaches on how the licenses can be obtained. Trying to keep the interaction with the server whilst allowing a simple communication with the user, we have decided to include in a web site a database with the licenses. There, the user can see the different contents available, and the various possibilities each license offers. As the license states the characteristics of the media content associated with it, the end-user can purchase different licenses depending on his/her choices.

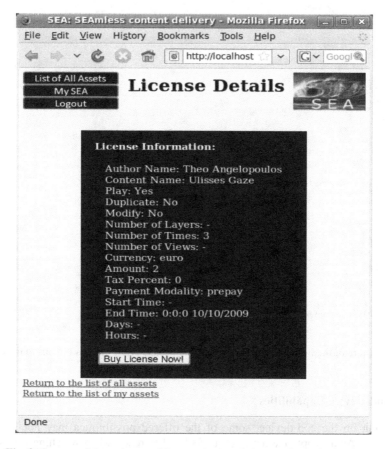

Fig. 2. Capture of the web page. License characteristics are shown in this figure.

The creator of the content will create a license associated with it. We have implemented a simple graphic interface to facilitate this task (see figure 3). On it, the content creator specifies the rights he grants to the users that buy the license. The possible rights he can offer are the number of layers, views and times a user can play the content. Moreover, we include payment and date/time requirements to indicate the license price and to complete the permissions given.

Once the content and license are created, the license should be uploaded to the web page, where the end users can easily navigate though the songs, videos and media contents offered, and the options given in each license. End-users should then purchase the desired license, start receiving the streaming data and enjoy the content they have bought.

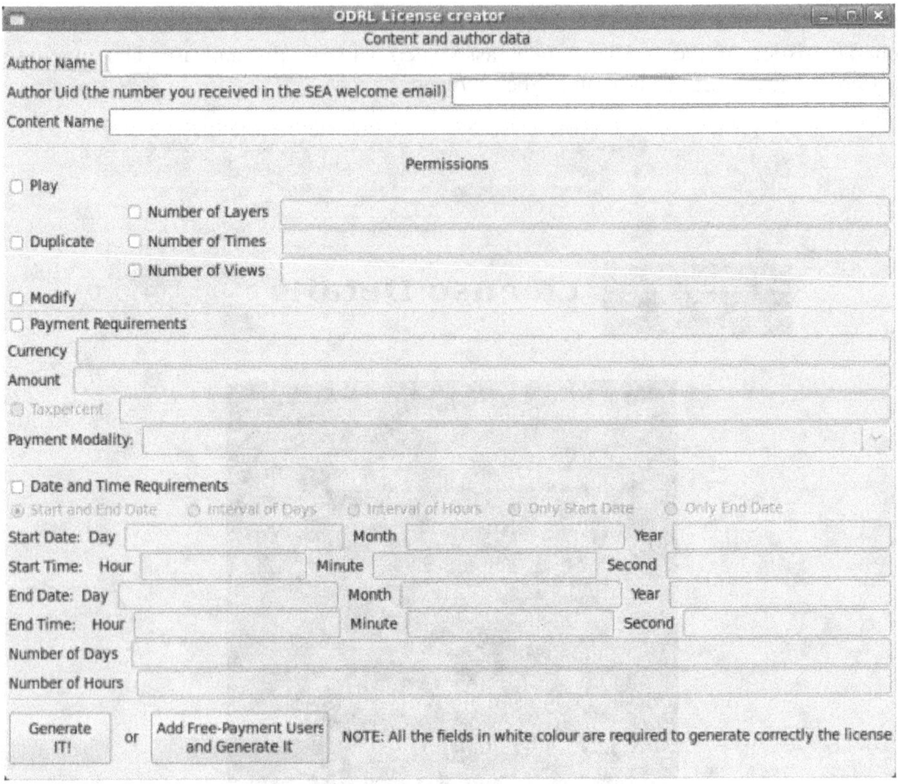

Fig. 3. Capture of the license creator program. The license offers various options to the content creator.

2.2 End-Device Capabilities

Depending on the end-device, some of the offered possibilities may not be really achievable. As a simple example, users feel a difference when watching a film in a 42" TV screen or in a 4" PDA screen. For instance, if high resolution content is delivered, the user will find an improvement in the video resolution when watching it on the TV, but may not notice a clear difference in the PDA. In addition, we have to consider that different machines have different maximum decoding bit rates. Furthermore, if the viewing device is a mobile device we have consider other issues, as would be batteries consumption, mobility (including service continuity while the user position moves and forces a change in the base station that relies it to the main network), etc.

In our network, we consider these constraints. A module aware of the terminal characteristics is incorporated in our network [3]. This module is aware of the terminal properties, its processing constraints, codec capabilities, etc. and is in charge of the management of any possible restriction the end-device may impose. If an end-device does not have the suitable characteristics to receive an enhanced video layer or multiple views, these streams should not be sent. Our objective is to avoid the transfer over the network of data that would be useless to the end-user and may result in network congestion. This network saturation produce several problems, as the one we will explain in the next section.

2.3 Network Parameters

Once the end-user have purchased the adequate license and the terminal awareness module has assured that the devices involved have the suitable characteristics to send, receive and play the media content chosen, the data streaming are delivered over the P2P network. However, one last problem may occur. If we want the terminal to be capable of displaying the content, the reception of the media content stream needs to fulfill a certain QoS (Quality of Service), especially regarding the maximum tolerable jitter in the packet arrival. If the network is congested, the general QoS may not be sufficient to permit the terminal to reproduce the media content as desired. Nevertheless, the network may have the necessary bandwidth to transmit correctly the base layer of the content. A module aware of the network characteristics is included in our system [3]. If this module realizes that the network is congested, it will inform the sender and the enhanced layers and additional views will not be sent.

3 Conclusion

In this paper we have analyzed further our proposed seamless P2P network to deliver media. We have identified the different causes that may force an adaption in the delivery of the media content to the end-user. Bearing in mind our interoperability aim, this personalization can take place for mainly three reasons: end-user preferences, devices restrictions and network proprieties. We have proposed three solutions that allow to manage the adaptation for each one of the tree cases presented. The end-user preferences can be easily considered as we enhance the network security using licenses where these preferences are described. In addition, our network model includes the necessary device capabilities aware and network aware elements to seamlessly perform the media delivery.

Acknowledgments. This publication presented the authors opinion. Yet, it is based on work performed in the framework of the Project SEA IST-214063, which is funded by the European Community. The authors would like to acknowledge the contributions to the SEA Project of colleagues from: STM, Synelixis, Thomson, Philips, Vodafone, Nomor, Fraunhofer HHI, Politechnico di Torino, Universidad Politécnica de Madrid, University of California, Los Angeles.

References

1. SEA, SEAmless Content delivery project, http://www.ist-sea.eu
2. Zahariadis, T., Negru, O., Álvarez, F.: Scalable Content Delivery over P2P Convergent Networks. In: 12th IEEE International Symposium on Consumer Electronics (ISCE 2008), Vilamoura, Portugal, April 14-16 (2008)
3. Zahariadis, T., et al.: D4.1 - Cross layer control, adaptation modelling and metadata specification. Deliverable of the SEA project, September, 30 (2008)

Shopping Assistant

Bernd Klasen, Alexander Vinzl, and Takeshi Martinez

SES ASTRA TechCom
Chateau de Betzdorf, L-6815 Betzdorf, Grand Duchy of Luxembourg
in Cooperation with University of Trier and University of Luxembourg
supported by the Fonds National de la Recherche, Luxembourg
{bernd.klasen,takeshi.martinez}@ses-astra.com,vinzl@syssoft.uni-trier.de

Abstract. This paper presents the Shopping Assistant, the prototype of a platform which provides personalized advertisements, ontology based product recommendations and user support to find a (non–web) store selling desired products. The benefits are satisfied consumers, better advertising revenues and fine grained TV usage statistics which enable broadcasters to provide a more user centric program composition.

Keywords: personalized advertisement, location based services (LBS).

1 Introduction

In todays media, advertisements have become an omnipresent part which is not uncommonly considered as very annoying. While technical advances make it easier to skip commercials, broadcasters try to prohibit that by means of technical restrictions. Instead of trying to force customers to see commercials, the Shopping Assistant – developed by SES ASTRA TechCom, Inverto and CRP Henri Tudor in the scope of the ITEA 2 WellCom Project [2] – aims at making advertisements more interesting and valuable for customers by offering personalized advertisements. This is a long–needed feature of marketing experts and many other global players are going the same direction (e.g. Google [1]).

The Shopping Assistant can be subdivided into two parts that differ in time and place. These parts, the Home–Fraction and the Shopping–Fraction, are briefly described in section 2 and 3.

2 Home–Fraction

The Home–Fraction assumes the following environment: TV, *Set–Top–Box*[1] (STB), WiFi capable mobile phone, a WiFi access point and a permanent internet connection. A new development is the *Home Gateway* (HGW), which is running an HTTP–Server on a Linux OS, providing an NFS share and an event notification interface listening for UDP packets. The HGW is a seperate device in this prototype but is proposed to be integrated into the STB later. All devices except of the TV are interconnected via WiFi or Ethernet and form the home network.

---

[1] Assembled by Inverto.

P. Daras and O. Mayora (Eds.): UCMedia 2009, LNICST 40, pp. 214–217, 2010.

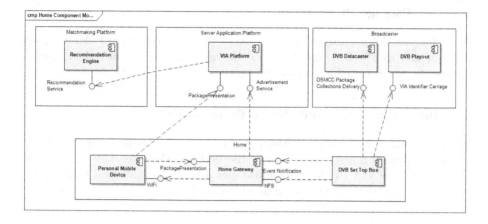

Fig. 1. Component model for the Shopping–Assistant

The following steps form the Home–Fraction:

2.1   Ad–Package Distribution and Identification

The registered user in this environment can receive more fine grained advertisements (*Ad–Packages*) on his mobile phone. These Ad–Packages are bundled to Ad–Collections – which correlate to one commercial–spot each – and broadcasted in advance using a *digital storage media command and control* (DSM–CC). The STB receives and stores them on the NFS share on the HGW. Ad–Collections have a unique ID (Ad–ID). The same ID is added as a subtitle segment[2] to the transport stream of the corresponding commercial.

2.2   Ad–Package Profile Matching

The STB inspects the subtitle segment and sends the notification to the HGW, including the ID. The HGW sends a SOAP–Message via Internet to the *Video Interactive Application (VIA) Platform* – where all user profiles are stored[3] – including the Ad–ID and the IDs of all connected users. This returns – after consulting the *Recommendation Engine* (RE) – the ID of the Ad–Package which matches the users interests best.

---

[2] The DVB Subtitling Systems standard (ETSI EN 300 743 v010301) has provisions for including stream synchronized private data. The VIA Identifier is encoded in a private subtitling segment with the segment type 0xAD.

[3] The centralized data storing on the VIA Platform facilitate an easy and quick recovery after a device failure, since nothing but the login credentials are stored on the end user device.

2.3 Ad–Package Presentation

According to this return values, the HGW prepares the correct Ad–Package for each user as a website on its HTTP–Server. Each user is offered the choice whether he is interested in this product or not.

2.4 Input Processing and Profile Enhancement

The input is sent to the VIA Platform and added to his profile. This enables an increasing precision in the selection of the best Ad–Package. The information that reaches the VIA Platform facilitate user acceptance statistics, that can be used to enhance the general design of commercials.

3 Shopping–Fraction

The user on the move must have the Shopping–Fraction application running on his mobile device in order to use the service. This application is implemented using JavaME. Its functions are to submit the users position[4] in individually predefined intervals to the VIA Platform and to listen for replies from the VIA Platform containing shop and product information. On reception of a postion update of a certain user, the VIA Platform generates a list of shops nearby that are selling products the user stated his interest in (see Shopping–Fraction in section 3). This is achieved by simple database requests, which select all shops whose euclidian distance to the user's position is less than a certain value

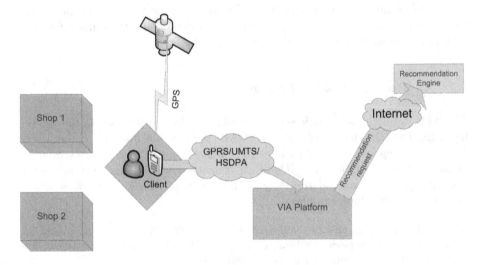

Fig. 2. Data exchange in the Shopping–Fraction

---

[4] The current position is determined using GPS.

(depending on the density of shops in the area) AND which are selling products that are in the user's interest–list. The interest–list is also the result of a database request. In this prototype, the VIA Platform only sends the user profile and the interest–list containing product profiles to the Recommandation Engine. Product profiles are given as OWL files. The corresponding ontologies have been designed by CRP Henri tudor. The request uses the same web service that has been used for Ad–Package recommendations described in section 2. This is proposed to be enhanced in future implementations as shown in section 4. At the time given, the return value from the RE contains the ID of the product that the user – according to his profile – is supposed to desire most. The VIA Platform selects the closest shop that is selling this product and sends the information to the user's mobile phone, where an alert is raised.

4   Future Prospects

The aim is to provide a closed system, where the final buying decision is recognized and used to enhance user profiles and future product recommandations automatically. As mentioned in the preceding section, some enhancements are proposed concerning the Shopping–Fraction. A new web service is to be implemented, which takes list of products, lists of shops and a user profile including his current position. This will enable more complex calculations and better recommendations (e.g. return one shop which sells most of the products the user desires, considering also the distance to the user).

In this context a lot of user data like interests, desires and current location is exchanged, which is highly confidential information. Even if security has always been taken into account during the design and implementation of the Shopping Assistant, this has to be reviewed before it can be officially launched. To achieve the best security and privacy, existing and sophisticated techniques to provide security will be employed and all user data will be anonymzed whenever it is communicated.

Another aspect is a completely new advertising paradigm, where Ad–Packages are injected directly into the transport stream of movies whenever they fit to the content and the users desires and only when they fit.

Finally, social networking functionality should be integrated, to enable suggestions for gifts for the upcoming birthday of a friend as well as collaborative filtering for the best possible product recommendations.

References

1. Wojcicki, S.: (VP, Google Product Management): Making ads more interesting (2009)
2. ITEA 2 WellCom Project, http://www.itea-wellcom.org

Streaming Content Wars: Download and Play Strikes Back

Charalampos Patrikakis[1], Nikos Papaoulakis[2], Chryssanthi Stefanoudaki[3], and Mário Nunes[4]

[1] National Technical University of Athens
bpatr@telecom.ntua.gr
[2] National Technical University of Athens
npapaoul@telecom.ntua.gr
[3] National Technical University of Athens
xrussanthi@telecom.ntua.gr
[4] INOV - INESC Inovação / IST, Lisbon
mario.nunes@inov.pt

Abstract. In this paper, the latest developments in the provision of streaming multimedia over the internet are discussed. We emphasize on the newly appearing HTTP adaptive streaming approach that threatens the reign of RTSP, as regards the support of real time media streaming, and we examine the possible uses of each of the two protocols with respect to the different transmission needs, as these are seen from the transport protocol view point.

Keywords: download, progressive download, streaming, HTTP adaptive streaming.

1 Introduction

In response to the technological developments in the technologies available for enhancing the user's viewing experience, media consuming society is driving more and more demanding. The fact that consumers can choose and switch among many different but allied media contents (i.e. when visiting a site on the Internet), reveals the increasing need of streaming services development and the corresponding support by networking protocols and media playback applications.

To this direction, several alternatives have been proposed, as regards the use of protocols, architectures and media distribution models. In the meanwhile, the diffusion of Content Delivery Networks (CDNs) has lead to the increase of the available bandwidth for clients. This fact has an impact on the technologies used for media streaming, as the availability of extra bandwidth has permitted the consideration of protocols traditionally considered as inappropriate for streaming, due to the large overhead that they incur [2]. As a consequence, we have to re-evaluate the standard streaming technologies under the light of this new situation, and also to take into account extra parameters, such as compatibility with web browsers, popularity of technologies for accessing short clips (i.e. videos over YouTube) and current conditions in the media market.

P. Daras and O. Mayora (Eds.): UCMedia 2009, LNICST 40, pp. 218–226, 2010.

In this paper, we make an evaluation of the available technologies, having in mind the most challenging scenario: that of live media streaming. In this evaluation, both the main rivals: HTTP and RTSP are considered, and the advantages of the use of each protocol, with respect to the particular distribution environment (unicast, multicast, broadcast) are highlighted. The rest of the paper is organized as follows: In section 2, a report on the technologies used for media distribution is provided. In section 3, an analysis on the selection of the appropriate protocol based on the transmission mode is given. Finally, section 4 closes the paper with conclusions and a use case scenario.

2   History and State of the Art in Media Streaming Technologies

Before continuing with our evaluation, we will go through a general overview of the different technologies used for accessing media content streams over the web. Here, it should be noted that the term of media content is considered to cover information that has a temporal dimension (video and sound) and is therefore subject to streaming. Also both stakeholder sides, namely content consumers and content producers are taken into consideration as regards their needs and requirements from a steaming platform.

From the streaming media consumers' side, it is really important to choose the latest technology for streaming/download, for two main reasons: except from the quality of the result on client's screen which evidently increases as technology advances, each technology affects the number and type of plug-ins that have to be installed on someone's PC so as to view content. The latter introduces issues of compatibility and stability of the software and application components used. On the other side, streaming producers, most of them being media and entertainment companies, focus on the use advertising-based models that depend on user's viewer satisfaction [6]. Therefore, user satisfaction here is important from an economic perspective, as it affects the audience of a particular streaming service. As a conclusion, the selection of tools for supporting the download and play or direct streaming is very important, as it may have a direct effect on the expected revenues and the user's experience and consequent use of the corresponding technologies.

Following this brief analysis, we will continue with the presentation of the technologies available for direct access to multimedia information, starting from the traditional download and play, and going to the dynamically adaptive media streaming.

2.1   Download and Play

This is simplest version of accessing multimedia information: download and play. It can be applied to both static media such as images, but also to non static such as video and sound. In this method, the client has to download and store first the whole media content on his PC and only then he/she will be able to view and use it. However, the media has to be available in the form of a file, excluding the use for covering live events, as the multimedia information that is to be transmitted cannot be "a-priori" available. Download and play is quite an old method that is unable to meet the needs of contemporary media market. It was the first attempt of getting media content on a personal computer from a server, which, though very simple to be supported even over a web browsing session is no longer considered as an efficient solution for video access.

2.2 Traditional Streaming

In streaming, multimedia data is transferred in a stream of packets that are interpreted and rendered in real time by a software application, as the packets arrive in the user's terminal. For this reason, we need a protocol that is capable of keeping track of the client application state and the progress of the streaming process, from the initial client request for multimedia access, up to the point of connection teardown.

RTSP is the traditional and standardized streaming protocol that meets the above needs [10]. In RTSP, packets can be transmitted either over UDP or TCP; the latter is preferred due to its guaranteed delivery, in important aspect in signaling It should be noted that RTSP is used for transferring the control information as regards the handling of the stream, while the actual streaming data are sent using the Real-time Transport Protocol, RTP. Here, it should be noted that since the use of RTP is followed by the use of UDP for delivery of the stream, in cases where there is use behind firewalls, this may lead to blocking of the download stream packets. Furthermore, as UDP does not provide any congestion monitoring mechanism, diagnosis of congestion problems may take place too late, after the problem has already appeared and has affected the user experience, even lead to disconnection of the stream. The following figure provides a graphical representation of the traditional streaming approach.

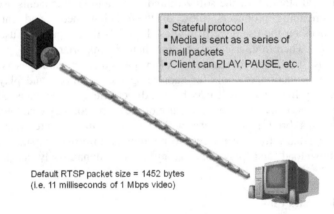

- Stateful protocol
- Media is sent as a series of small packets
- Client can PLAY, PAUSE, etc.

Default RTSP packet size = 1452 bytes
(i.e. 11 milliseconds of 1 Mbps video)

Fig. 1. Traditional streaming [1]

2.3 Progressive Download

Alternatively to the use of RTSP and RTP, the use of the HTTP protocol has been proposed, in a near streaming like approach, called progressive download. Progressive Download is a hybrid approach between the download and play and the streaming, in which the client does not need to download the whole file before being able to locally reproduce it. Contrary to the download and play approach, content download is not downloaded as a whole but in segments. In this way, the client is able to begin playback of the segment that has already been downloaded locally, before the full download is complete. In the case of any problem, download is paused. One basic difference between streaming and progressive download is in how the media data is received and stored at the end user's terminal. In progressive download, the media data is stored on the terminal's hard drive, in contrast with (real-time) streaming, where media is only delivered, but not stored by the client. A visual presentation of the procedure described above follows in Figure 2.

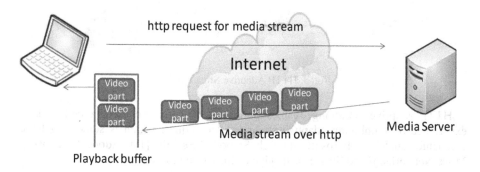

Fig. 2. Progressive Download

2.4 HTTP Adaptive Streaming

Advancing the idea of progressive download of chunks of media data so as to emulate a streaming behavior, the latest approach that seems to incorporate both the advantages of using a standard web protocol and the ability to be used for streaming even live content is the adaptive (HTTP-based) streaming. Adaptive HTTP streaming is based on HTTP progressive download, but contrary to the previous approach, here the files are very small, so that they can be compared to the streaming of packets, much like the case of using RTSP and RTP. To achieve this, the streaming server first divides video/audio file into many small segments (of various sizes) and then encodes them to the desired format. After that, according to network conditions at the time of real streaming (i.e. bandwidth), server sends to each client segments of the client requested content with appropriate size, as depicted in Figure 3. For example, when bandwidth is reduced, in order to avoid the pause of the file download, server sends segments of smaller size, reducing by this way the bit rate of the downloading process. The result is momentarily lower in quality, but uninterrupted viewing experience. Of course, after the recovering of the problem, bit rate increases. So, we get the best quality we can receive each time. The use of HTTP for conveying the streamed information introduces the advantage of having the underlying TCP protocol monitor the network conditions, and through its inherent congestion control mechanism provide the means for early detection of congestion problems.

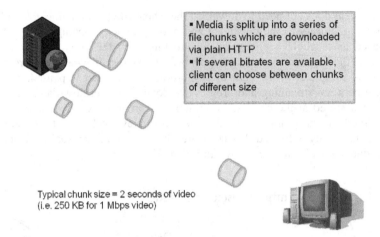

Typical chunk size = 2 seconds of video
(i.e. 250 KB for 1 Mbps video)

Fig. 3. HTTP Adaptive Streaming [1]

HTTP adaptive streaming enables publishers to offer the users a better video experience by employing adaptive streaming techniques, which is something large companies such as Microsoft (through Smooth Streaming[1]), Adobe Systems[8], Move Networks[5] and Swarmcast[7] have already embraced.

3   Selecting the Best Method Based on User (Transmission) Modes

Following the brief presentation of the different methods for supporting media streaming, in this section we identify the suitability of each method, based in the particular parameters that exist as regards the needs for transmission. The analysis that will follow is focused on the support of real time streaming information, targeting at the live coverage of event. This is considered the worse scenario, which by far covers the needs for streaming information in general, as it has to take into account the direct processing and transmission of the streams, upon production based on live feeds. This automatically allows only two candidates: RTSP and HTTP adaptive streaming.

Furthermore, emphasis is given in the aspect of personalization, as this is an innovation that IPTV or Web-TV services can offer compared to the traditional TV services. By the term personalization, we refer to the specialization of the media streams offered to the end users, so as to match the personal preferences as regards both the content and the specific device parameters (size of screen, processor power e.a.). A detailed analysis as regards the personalization of the media streams can be found in the corresponding report of my-eDirector 2012 project [3] that addresses these issues. It should be noted here that as regards the selection of network access, taking into account the plethora of technologies, all wired, wireless and mobile technologies should be taken under consideration (xDSL, Ethernet, WiMax, WLAN, 3G, and Bluetooth). As many of the above cases present similarities regardless of the status of the physical media involved (i.e. Ethernet and WLAN), evaluation should not be performed based on the particular characteristics of the physical medium, but rather based on the characteristics of the transmission method. Based on this, the three transmission methods, Unicast, Multicast and Broadcast, are used in the following sections of the paper to present the suitability for use of each protocol.

3.1 Supporting the Unicast Transmission Mode

Here, the use of HTTP adaptive streaming is considered the best candidate, since through the use of the appropriate technology at the server (i.e. Smooth Streaming [1]) it is able to adapt the stream to the available bandwidth. Furthermore, it is able to seamlessly address issues arising both during mobility across domains, but also when switching between network technologies, with immediate redirection of the streaming packet routes. Based on the use of HTTP adaptive streaming, seamless switching between bitrates and encoding levels in order to adapt to the network conditions, and seamless switching across network technologies can be supported. Furthermore, seamless switching across devices is also possible, to the level that this can be supported by the operating systems and the application frameworks available in each device.

3.2 Supporting the Multicast Transmission Mode

As regards the use of multicast, this is mainly of interest to the case where wireless access network technologies are involved, in order to provide a more efficient use of the scarce available bandwidth. Use over wired technologies of course presents no challenges and can be provided in the same way. In the case of multicast use case scenarios, one major challenge is the ability to support seamless switching between transmission modes (unicast to multicast and vice versa) within the same access network technologies, and adaptation to the network conditions based on selection of transmission scheme. To support the above, the use of selection of RTP based adaptive streaming with the use of SIP-RTSP protocols is the best option, as it inherently supports multicasting. The use of HTTP here is not an option due to its unicast oriented nature.

It should be noted that an added value feature in the case of multicasting is the ability to provide personalization of the media streaming experience. This can either be supported through the dynamic selection of multicast channel according to the personal preferences of users, or through the switching from multicast to unicast in cases of reduced interest to particular streams.

3.3 Supporting the Broadcast Transmission Mode

As regards the use of broadcast, this is best suited to TV like use case scenarios, and for this DVB-H and DVB-T technologies are the most appropriate. Here, one major challenge is the personalization of the streaming experience. Two different approaches can be considered here: the first regards personalized coverage over DVB-H, bypassing the issue that use of broadcasting does not allow the provision of a separate per user stream at the download, while the second addresses the issue of the effective use of DVB-H in order to offer the user a more rich experience that will allow him/her to personalize the viewing experience.

To achieve the above, extensions to the traditional DVB-H support are needed, according to which the terminal is considered to be operating as a plain TV-like receiver will be able to receive extra recommendation information and real-time metadata about the received streams, while feedback from the user's selections will be sent back to the system. This can be available through the use of extra communication channels based on multicast and unicast transmission through the use of 3G or WiFi technologies, in parallel to the DVB-H streams. Use of SIP protocol can be deployed here in order to support the return channel and to convey the user's selections, enhancing the end user experience over the DVB-H transmitted stream and introducing full interactivity capabilities to the media distribution model.

In the context of this scenario, switching (not necessarily seamlessly) between unicast and broadcast can be performed according to the personal preferences of the users, so as to better accommodate the needs for personalization of the transmission.

4 Use Case Scenario and Conclusions

From the above analysis, we can see that the selection of the best way to support the media streaming services depends on the transmission mode that will be applied. Though HTTP offers more capabilities for personalization of the offered streams, it is oriented towards unicast transmission, which is not as effective as alternative methods such as multicast and broadcast. Furthermore, since HTTP adaptive streaming is making use of standard protocols (the exact same protocols used for access to web services) their deployment over a wide range of platforms and network environments is seamless, creating minimum implications as regards the use behind firewalls and NATs. Finally, the ability to deploy TCP transmission for the media streams and the control information, offers an advantage as regards early detection of network congestion and also fast adaptation of the transported stream rate.

On the other hand, use of RTSP is closely related to the efficient use of the network resources, as it can be easily deployed over all types of transmission. Therefore, bulk distribution of streams using multicast and broadcast can be supported, while for accommodating the return channel needs with user selections and feedback can be still be supported over unicast. Furthermore, RTSP protocol is designed to accommodate streaming of media and therefore inherently supports media handling capabilities.

The selection of the most appropriate method therefore lies in the hands of the engineers that design the end to end distribution architecture. Though there is no golden rule as regards the selection of the protocol, several parameters can be evaluated so as the reach a decision: Need for multicast/broadcast, need for operation behind firewalls and NATs, switching among devices and networks. An example for selection is presented in the following table, taken from the access network technologies selected in my-eDirector 2012 project and the corresponding support in terms of protocols.

Table 1. Selection of protocols and access network technologies in My-e-Director 2012[9]

	Access Network Technology	Data Transfer	Streaming Protocol	Transfer Protocol	Return channel feedback
	Ethernet	IP	HTTP	TCP	HTTP
	ADSL	IP	HTTP	TCP	HTTP
UNICAST	WiFi	IP	HTTP	TCP	HTTP
	3G	IP	HTTP	TCP	HTTP
	WiMAX	IP	HTTP	TCP	HTTP
	WiFi	IP	RTP	UDP	SIP - RTSP
MULTICAST	WiMAX	IP	RTP	UDP	SIP - RTSP
	Ethernet	IP	RTP	UDP	SIP - RTSP
BROADCAST	DVB - H	TS	MPE 2-TS	MPE-FEC	HTTP

In the above table, we see that following the analysis presented earlier in the paper, the corresponding use of HTTP and RTSP has been selected so as to match the different needs for transmission of the streams. As one of the key aspects of the project is personalization of the streams, provision has been taken in order to be able to offer personalized services even in the case where multicast and broadcast is used. The results clearly indicated that no single universal can be adopted.

As a conclusion, the return of the progressive download has been remarkable, as it has been accompanied by the corresponding capability for adaptive transformation of the stream, even in conditions of real time coverage of events. However, the use of RTSP is still the only solution when it comes to the use of multicast and broadcast, while the forthcoming developments with the introduction of IPv6 and the next generation of RTSP may create new scenery in the field of media streaming over the internet.

Acknowledgments

The work presented here has been performed in the context of the European My-eDirector 2012 project. The authors would like to thank all project partners for their support.

References

1. Zambelli, A.: IIS Smooth Streaming Technical Overview (2009)
2. AT&T, Choosing a Content Delivery Method (2008)
3. My-e-Director 2012, Real-Time Context-Aware and Personalized Media Streaming Environments for Large Scale Broadcasting Applications, D2.1-081002-End-User Requirements (2009)
4. Apple Inc. QuickTime streaming, http://www.apple.com/ (accessed on September 14, 2009)

5. Move Networks, MOVE ADAPTIVE STREAM On-the-fly Stream Selection for an Uninterrupted Experience, http://www.movenetworks.com/ (accessed on September 12, 2009)
6. Ozer, J.: Streaming Gets Smarter: Evaluating the Adaptive Streaming Technologies, http://www.streamingmedia.com (accessed on September 12, 2009)
7. Hinchey, M.G., Sterritt, R., Rouff, C.: Swarms and Swarm Intelligence. Computer 40(4), 111–113 (2007)
8. Adobe Systems Inc. Flash Video Streaming Service, http://www.adobe.com/ (accessed on September 10, 2009)
9. My-e-Director 2012, Real-Time Context-Aware and Personalized Media Streaming Environments for Large Scale Broadcasting Applications, D 6.1: My-e-Director 2012 Heterogeneous internetworking and mobility architecture (2009)
10. RFC 2326, Real Time Streaming Protocol (RTSP), IETF (April 1998)

A Personalized HTTP Adaptive Streaming WebTV

Rui Santos Cruz[1], Mário Serafim Nunes[1], and João Espadanal Gonçalves[2]

[1] IST/INESC-ID/INOV, Lisboa, Portugal
mario.nunes@ieee.org, rui.cruz@ieee.org
[2] Instituto Superior Técnico, Lisboa, Portugal
espadanal@gmail.com

Abstract. This paper presents a HTTP based multimedia content streaming architecture able to provide large scale WebTV services to end users connected over several access network technologies. The architecture features smooth Rate Adaptation, context Personalization information (Channel Suggestions), fast program channel switching, Network Portability (Vertical Handover) and Terminal Portability capabilities. The prototypical solution is being developed under the scope of the *My-eDirector 2012* european project.

Keywords: WebTV, Smooth Streaming, Content Adaptation, Rate Adaptation, HTTP Progressive Download.

1 Introduction

An increasing large number of people is already having access to all kinds of multimedia content (video, television, audio, graphics or simply data) from all kinds of networks using a large variety of terminal devices, such as computers, Set-Top-Boxes and televisions, networked portable media players, Internet-enabled smartphones or just mobile phones. The number of multimedia contents accessed by users over IP networks is also reaching very high levels, mostly due to video storage availability and live streaming of videos or television programs.

On the content side the worldwide trend is clearly towards user-generated or personalized contents. This trend, associated with the maturity level of perceptual technologies for context-awareness and the need and demand from operators, service providers and broadcasters for novel revenue generating networked services is the main driver for the development of innovative personalized media streaming services that enable new forms of rich user interaction with live or archived contents.

Streaming producers, most of them being media and entertainment companies, have a particular interest in personalization capabilities as they can affect the audience of a particular streaming service and also because their business models are mainly based on advertisement, depending ultimately on the user's viewing satisfaction.

P. Daras and O. Mayora (Eds.): UCMedia 2009, LNICST 40, pp. 227–233, 2010.
© Institute for Computer Sciences, Social-Informatics and Telecommunications Engineering 2010

These personalized media streaming services aim to provide a novel viewing experience on the end user's device by means of recommendation services that are used for the selection and viewing of particular streams matched to the user preferences and profile.

But these personalization capabilities require adaptation of contents at different levels (media production, distribution and viewing) and adaptation of the viewing experience to the status of the environment (network capabilities, device type capabilities, physical environment).

This paper shares the experience on the development of a personalized HTTP Adaptive Streaming architecture for WebTV services, realized within the scope of the european project *My eDirector 2012* [1].

This WebTV oriented solution uses standard HTTP transactions for multimedia content streaming that are capable of traversing any firewall or proxy server that lets through standard HTTP traffic.

The HTTP Adaptive Streaming architecture is capable of providing video streaming to end users, connected either directly or through Content Delivery Networks (CDN) over several access network technologies, enabling an uninterrupted streaming experience, guaranteeing, for each user, the best possible video quality reception. With this architecture,, multimedia sources (video and audio) can be encoded in various bitrates, and a HTTP progressive download method provides the capability to react to both bandwidth and local terminal conditions in order to seamlessly switch the video quality that the user is receiving, maximizing the Quality of Experience (QoE).

One of the most important functionalities of the architecture is the seamless bitrate switching. The switching heuristics are implemented on the Client side, not involving any process at the Streaming Server. It is the responsibility of the Client to monitor network conditions, chunk download times, buffer fullness, rendered frame rates, and other factors, such as screen resolution, playing/paused status of the content, or the CPU load, in order to decide when to request higher or lower bitrates from the Streaming Server.

Other important features implemented in the architecture are the Network Portability (Vertical Handover) and terminal Portability capabilities, allowing a flexible user experience in terms of either a continuous uninterrupted playback during network handover with the same terminal or a suspend/resume session playback, either with the same terminal or with a different terminal (with same or with different capabilities like screen resolution or access network attachments).

In this paper, Section 2 describes the overall architecture of the HTTP Adaptive Streaming WebTV solution, Section 3 presents the Web Media Player Client architecture, Sections 4 and 5 the Portability functionalities of the solution prototype and Section 6 concludes the paper.

2   The HTTP Adaptive Streaming

The HTTP Adaptive Streaming architecture comprehends a Server component, a Distribution component and a Client component.

The Server component (Service Provider Encoder) is responsible for taking input streams of media from Content Provider Video Streamers and encode them digitally, to a format suitable for delivery, and preparing the media for distribution.

The Distribution component (Smooth Streaming Server) consists of Web Servers responsible for accepting Client requests and delivering prepared media and associated resources to the Client.

The Client component consists of a Web Media Player responsible for determining the appropriate media to request, download those resources, and then reassemble them to be presented as a continuous stream.

The architecture also uses HTTP for personalization and program channel information (Channel Suggestions/Recommendations) and associated Metadata, with exchange format based on eXtensible Application Markup Language (XAML), offering Camera Selection suggestions, fine control over presentation of closed captions and other timed text content.

The HTTP Adaptive Streaming architecture uses a streaming technology based on a HTTP progressive download method.

Both Microsoft, with "Smooth Streaming" [6], and Apple Inc., with "HTTP Live Streaming" [5], propose similar technologies based on HTTP progressive download.

For the initial implementation of the WebTV architecture the choice fall on Microsoft's Smooth Streaming.

Smooth Streaming: The Smooth Streaming technology from Microsoft uses MP4 Media File Format specification, MPEG-4 Part 14 (ISO/IEC 14496-12) [2], for both the storage (*disk*) and transport (*wire*) formats, architected with H.264 [3] video codec support.

The *wire* format defines the structure of the chunks that are sent to the Client, for both stored or live contents, whereas the *disk* format defines the structure of the contiguous files on disk for stored content.

For the Smooth Streaming method, the multimedia source at the Server is segmented into "chunks" from 2 to 4 seconds long, cut along a video GOP (Group of Pictures) boundary (starting with a key-frame) and have no dependencies on past or future chunks/GOPs. The chunks are encoded to the desired delivery format (H.264) and made available at the Distribution server to be delivered upon Client requests. The Client uses the Smooth Streaming method to request the chunks. Each chunk can be decoded independently of other chunks.

3   The Web Media Player Client

The Web Media Player Client architecture, a cross platform, cross browser and cross device Rich Interactive Application (RIA) based on Microsoft *Silverlight* technology [4], takes the form of a "plug-in" with a modular design, as illustrated in Figure 1, facilitating the implementation of enhanced/new functionalities.

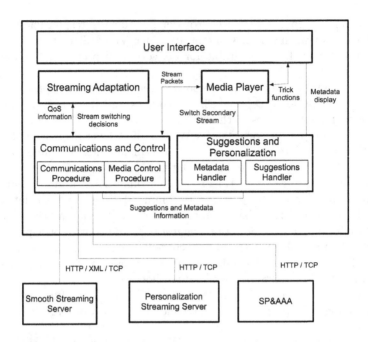

Fig. 1. Web Media Player Client Application Modules

The Communications and Control Module: The Communications and Control Module implements the Smooth Streaming method and is responsible for Session Setup, Session Termination, content negotiations with the Smooth Streaming Server, polling the Personalization Streaming Server (PSS) for Channel Suggestions and stream Metadata and for monitoring network conditions.

A typical content negotiation message sequence is illustrated in Figure 2. In this module a Communications Procedure handles all session control while a Media Control Procedure handles network Trick Functions, in order to control the media stream status (PLAY, PAUSE) as well as the session SUSPEND/RESUME feature for terminal Portability.

The Suggestions and Personalization Module: The Suggestions and Personalization Module is used for the treatment of textual information on Personalization and content Metadata, including Camera Selection, Event comments, subpicture information or news about some other events.

A polling method with a resolution of 1 second has been chosen for all requests to the Personalization Streaming Server (PSS), through port 80, preventing blocking/filtering conditions on traversing network boundaries.

The Streaming Adaptation Module: The Streaming Adaptation Module implements the Network Quality and Host Capabilities adaptation strategies for the selection of the most appropriate bitrate at any moment in time.

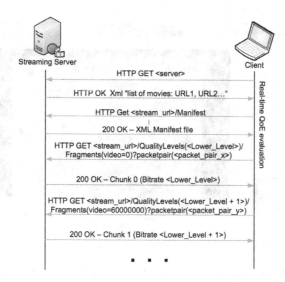

Fig. 2. Example of Content negotiation with the Smooth Streaming Server

The Network Quality strategy heuristic is responsible for the network logic on determining the adequate bandwidth for an initial fast bitrate ramp up and for a subsequent sustained buffer fullness within lower and upper bitrate boundaries, constrained by a validation of supported bitrate determined by the Host Capabilities strategy.

The Host Capabilities strategy heuristic is responsible for the environment logic. Essentially, it keeps a cumulative playtime validation, on a per bitrate basis, and continuously analyses the dropped frames to infer the bitrates supported by the Client.

The Media Player Module: The Media Player Module corresponds to the media player itself, responsible for the reproduction of streams received from the Communications and Control Module. The media player can reproduce content from a variety of sources, including multiple bitrate streaming of live, video on demand H.264/AAC/MP3 and unicast HTTP Windows Media Services streams.

The User Interface Module: The User Interface Module implements the Application Graphical User Interface (GUI), dynamically generated, based on the type of terminal the Client is running.

Figure 3 illustrates the prototype UI for a desktop/laptop browser and Figure 4 for a smartphone or Personal Digital Assistant (PDA) browser. Via the GUI the user is able to authenticate in the Service Platform (SP&AAA) and also to select contents, or events to watch. The GUI also presents program Suggestions to the user based on the Personalization information.

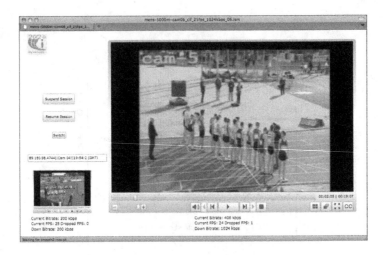

Fig. 3. Web Media Player Client Application in a standard browser

Fig. 4. Web Media Player Client Application in a PDA browser

4 Network Portability

The Network Portability, or Vertical Handover, is assured within this architecture in a transparent manner, as long as the Client terminal provides adequate networking capabilities (multiple active interfaces, like WiFi/WiMax, 3G, etc.).

During the handover, the Web Media Player continues the playback with the "chunks" still in buffer. After the IP connection re-established (and the HTTP/TCP session automatically re-established), the Web Media Player starts requesting new "chunks", from the new connection, completely unaware of the network handover (unless the handover process takes longer than the playback of the buffered content).

5   Terminal Portability

The terminal Portability feature allows the user to suspend the current session on one terminal and resume it later on the same, or on a different terminal and independently of its type (e.g., transfer from a personal computer to a PDA).

By means of two buttons, a SUSPEND and a RESUME, the user is requested to type a code (with duplicate verification) that will be used later for resuming that session.

The Web Media Player Client exchanges XML files with the server, by HTTP POST methods, that includes the chosen code and/or the URL of the stream and the time of playback for that stream.

These RESUME requests always start with the lowest resolution/bitrate, in order to start re-playing the video as quickly as possible.

6   Conclusion and Future Work

This project is currently under development. Almost all components are implemented, but not fully tested.

The current roadmap includes both stored (video-on-demand) and live streaming media player capabilities and a full graphical design of the User Interface, the fix of eventual bugs that may exist and an exhaustive test plan (with both Functional and Performance tests), to prove the effectiveness of this solution.

Future work on this project will include the deployment of objective quality estimations at the Client and enhanced Streaming Adaptation heuristics related with physical environment status (like adaptation of resolution and zooming capabilities to light brightness, sound level and movement of user/device or display orientation of the device).

References

1. My eDirector (2012), http://www.myedirector2012.eu
2. ISO/IEC: Information technology – Coding of audio-visual objects – Part 14: MP4 file format. Tech. Rep. ISO/IEC 14496-14:2003 (2003)
3. ISO/IEC: Information technology – Coding of audio-visual objects – Part 10: Advanced Video Coding. Tech. Rep. ISO/IEC 14496-10:2005 (2005)
4. Moroney, L.: Introducing Microsoft Silverlight 3. Microsoft Press, Redmond•(2009)
5. Pantos, R.: HTTP Live Streaming. IETF (2009)
6. Zambelli, A.: IIS Smooth Streaming Technical Overview. Microsoft Corporation (2009)

PerMeD 2009

Session 2

Scalable IPTV Delivery to Home via VPN

Shuai Qu and Jonas Lindqvist

Acreo, NetLab,
Håstaholmen. 4, 82412 Hudiksvall, Sweden
{shuai.qu,jonas.lindqvist}@acreo.se

Abstract. The significant interest in IPTV drives the needs for flexible and scalable IPTV delivery way, especially when distributing IPTV service to end-users who are in a separate network or not in an IPTV enabled network. The recent popularity of VPN has made scalable distribution of IPTV possible. VPN can provide global IPTV networking opportunities and extended geographic connectivity. Additionally, the native traits of VPN also provide secure and controllable service features to IPTV. This paper addresses one important area related to IPTV distribution, namely scalability. We present a novel solution to distribute IPTV via VPN to remote end-users over public networks. The solution allows end-users over a wider geographical area to get IPTV service, and it also reduces operating costs. Traffic measurements and evaluation of services performance are also illustrated and discussed in this paper.

Keywords: IPTV, VPN, scalability.

1 Introduction

1.1 Background and Problem Motivations

Internet Protocol Television (IPTV) [1], [2] is a system where a digital TV service is delivered to end-users by using IP over a network infrastructure. IPTV is now gaining popularity very rapidly, Informa Research [3] state that the market will grow by a factor of seven by 2011 based on the 2006 numbers. The significant interest in IPTV services and wholesale business models are driving the need to consider more scalable ways to deliver multicast services[4]. Generally, IPTV platform has been physical platform: leased lines connecting a limited set of locations. The coverage areas of IPTV service are dedicated and depend on network infrastructure built for IPTV distributions. It is therefore difficult to make IPTV service globally available for remote users who are in a separate network or not part of IPTV enabled network. In addition, it is also quite expensive to extend and operate IPTV at very large scale. Therefore, traditional IPTV scheme do not address the challenges that will be faced in the future and that will drive the need of flexible IPTV delivery.

One IPTV platform is in Acreo's National Testbed (ANT) for broadband [5], which is physically built on the fiber infrastructure of the local municipality network in Hudiksvall in Sweden, Fibstaden. There are around 60 households

P. Daras and O. Mayora (Eds.): UCMedia 2009, LNICST 40, pp. 237–246, 2010.
© Institute for Computer Sciences, Social-Informatics and Telecommunications Engineering 2010

comprising end-users living in Hudiksvall, and they are supplied with IPTV via Fiber to the Home (FTTH). As a result of geographic limitation, IPTV service in ANT is only locally accessible. It is also costly to extend and operate ANT a wider geographical area. Thus, IPTV service in ANT is typically of small geographic extent and cannot meet the scalability requirements in future.

To address the problems mentioned above, IPTV VPN is proposed to addresses one area related IPTV distribution, namely scalability. And this solution has been implemented and tested in a small scale field trial. With the help of this novel solution, IPTV is distributed to remote end-users who are not part of ANT network via VPN over public networks, and to therefore provide a path for scalable IPTV service to be globally delivered. VPN is a generic term that covers the use of public or private networks to create groups of users that are separated from other network users and that may communicate among them as if they were on a private network [6]. VPN can extend geographic connectivity, provide global networking opportunities, reduce operational costs versus traditional WAN and transit time and transportation costs for remote users. These main VPN benefits can facilitate connections to an IPTV platform, and remote end-users can enjoy IPTV in a scalable way and at a low cost. Therefore, IPTV VPN is an ideal way to tackle the scalability issue of IPTV distribution.

1.2   Related Work

Some standards and specifications about IPTV VPN have been designed and released. "ITU-T IPTV Focus Group Proceedings" [7] promotes the global IPTV standards. In other aspect part of the standards, the Work Group (WG) 3 has identified some requirements on Multicast VPN in IPTV network Control and Multicast VPN Group Management aspect. The Internet Draft "Multicast in MPLS/BGP IP VPNs" [8] was written by engineers at Cisco and describes the MVPN (Multicast in Border Gateway Protocol (BGP)/Multi-Protocol Label Switch (MPLS) IP VPNs) solution of Cisco Systems. The "MPLS and VPN Architectures Volume II" [9], in Chapter 7 Multicast VPN, defines a few multicast VPN concepts and introduces some detailed examples. For these VPN solutions, most standards focus on MPLS VPNs which need in distribution and core networks to support MPLS. However, delivery of IPTV over an MPLS-enabled network cannot be done in an especially scalable way. To ensure interoperability among systems that implement this VPN architecture using MPLS label switched paths as the tunneling technology, all such systems MUST support Label Distribution Protocol (LDP) [MPLS-LDP] [10]. The scheme presented in this paper is built on a variety of networks using IP, which is much easier to implement and distribute IPTV to remote end-users.

1.3   Contributions

The contributions in this paper are threefold: 1) One novel solution - IPTV VPN is proposed and implemented to provide a scalable IPTV delivery way. As long as the bandwidth is sufficient, it is possible for people who have broadband connections to get IPTV service via the Internet all over world. 2) The traffic measurements had been performed, and the results showed that a VPN solution can provide IPTV with acceptable Quality of Service (QoS) to remote end-users. 3) All implementations are built upon different kinds of open source software, which makes the service more scalable and extendable. The rest of this paper is organized as the follows. The proposed scheme is presented in Section 2. Section 3 describes experiments designed to implement proposed scheme. Section 4 presents the performance evaluations and test results. Concluding remarks are made in Section 5.

2   Proposed Scheme

2.1   OpenVPN

In our proposed scheme, OpenVPN [11] is used to provide VPN tunnels from ANT network to remote end-users, and then IPTV is delivered to remote end-users over the VPN tunnels.

OpenVPN is a full-featured open source Secure Socket Layer (SSL) VPN solution that accommodates a wide range of configurations, including remote access, site-to-site VPNs, Wi-Fi security, and enterprise-scale remote access solutions with load balancing, failover, and fine-grained access-controls [11]. It's a real VPN in the sense that IP or Ethernet frames from a virtual network interface are being encrypted, encapsulated in a carrier protocol (TCP or UDP), and tunneled [12]. OpenVPN provides VPN connections via TUN/TAP virtual devices which allow for creating numerous endpoints through scripted interactions that work with "push"or "pull"options. OpenVPN uses the widespread and mature industry standard SSL infrastructure to provide secure communications over the Internet with encryption of data packages and control channels. There are some benefits for using OpenVPN. With OpenVPN, you can [13]:

- tunnel any IP sub-network or virtual Ethernet adapter over a single UDP or TCP port [13],
- multiple load-balanced VPN servers farm which can handle thousands of dynamic VPN connections,
- use security features of the OpenSSL library to protect network traffic,
- use real-time adaptive link compression and traffic-shaping to manage link bandwidth utilization[13],
- tunnel networks over NAT [13].

2.2   IPTV VPN

Figure 1 illustrates an example of basic IPTV VPN. The main office offers IPTV service to different types of end-users over VPN connections. The IPTV distributions are not constrained by geographic locations, e.g., the main office offers IPTV service to remote office with connected IPTV VPN network, and the remote office could locate anyplace in the world. In addition, IPTV VPN is able

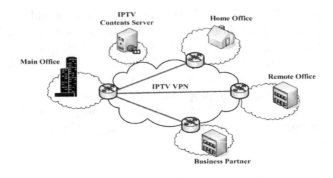

Fig. 1. An example of IPTV service via VPN [7]

to reduce operation costs, transportation costs, provide improved security and better control due to native traits of VPN. IPTV VPN can also provide classified IPTV service features according to geographical groups and customers' demands [7], classified IPTV group services features [7], etc.

3   Experiment Setup

The implementation is based on ANT, which provides different access networks and network applications to support the related research and test activities. The infrastructure of ANT is shown in Figure 2. Based on ANT infrastructure, IPTV VPN network layout was designed as shown in Figure 3.

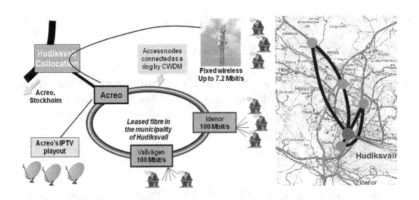

Fig. 2. The ANT network infrastructures

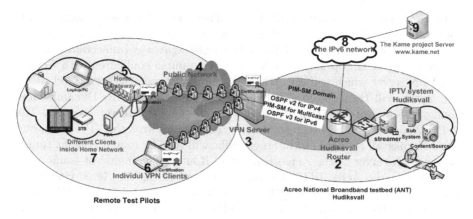

Fig. 3. IPTV VPN network layout. The different components labeled with numbers are described in section 3.1: *IPTV VPN network layout description*.

3.1   IPTV VPN Network Layout Description

The following descriptions all are related to Figure 3.

- Number 1, IPTV system Hudiksvall: The IPTV system is in ANT network and includes the content source, streaming server, sub-systems and the other components.
- Number 2, Acreo Hudiksvall Router: the core router in ANT Hudiksvall.
- Number 3, VPN Server: the VPN Server is linked up together over a VPN tunnel with the VPN individual clients or home gateway. Different open source software was installed on this server. Together with the core router, the VPN server provides VPN and multicast services to VPN clients.
- Number 4, The Public Network.
- Number 5, Home Gateway: the home gateway is physical placed in-between the link network of the VPN server and home network. The home gateway acts as vpn client for vpn connections, Internet Group Management Protocol (IGMP) proxy[14] for multicast routing besides the roles of normal gateway with the route and DHCP functions. The home gateway is running on an open source routing platform – OpenWRT [15], based an embedded Linux box.
- Number 6, Individual VPN clients: the laptops installed the VPN client program.
- Number 7, Different clients inside home network.

The main implementation is IPTV VPN implementation. In the implementation, OpenVPN was set up to provide VPN services; Open Shortest Path First version 2 (OSPFv2) was implemented to provide unicast routing; Protocol Independent Multicast - Sparse Mode (PIM-SM) was built up to provide multicast routing; Home gateway was developed to support gateway-to-gateway VPN connections. The home gateway was built on embedded Linux box with different open source software to establish an automatic VPN connection to VPN server and provide home network connectivity for different clients inside home network.

The IPTV VPN starts up as follows. For host-to-gateway connections, an end-user starts up a laptop and a VPN client programme configured with Acreo's own VPN server which will set up a VPN-tunnel between the server and client. The laptop will then obtain a public VPN IP address via the Dynamic Host Configuration Protocol (DHCP) service which the VPN server provides. The OSPFv2 and PIM-SM routing protocol are running between the VPN server and Acreo Hudiksvall Router. The internet traffic will then be routed over the tunnel via the VPN server to the Acreo Hudiksvall Router. The multicast traffic from the source in the Acreo IPTV system will be routed via the Acreo Hudiksvall Router (the Rendezvous Point (RP) in the PIM-SM domain) to the VPN server (PIM-SM enabled) over a VPN-tunnel to the client. The difference between the gateway-to-gateway and host-to-gateway VPN connection is the home gateway acts as a VPN client and IGMP proxy, besides playing normal gateway role with the route and DHCP functions for clients inside the home network.

4   Measurement and Analysis

4.1   Test Methodology

Various measurement instruments and methods were used to evaluate the QoS of IPTV VPN service. Most used for the IPTV testing was one professional IPTV measurement system - Agama Analyzer [16]. Other tools were also used to test network delay, network connectivity, network capacities, etc. In addition, end-users' perceived inspections are also a common method and used to measure IPTV visual quality. The main test activities are as follows.

– Evaluate the VPN services qualities.
– Compare IPTV service qualities between the VPN and normal wired line connection.

4.2   VPN Service Qualities Measurements

As carrier tunnels to deliver IPTV service, the QoS of VPN connections will determine IPTV service qualities. Therefore, VPN connections qualities (network delay, network connectivity, capacities loss, etc) were quantified under different VPN server configuration options. These options have some influences on VPN connectivity performance, such as some security options for different encryption algorithms, keyed-Hash Message Authentication Code (HMAC) for integrity check, data compression with "comp-lzo" option, etc. The VPN connectivity was measured with two test cases shown in Figure 4 and Figure 5.

The test case 1 and test case 2 were performed five times in a row with different VPN server configuration options. The measurement values are presented in Table 1. There are six different options in the first row of the table, for example, option 1 is original network bandwidth test without VPN connections; option 5 is VPN bandwidth test with data compression enabled.

skicka **32.59 Mbit/sek**

ta emot **50.55 Mbit/sek**

Svarstid: **7 ms**   Mätserver: **Sundsvall**

Fig. 4. The test case 1, network bandwidth check against www.bredbandskollen.se. skicka=send,ta emot=receive, Svarstid= response time, Mätserver=mesurement server.

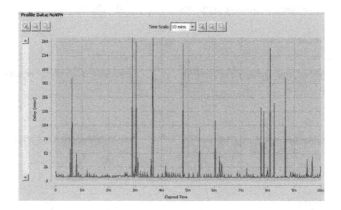

Fig. 5. The test case 2, network delay check against www.bredbandskollen.se with "Anue Network Profiler" [17] test tool in ten minutes

Table 1. Bandwidth connectivity test results for VPN with different VPN server options

	option 1	option 2	option 3	option 4	option 5	option 6
VPN connections		•	•	•	•	•
Encryption		•	•			
Integrity check HMAC		•		•		
Secret Key for HMAC		•		•		
Data compression		•	•	•	•	
Average Sent	32.59Mb/s	22.58Mb/s	22Mb/s	24.77Mb/s	22.40Mb/s	23.37Mb/s
Average Received	50.55Mb/s	36.34Mb/s	36.4Mb/s	37.49Mb/s	36.86Mb/s	37.96Mb/s
Average Network Delay	13.01 ms	36.07 ms	21.27 ms	24.97 ms	14.68 ms	17.30 ms
Maximum Sent	33.17Mb/s	23.55Mb/s	24.3Mb/s	26.03Mb/s	26.96Mb/s	27.12Mb/s
Maximum Received	55.34Mb/s	38.56Mb/s	38.6Mb/s	39.96Mb/s	39.50Mb/s	39.60Mb/s
Shortest Network Delay	7ms	7.95ms	7.94ms	7.94ms	7.91ms	7.76ms

4.3   IPTV VPN Service Qualities

IPTV service qualities comparisons between the VPN and normal wired line connections had been done with an Agama instrument and from user perspectives. Below Figure 6 and Figure 7 are shown that represent test results that one IPTV channel from same streamer was measured by Agama Analyzer during 72 hours (from 2009-05-29 8:00 to 2009-06-01 8:00). In the context of computer networks, the term jitter is often used as a measure of the variability over time of

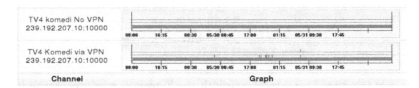

Fig. 6. SVT TV4 Komedi channel measuring graph from Agama Analyzer. Green=OK, Blue=minor distortion, Yellow=major distortion, Red=Packet loss.

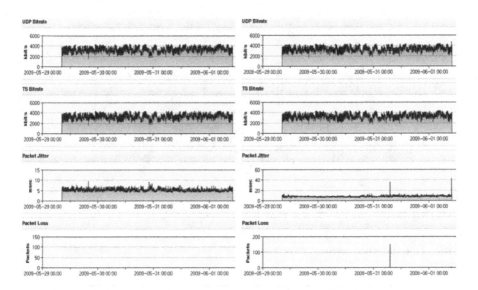

Fig. 7. SVT TV4 Komedi channel Real-Time measurement graphs from Agama Analyzer. The left chart is the measurement results with no VPN connections; the right chart is measurement results with VPN connections.

Table 2. The Packet Jitter measurement results from Agama Analyzer

	normal wired lines	VPN
Average Packet Jitter	5.9 ms	9.8ms
Maximum Packet Jitter	9.8ms	41.5ms

the packet latency across a network [18]. A bigger number of packet jitter value means larger packet latency. The Packet Jitter measurements were performed and test results are presented in Table 2. From user perspectives, normal quality TV and High Definition (HD) TV were measured in terms of zapping time in a small scale trial. In the measurement, a normal quality TV's bitrate is around 4Mb/s and the bitrate for a HD TV is above 7Mb/s. The zapping time for normal quality TV channels between VPN and normal wired line connections are almost same. For HD TV channels, there were comparative long time delay in terms of zapping time.

4.4   Discussion

The VPN service connectivity benchmark results can be summarized as follows.
1) The VPN network bandwidth loss rate is approximate 26%–32% comparing
to original network bandwidth. 2) The VPN network bandwidth is nothing with
the security options (encryption algorithm, session authentication, HMAC, etc).
3) For network delay, the data compression "comp-lzo" option can reduce VPN
network delay while the security options worsen the network delay. In table I,
the VPN connection without security options but with data compression enabled
is the winner in all tests. The VPN connection with all security options shows
rather larger network delay (average 36.07ms). The mission-critical IPTV service requires low network delay and high real-time multicast traffics. However,
securing multicast streaming will consume system resource and give negative
impact on the service performance, bandwidth, QoS, etc. If no confidentiality
requirement for multicast streaming, to some extent, authentication of both communication parties can ensure IPTV security. In this way, the consumption of
system resource is reduced and the services performance is improved.

For IPTV VPN, the measurement results show that the qualities of IPTV
VPN service is acceptable both from IPTV measurement instrument and user
perspectives. By comparison, there was no obvious difference for normal quality
TVs between the VPN and normal wired line connections. However, for HD TVs
usually with bitrate above 7Mb/s, VPN connections gave a comparatively poor
Quality of Experience (QoE) [19].

5   Conclusion

Previously Acreo only had access to end-users in "its own" networks in terms
of IPTV. In this paper, a VPN solution is designed and implemented to realize
a scalable IPTV delivery way, which can allow remote end-users over a wider
geographical area to access IPTV service at a lower operation cost. The evaluations of proposed schema show that the qualities of IPTV service via VPN
are acceptable. Although there is a network capacity reduction of VPN due to
network management traffic overhead, VPN is still a good way or in some case
the only solution of scalable IPTV distributions. Additionally, the VPN solution
supports certificate infrastructure and can provide a flexible way for test pilots
control simply by creating or revoking different certificates for different groups
of users. Besides, this solution is able to reduce operation costs, transportation
costs, provide improved security and better control due to native traits of VPN.
Finally, almost all implementations are based on open source software, which
makes the whole system more scalable and extendable.

References

1. Walko, J.: I love my IPTV. IEEE Communications Engineer 3(6), 16–19 (2005)
2. Yarali, A., Cherry, A.: Internet protocol television(IPTV). In: TENCON 2005 IEEE
   Region 10, pp.1–6 (2005)

3. Information Telecom & Media: IPTV: a global analysis (2nd edition). Information Telecom & Media, Tech. Rep. (2006)
4. WHITE PAPER - Emerging Multicast VPN Applications. Juniper Networks, 1–2 (2009)
5. Larsen, C.P., Andersson, L., Berntson, A., Gavler, A., Kauppinen, T., Lindqvist, C., Madsen, T., Mårtensson, J.: Experiences from the Acreo National Broadband Testbed. In: OFC/NFOEC 2006, paper NThF2, Anaheim, CA, USA (2006)
6. Andersson, L., Madsen, T.: Provider Provisioned Virtual Private Network (VPN) Terminology. Intenet Request For Comments RFC 4026 (2005)
7. ITU-T: ITU-T IPTV Focus Group Proceedings, pp. 389–390 (2008)
8. Rosen, E., Cai, Y., Wijsnands, J.: Multicast in MPLS/BGP VPNs. Internet Draft (2009)
9. Pepelnjak, I., Guichard, J., Apcar, J.: MPLS and VPN Architectures, vol. II, pp. 333–387. Cisco Press (2003)
10. Rosen, E., Rekhter, Y.: BGP/MPLS IP Virtual Private Networks(VPNs). Internet Draft RFC4364 (2006)
11. OpenVPN, http://www.openvpn.net
12. Yonan, J.: OpenVPN and SSL VPNs (January 26, 2005), http://www.mail-archive.com/cryptography@metzdowd.com/msg03333.html (accessed on August 18, 2009)
13. Yonan, J.: Open Source Overview, http://www.openvpn.net/index.php/open-source.html (accessed on August 18, 2009)
14. Cho, C., Han, I., Jun, Y., Lee, H.: Improvement of Channel Zapping Time in IPTV Services Using the Adjacent Groups Join-Leave Method. In: 6th International Conference on Advanced Communication Technology, vol. 2, pp. 971–975 (2004)
15. OpenWrt–Wireless Freedom, http://www.openwrt.org
16. Agama Analyzer, Agama Technologies AB, Box 602, SE-581 07 Linköping, Sweden (2009)
17. Anue Network Profilerp, Anue Systems Inc., 9737 Great Hills Trail, Suite 200, Austin, TX 78759 (2009)
18. Wolaver, D.H.: Phase-Locked Loop Circuit Design, pp. 211–237. Prentice-Hall, Englewood Cliffs (1991)
19. Siller, M., Woods, J.: Improving quality of experience for multimedia services by QoS arbitration on a QoE Framework. In: Procceedings of the 13th Packed Video Workshop (2003)

Location-Aware and User-Centric Touristic Media

Michael Weber, Alexandru Stan, and George Ioannidis

IN2 search interfaces development Ltd,
Fahrenheitstrasse 1, D-28359 Bremen, Germany
{mw,as,gi}@in-two.com

Abstract. This paper presents an approach that uses geo-temporal media tagging and Web 2.0 ratings to demonstrate the true potential of location-based applications for the tourism and heritage industries. It presents a framework, associated services and a general user workflow, which will allow end-users to automatically retrieve additional information about objects they capture with their GNSS-enabled mobile device, document their route, exchange and acquire geo-based information, connect with peer visitors and receive route recommendations. It provides therefore a new, integrated scenario for accessing on-the-spot information about objects having a cultural value and supporting users on the move to share and retrieve personalised information regarding cultural attractions and routes, all with one click.

Keywords: Geo-temporal coding, location-based services, tourism, media tagging, geo-tagging, content-based image retrieval.

1 Introduction

This paper presents an approach that exploits accurate geo-temporal coding in photos, which will be available in Galileo [1], to develop new ways to retrieve and present location-based media. Its prospect is to transform the use of geo-temporal media tagging and demonstrate the true potential of user-centric location-based applications for the tourism and heritage industries. The proposed scenarios will be used to transform the way current tagging processes are organised, especially how geo-temporal tagged media content is offered and searched for.

The approach presented is based on an open server-side architecture with integrated services and advanced solutions for experiencing, searching, and accessing geo-temporal media content, enabling intuitive presentation and delivery both for professional and non-professional users. Doing this, users move away from today's widely prevalent text search paradigm, towards unobtrusive mixed-media queries and more efficient geo-tagged and user-centric visual content presentation. Through careful consideration and focus on the information interaction needs of users on the move (e.g. travellers, cyclists, mountain bikers, hikers), it enables new user-centric ways of describing and retrieving visual information and significantly enhances travel and leisure activities.

The rest of this paper presents in the next section the state of the art in the areas concerned, the proposed system architecture and the general user workflow as well areas of applications in the field of tourism.

P. Daras and O. Mayora (Eds.): UCMedia 2009, LNICST 40, pp. 247–254, 2010.

2 State of the Art

"The increasing power and ubiquity of smart devices such as mobile phones and PDAs means that a visitor to a city now carries with them a device capable of giving location-specific guiding and routing information." [2]

In information technology, developments are rapid and ubiquitous. The availability of global navigation satellite systems (GNSS) such as the global positioning system (GPS) and the emerging European Union's Galileo allows pedestrians to precisely position themselves in an 'absolute' earth reference frame without any knowledge of the body motion, while also providing highly accurate time and velocity, thanks to rapid advances in low cost and low power microelectronics that have produced highly portable miniature user equipment. It is only since a few years that consumers are using GNSS-based devices on a day-to-day basis. The advent of GPS-enabled route planning devices (e.g. TomTom [3], Navigon [4], GoogleMaps [5]), geo-enabled PDA's and cellular phones have instigated a host of new services and devices. The presented approach utilises three basic technologies as a backbone of developments: GNSS-enabled devices, geographical mapping technology, and the forthcoming Galileo satellite information.

Mass-market applications of devices such as GNSS-enabled cameras, navigators, and smartphones, which started with the above mentioned car navigation over the last five years, contributed to the widespread use of GNSS receivers. Multi-modal navigation, local search and social networking are just a few examples of the wide range of emerging applications. The rollout of most accuracy-critical applications, as in e-tourism for instance, has been slowed down by the difficulties of GNSS-based positioning for pedestrians in urban areas. Multipath and masking effects of urban canyons degrade the accuracy of GNSS ranging and increase geometric dilution of precision in receivers that operate in dense urban areas. In the case of GNSS applications designed for vehicles, the effects of these phenomena on accuracy can be reduced, thanks to the velocity of the user that contributes in averaging multipath and the use of map matching. However, pedestrians do not benefit from the same circumstances, and GNSS-based positioning for pedestrians in dense urban areas suffers from inadequate accuracy and integrity. Tests performed in down town urban areas over a variety of mass-market terminals with integrated GNSS receivers show 95 percent circular error probable (CEP) performances between 50 and 100 meters [6]. The presented approach in this paper is motivated by the fact that Galileo will provide more accurate positioning than it is possible nowadays and GNSS receivers have already started to be integrated into mobile phones (e.g. the N95 and E90 mobile devices from Nokia) and digital cameras (e.g. the Ricoh 500SE camera [22]). Manual geo-tagging is possible but not very widespread because it is cumbersome and requires two different devices with synchronised time.

A GPS unit showing basic waypoints and tracking information is typically adopted for outdoor sport and recreational use. Bicycles for example often use GNSS in racing and touring to allow cyclists to plot their trips in advance and follow the defined course without having to stop frequently to refer to separate maps. Hikers, climbers, and even ordinary pedestrians in urban or rural environments increasingly use GNSS integrated in their smartphones or in stand-alone devices to determine their position. In isolated areas, the ability of GNSS to provide a precise position can greatly enhance the chances of rescue when climbers or hikers are disabled or lost (if they have a means of communication with rescue workers). For specific categories of users with disabilities (e.g. for the visually impaired) dedicated GNSS equipment is also available.

The segment of mobile devices is still growing and new devices are released all the time. Nowadays most of the mobile phones are equipped with digital cameras by default and often they have special abilities like accelerometers, digital compasses, and RFID (Radio Frequency Identification). They offer new ways of interaction with the environment to the user. A report of Harris Interactive [7] showed that 64% of the interviewees, who consisted of young teenagers in the age range of 13-19, use their mobile devices frequently to take pictures. Though it is not only the young people who adopt to work with these technologies, but older users as well. Having a mobile phone of great functional range is still a status symbol. Especially with the release of the iPhone™ and the G1™ the awareness of the greater public for mobile applications increased enormously. They already come with a built-in GPS functionality. Due to the previously named developments the field of location-based applications is of growing interest. Since many people carry around their mobile devices all the time they are interested in using its functionality at each time, wherever they are.

Recently many interesting location-based services and applications have been published and become known to a greater public than before. This year for example Google released an application called Latitude [8] which allows people to access information on where, when and what their friends are doing. Another currently published example is Wikitude [9], which can provide information on prominent locations that are captured with the camera of the G1™.

Beyond the field of fast developing mobile devices the role of the Internet is becoming more and more ubiquitous. The usage of the Internet has become part of everyday life for most of the population, and also the amount of actively contributing users is growing steadily. This is mainly dedicated to the development of the so called Web 2.0, which stands for innovative web applications that allow their users to become an active part of a community and to create their own content. A Web 2.0 platform usually consists of a web space where users can upload, share, and comment on the available content that has been provided by the users. All users generate content and thus turn into producers and consumers (referred also as prosumers), instead of being a pure consumer of the given information. Furthermore, the increasing numbers of so-called mash-ups, even based on simple interfaces, illustrate how content can be produced and processed in a more efficient manner in the future.

In the last few years tagging became increasingly important as well. Usually associated with the above-discussed Web 2.0, tagging describes the process of adding keywords to content that can be text, images, audio or video. These keywords are used to improve the findability of content in the Internet. The advance of tagging started with social bookmarking services like Delicious [10], that allow users to save all their bookmarks in one place and to classify and sort them with the help of keywords/tags. The current state-of-the-art of accessing, searching, and retrieving multimedia information, requires to manually attaching metadata, keywords, and other information to the content though. This is a highly subjective, non-scalable, and erroneous process. Moreover, while the amount of geo-tagged visual information is increasing, the users on the move do currently lack the appropriate interfaces to have efficient and flexible access to visual data in changing contexts, mobility requirements, edutainment, and travel settings. Such shortcomings can hinder the evolution of location-based applications. In the presented approach we will exploit the more accurate GNSS positioning and combine it with innovative visual features and image header information data to enable more robust object and scene recognition and therefore provide the potential to deliver instant information about cultural objects/scenes captured. This will revolutionise many information push scenarios because it can provide instantly the right information with just on photo-click and demonstrate, therefore, the true potential of the Galileo infrastructure to industry, government and consumers alike.

3 System Architecture and User Workflow

The presented approach builds its developments on an open service-oriented architecture (see Fig. 1) where integrated services and advanced solutions for experiencing, searching, and accessing geo-temporal media content will be implemented. The main aims of the system are to provide:

- A natural interaction with information related to objects, which tourists are capturing with their mobile device or camera, and
- An increased usability of a web platform where users can share, display and retrieve user-centric geo-information with one click.

As seen in Fig. 1, the proposed system consists of several components, which are provided as services and which are connected by the Enterprise Service Bus (ESB). These services contain location- and content-based multimedia analysis, for extracting colour and texture features of the captured images to enable search and retrieval. Other services, take care of the user management, which incorporates individualised storage of user content, as pictures, tracks, descriptions, etc., that are able to be shared with other users. Furthermore the architecture covers presentation interfaces, which build the front-end of the platform. These front-end layers, with which the users interact with the common service of the back-ends, are:

- A mobile client, which will allow end-users to document their route, exchange and acquire geo-based information, just by using their GNSS-enabled device and interfaces to search and retrieve geo-sensitive user-centric media

- A Web 2.0 platform where touristic regions advertising their specific destinations and users visiting those specific locations can share, exchange, search, explore and browse user-generated content. This platform will support pre-travel information gathering and decisions about visiting a specific destination, as well as providing individual users the options to track and present their specific travel routes and sight-seeing tours. During travelling, the platform will act as the back-end for providing information on objects that the user captures with his mobile phone camera. Through this platform it is expected that the adoption of GNSS applications and services will be increased and the transition to Galileo will provide immediately economic and social benefits.

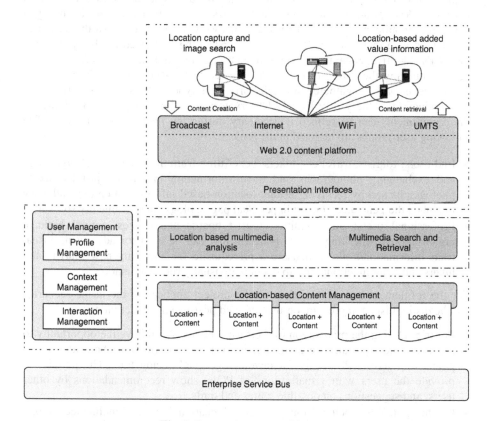

Fig. 1. Proposed system architecture

Since the proposed system is partly envisioned as a Web 2.0 travel platform, it will extensively employ mapping technologies. Both, commercial and open source GEO-API's, will be considered. Non-commercial API's are the ones from OpenStreetMap [11] and OpenRouteService [12]. Commercial API's are developed by the major players in the industry, namely Google (Google Maps [13]), Microsoft (Virtual Earth, including Bing Maps [14]), and Yahoo! (Yahoo! Maps [15]).

Consumers and organisations are more likely to adopt technologies if they can ensure ownership of their own submitted information. Following this observation the proposed scenarios reflect the possibility to tie in personal collections of photographs and thereby generated route information in their personal web presence. Therefore it is also planned to tie in seamlessly with popular content management systems (CMS) or communities (Facebook [16], Hyves [17], LinkedIn [18], etc.). Simple functionality similar to YouTube's methodology [19] could suffice to meet this initial requirement.

In the system development we follow an iterative approach utilising scenarios to portray the experiences that users will be able to have with the proposed (mobile) applications and services. In this context, the word "scenario" is used to mean a narrative description of what the user does and experiences while using a computing/mobile system [20]. To achieve this we will make sure, that the designers understand who their users will be. This understanding is derived partly by studying the user's cognitive, behavioural, anthropometric, and attitudinal characteristics and in part by studying the nature of the work expected to be accomplished. The end user requirements will be gathered through early prototype tests, scenario testing, and the conduction of the yearly Utrecht Observatory of online behaviour. By doing so, the requirements can be refined if necessary during the iterative design cycle. The following user groups are targeted in the project:

- End users: tourists, travellers, hikers; benefitting from the simplicity of organising and exchanging photographs and travel-information in one go just by taking photographs and by receiving instant location-based information on-demand, easy to embed the generated content in personal pages, blogs, and community software
- Professional users: municipalities, tourist boards and offices, natural heritage organisations; benefiting from a platform that can be used for advertising visual route-information/trails, which are a prerequisite for attracting new tourists and hikers and from functionalities to document visually rural landmarks.

For the refinement and development of the project application the usage scenarios will be distinguished in pre-trip, on-the-spot and post-trip travel activities:

- Pre-trip: During the preparation of their trip, the users will have the opportunity to use the project portal. The main aim is to gather information about an area or points of interest (POI), e.g. for developing a sightseeing tour. The portal will provide the users with visualisations of POIs, show recommendations by other users, and suggestions on possible routes and trails.
- On-the-spot: While being at the actual travel location, the use of mobile technology is central to the scenario. The user receives on-spot directions from other trusted users on the platform and can also receive information of captured objects as well document and share routes to his/her community, simply by capturing photographs.
- Post-trip: The main aim is to allow the user to mentally re-visit a trip or parts of a travel by accessing the original photographs placed in the saved route. The portal will improve the way users organise their collection of photographs geographically and time-based.

The general user workflow and processing of the proposed platform is shown in Fig. 2.

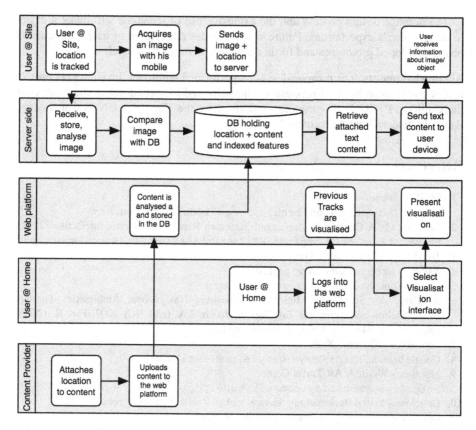

Fig. 2. Typical user interaction with the proposed system and the related processes

4 Conclusions

This paper presents a novel approach to combine location-based image retrieval technology with a web platform and mobile clients for the tourism and heritage sector, the proposed plan to develop the system and the related use cases and scenarios. The open service-oriented architecture, on which the proposed system is based, allows an easy access and improved retrieving methods for multimedia content produced by the users on the move.

By allowing users to contribute to the Web 2.0 platform, a fast-growing community will be initiated to help find and share relevant information, e.g. on a bike trip and stimulate the communication and exchange of experiences with other fellow travellers. Accessing and exchanging information through the platform facilitates the pre-trip, on-the-spot and post-trip activities. During their trip users have the opportunity to immediately contribute to their community and retrieve relevant information.

The iterative design process and the extensive use of scenarios will make it easier to achieve user's expectations. Future work includes elaboration of user requirements on the developed prototypes and further iteration on the system level.

Acknowledgments. The presented work is performed within the ImaGeo [21] project that is co-financed by the European Community and is carried out in the context of the Galileo FP7 R&D programme supervised by the European GNSS Supervisory Authority under Grant Agreement No: 228341.

References

1. EC Programme – Galileo,
   http://ec.europa.eu/transport/galileo/index_en.htm
2. Lejnieks, C.: A Generation Unplugged. Research Report from Harris Interactive (2008),
   http://files.ctia.org/pdf/HI_TeenMobileStudy_ResearchReport.pdf
3. TomTom, http://www.tomtom.com/
4. Navigon, http://www.navigon.com/
5. Google maps, http://maps.google.com/
6. Beeharee, A., Steed, A.: Minimising Pedestrian Navigational Ambiguities Through Geoannotation and Temporal Tagging. In: Jacko, J.A. (ed.) HCI 2007, Part II. LNCS, vol. 4551, pp. 748–757. Springer, Heidelberg (2007)
7. Lejnieks, C.: Op. cit. (2008)
8. Google latitude, http://www.google.com/latitude
9. Mobilizy - Wikitude AR Travel Guide,
   http://www.mobilizy.com/wikitude.php
10. Delicious – Social Bookmarking Service, http://delicious.com/
11. OpenStreetMap, http://www.openstreetmap.de/
12. OpenRouteService, http://data.giub.uni-bonn.de/openrouteservice/
13. Google maps, op. cit
14. Bing Maps, http://www.bing.com/maps/
15. Yahoo! Maps, http://maps.yahoo.com/
16. Facebook, http://www.facebook.com/
17. Hyves, http://www.hyves.nl/
18. LinkedIn, http://www.linkedin.com/
19. YouTube, http://www.youtube.com/
20. Caroll, J.M.: Scenario-Based Design: Envisioning Work and Technology in System Development. Wiley & Sons, Indianapolis (1995)
21. ImaGeo project, http://www.imageoweb.eu
22. Ricoh 500SE camera,
    http://www.ricoh-usa.com/solutions/geoimaging/brochures/
    500SE_brochure.pdf

Optimal Ranking for Video Recommendation

Zeno Gantner, Christoph Freudenthaler, Steffen Rendle,
and Lars Schmidt-Thieme

Machine Learning Lab, University of Hildesheim
Marienburger Platz 22, 31141 Hildesheim, Germany
{gantner,freudenthaler,srendle,schmidt-thieme}@ismll.de

Abstract. Item recommendation from implicit feedback is the task of predicting a personalized ranking on a set of items (e.g. movies, products, video clips) from user feedback like clicks or product purchases. We evaluate the performance of a matrix factorization model optimized for the new ranking criterion BPR-OPT on data from a BBC video web application. The experimental results indicate that our approach is superior to state-of-the-art models not directly optimized for personalized ranking.

1 Introduction

Recommendations are an important feature of many websites. For example, online shops like Amazon provide customers with personalized product offers. In media applications, personalization is attractive both for content providers, who can increase sales or views, and for customers, who can find interesting content more easily. In this paper, we focus on item recommendation from implicit data. The task of item recommendation is to create a user-specific ranking for a set of items. User preferences are learned from the users' past interaction with the system, e.g. their buying/viewing history.

BPR was recently [3] proposed as a generic optimization method for item prediction from implicit feedback. To our knowledge, models optimized for BPR-OPT are among the most competitive methods for item prediction from implicit feedback. BPR has also been adapted to tag prediction; a factorization model using BPR won the ECML PKDD Discovery Challenge 2009 [4].

Its advantages make BPR an obvious choice for video recommendation from click data. In this paper, we evaluate the performance of a matrix factorization model optimized for BPR-OPT on data from a video web application ran by the British Broadcasting Corporation (BBC).

2 Bayesian Personalized Ranking (BPR)

We briefly introduce BPR, which consists of the objective criterion BPR-OPT, for which we sketch the underlying idea, and the learning algorithm LEARNBPR. For more details refer to the long version of Rendle et al. [3].

P. Daras and O. Mayora (Eds.): UCMedia 2009, LNICST 40, pp. 255–258, 2010.
© Institute for Computer Sciences, Social-Informatics and Telecommunications Engineering 2010

2.1 Motivation for BPR-Opt

Implicit feedback often consists only of positive observations. The non-observed user-item pairs – e.g. a user has not viewed a video – are a mix of real negative feedback (not interested in viewing it at all) and missing values (may want to view in the future).

Let U be the set of all users and I the set of all items. The known implicit feedback be $S \subseteq U \times I$. The task is now to provide a personalized total ranking $>_u \subset I^2$ for each user.

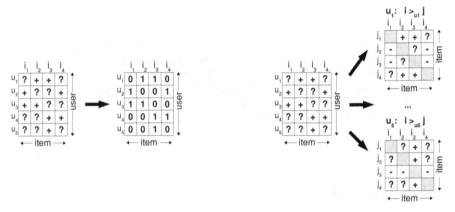

Fig. 1. *Left:* Usually, negative data is generated by filling the matrix with 0 values. *Right:* BPR creates user-specific preferences $i >_u j$ between item pairs. On the right side, + means a user prefers item i over item j; − means they prefer j over i.

Item recommenders [1,2] typically create the training data from S by giving pairs $(u, i) \in S$ a positive class label and all other combinations a negative one (see Fig. 1). Then a model is fitted to this data, which means is optimized to predict the value 1 for elements in S and 0 for the rest. The problem of this approach is that all elements to be ranked in the future are presented as negative examples to the learning algorithm.

We propose a different approach by using item pairs as training data and optimize for correctly ranked item pairs instead of scoring single items. This represents the problem better than just replacing missing values with negative ones. From S we try to reconstruct $>_u$ for each user. If an item has been viewed by a user, then we assume that the user prefers this item over all other non-observed items. See Fig. 1 for an example. For items that have both been seen by a user, we cannot infer any preference. The same is true for two items that a user has not seen yet.

$$D_S := \{(u, i, j) | i \in I_u^+ \wedge j \in I \setminus I_u^+\} \subseteq U \times I \times I$$

The interpretation of $(u, i, j) \in D_S$ is that user u prefers i over j. Our approach has two advantages: (1) The training data D_S consists of both positive and negative pairs and missing values. The missing values between two non-observed items are exactly the item pairs that have to be ranked in the future. (2) The training data is created for the actual objective of ranking, i.e. the observed subset D_S of $>_u$ is used for training.

2.2 Learning Algorithm and Application to Matrix Factorization

To optimize for BPR-OPT, [3] we use LEARNBPR, a stochastic gradient-descent algorithm (see Fig. 2). Using bootstrap sampling instead of full cycles through the data is especially useful as the number of examples is very large, and for convergence often a fraction of a full cycle is sufficient.

```
1: procedure LEARNBPR(D_S, Θ)
2:     initialize Θ
3:     repeat
4:         draw (u, i, j) from D_S
5:         Θ ← Θ + α ( (e^{-x̂_uij})/(1+e^{-x̂_uij}) · (∂/∂Θ) x̂_uij − λ_Θ · Θ )
6:     until convergence
7:     return Θ̂
8: end procedure
```

Fig. 2. Optimizing for BPR-OPT with bootstrapping-based gradient descent

Because we have triples $(u, i, j) \in D_S$, we decompose the estimator $\hat{x}_{uij}$: $\hat{x}_{uij} := \hat{x}_{ui} - \hat{x}_{uj}$. Now we can apply any model that predicts $\hat{x}_{ul}, l \in I$.

Matrix factorization (MF) models are known to outperform [5] many other models for the related task of rating prediction. They are also state-of-the-art for item prediction. MF approximates the target matrix X by the product of two low-rank matrices $W : |U| \times k$ and $H : |I| \times k$ by $\hat{X} := WH^t$. For estimating whether a user prefers one item over another, we optimize the parameters $\Theta = (W, H)$ for the BPR-OPT criterion using our algorithm. To apply LEARNBPR to MF, only the gradient of $\hat{x}_{uij}$ wrt. every model parameter has to be derived.

3 Evaluation

We compare learning the MF model with LEARNBPR to weighted regularized matrix factorization (WR-MF) [1,2]. We also report results for a baseline method that ranks the items by global frequency.

The BBC data was collected from one of BBC's online services during a ten day period in 2009. It contains 2,867,128 viewing events generated by 189,228 anonymous users on 5,125 different video clips. For comparability, we report results on a subset of the *Netflix* data.[1]

We randomly removed one user-item pair per user to create the training set. and repeated all experiments 10 times by drawing train/test splits in each round. You can see in Fig. 3 that on Netflix, BPR-MF outperforms all other methods in prediction quality on both datasets.

---

[1] The results are taken from [3]. The subset contains $10,000$ users, 5000 items, and $565,738$ rating actions, where each user and item have at least 10 ratings.

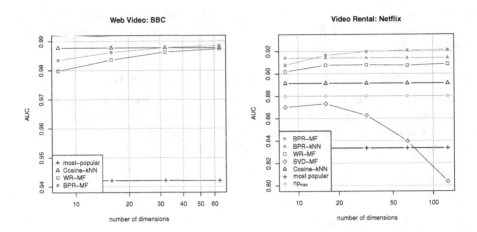

Fig. 3. Area under the ROC curve (AUC) prediction quality for BBC and Netflix

4   Conclusion and Future Work

We evaluated the performance of a matrix factorization model optimized for the recently proposed generic criterion BPR-OPT on data from a BBC video web application. The results indicate that the method outperforms other state-of-the-art methods. This is justified by the theoretical analysis in [3]. In the future, we will conduct a large-scale study on additional real-world datasets from the MyMedia project field trials.

Acknowledgements. We thank the BBC for providing the video dataset, and Chris Newell for fruitful discussions. This work was co-funded by the EC FP7 project MyMedia under the grant agreement no. 215006.

References

1. Hu, Y., Koren, Y., Volinsky, C.: Collaborative filtering for implicit feedback datasets. In: ICDM 2008 (2008)
2. Pan, R., Zhou, Y., Cao, B., Liu, N.N., Lukose, R.M., Scholz, M., Yang, Q.: One-class collaborative filtering. In: ICDM 2008 (2008)
3. Rendle, S., Freudenthaler, C., Gantner, Z., Schmidt-Thieme, L.: BPR: Bayesian personalized ranking from implicit feedback. In: UAI 2009 (2009)
4. Rendle, S., Schmidt-Thieme, L.: Factor models for tag recommendation in bibsonomy. In: Discovery Challenge at ECML PKDD 2009 (2009)
5. Rennie, J.D.M., Srebro, N.: Fast maximum margin matrix factorization for collaborative prediction. In: ICML 2005 (2005)

Gaming Platform for Running Games on Low-End Devices

Arto Laikari[1], Jukka-Pekka Laulajainen[1], Audrius Jurgelionis[2],
Philipp Fechteler[3], and Francesco Bellotti[2]

[1] VTT Technical Research Centre of Finland, Finland
[2] University of Genoa, Italy
[3] Fraunhofer Institute for Telecommunications, Germany
Arto.Laikari@vtt.fi, Jukka-Pekka.Laulajainen@vtt.fi,
jurge@elios.unige.it, Philipp.Fechteler@hhi.fraunhofer.de,
franz@elios.unige.it

Abstract. Low cost networked consumer electronics (CE) are widely used. Various applications are offered, including IPTV, VoIP, VoD, PVR and games. At the same time the requirements of computer games by means of CPU and graphics performance are continuously growing. For pervasive gaming in various environments like at home, hotels, or internet cafes, it is beneficial to run games also on mobile devices and modest performance CE devices such as set top boxes. EU IST Games@Large project is developing a new cross-platform approach for distributed 3D gaming in local networks. It introduces novel system architecture and protocols used to transfer the game graphics data across the network to end devices. Simultaneous execution of video games on a central server and a novel streaming approach of the 3D graphics output to multiple end devices enable the access of games on low cost devices that natively lack the power of executing high-quality games.

Keywords: distributed gaming, graphics streaming, video streaming, gaming architecture.

1 Introduction

Future home is considered to be an always-on connected digital home with various appliances. Computer gaming and other entertainment equipment has been utilizing networked home infrastructure already for some time. Modern games have become highly realistic and their requirements for CPU processing power and graphics performance are increasing. At the same time users have adopted to use various low execution power end devices for which modern 3D computer games are too heavy.

The Games@Large project is developing a novel system for gaming both for homes and for enterprise environments, like hotels, internet cafés and elderly homes [1]. In the Games@Large system the key concepts are execution distribution, audio, video and graphic streaming and decoupling of input control commands. According to the Cloud Computing concept, games are executed in one or more servers and the game display and audio is captured and streamed to the end device, where the stream

P. Daras and O. Mayora (Eds.): UCMedia 2009, LNICST 40, pp. 259–262, 2010.

is rendered and the game is actually played. Game control is captured at the end device, streamed back to the server and injected to the game. In comparison to other currently emerging systems, the Games@Large framework explicitly addresses support of end devices with rigorously different characteristics, ranging from PCs with different operating systems over set-top boxes to simple handheld devices with small displays and low CPU power, with adaptive streaming techniques.

There is a number of commercial Gaming on Demand systems, overviewed in [2], that have been presented to the market. More recently, there have been some new announcements about the upcoming systems such as Playcast Media Systems, Gaikai's Streaming Worlds technology, Onlive and GameTree.tv from TransGaming Technologies. However, there is very little detailed technical information publicly available about the commercial systems.

This paper presents briefly the novel Games@Large architecture and its key functionalities. Finally we present system performance conclusions based on our laboratory tests.

2 Games@Large Framework

The Games@Large system consists of three major element classes: servers, end devices and access points. Games are played on the end devices and executed on the servers. Games run on the Local Processing Server (LPS), which utilizes also the Local Storage Server (LSS). In the home version, these logical entities will be located in the same physical computer. In an enterprise version, the server entities are distributed into several physical computers. End devices, like STBs or Enhanced Multimedia Extenders (EME), Enhanced Handheld Devices (EHD) and notebooks are connected to the server either with wireless or wired connection. The system exploits an adaptive streaming architecture and uses Quality of Service (QoS) functionalities to ensure good quality gaming to a variety of devices connected to the wireless network. The details of each component have been presented in earlier publications [1,2,3,4,5,6,7,8] so only a brief summary is given here.

The objective of the streaming architecture is to support various end devices with an efficient and high quality game experience, independent of software or hardware capabilities. To meet these demands a streaming architecture has been developed that is able to support two streaming strategies to display the game remotely: graphics and video streaming.

Graphics streaming is used for end devices with accelerated graphics support, like computers or set-top-boxes, typically having screens of higher resolution. Here the graphics commands from the game are captured on the server, transmitted across the network using a protocol developed in the project, and rendered locally on the end device. In this way, high image quality is achieved, since the scenes are always directly rendered for the desired screen.

Video streaming is used mainly for end devices without a GPU, like handheld devices, typically having screens of lower resolution. Here the graphical output is rendered on the game server and the frame-buffer is captured and transmitted encoded as H.264 video stream. In this approach the bit rates are in general much more predictable in comparison to graphics streaming. However, H.264 encoding on server side as well as decoding on end devices is computationally demanding.

Audio streaming sub-system has been developed to produce sounds and music of the games at the end device. Since computer games typically produce their audio samples in a block-oriented manner, the current state-of-the-art audio encoder, the High Efficiency Advanced Audio Coding version 2 (HE AAC-v2) is used. UDP based Real-time Transmission Protocol (RTP) and Real Time Control Protocol (RTCP) are used for the synchronization at the playback.

Different end devices typically provide several different input modalities. In the initial discovery phase performed by the UPnP device discovery, the end device sends its properties and capabilities. During the game play, the input from the controllers are captured either using interrupts or through polling. The captured commands are then transmitted to the server. At the server side the input commands are mapped to the appropriate game control and injected in real-time to the game running at the server.

To ensure smooth game play, decent Quality of Service (QoS) is required from the wireless home network. In order to meet the requirement of using low cost components, the choice for the wireless home network has been to use IEEE 802.11 based technologies, because of their dominant position in the market. Priority based QoS is supported using Wi-Fi Multimedia (WMM) extensions at the MAC-layer and UPnP QoS framework for QoS management.

3 Games@Large Performance Potential

Lab experiments have shown that the G@L system is capable of running video games of different genres and streaming them with a high quality to concurrently connected clients via a wireless / wired network, in particular exploiting a QoS solution, which improves systems performance also in the presence of competing traffic.

In the case of graphics streaming the quality of gaming experience is typically correlated with the game frame rate, which in the G@L system is proportional to the network throughput. Video games that use high data rates per frame require very high bandwidth and thus, are problematic to be deployed via current networks.

For the video streaming, experiments have shown that the encoding time at the server enables the handling of multiple parallel streams on a single server. Further enhancements are expected by exploiting more information on the scene structure obtained from the graphics board. The performance of H.264 allows for satisfying visual quality at bit rates of several hundreds of kilobits [7].

With regard to device performance, the server must have enough CPU power and memory to run the game natively. Additionally, the amount of video memory that a game requires when running natively, must be available in the system memory when running in the Games@Large environment (because the graphic objects are emulated by the streaming module in the system memory). The most important hardware requirement for the client device is the video adapter (for 3D streaming). It should have hardware acceleration capabilities to enable fast rendering of 3D scenes, otherwise only the video streaming can be used. As on the server, the graphic resources that the game stores in the video memory should be available in the system memory to enable manipulation prediction and caching. So memory requirements for the client should be 200-300 MB available to the client application for fairly heavy games [8].

Another issue for a game to be playable on low-end devices is that it has to be compatible with the device screen size (in particular concerning resolution) and controller capabilities (e.g., some games are difficult to control with a gamepad, a PDA keypad or a remote control for TV). Thus, our research argues for new generation mobile devices to have an increase in their Human-Computer Interaction capabilities in order to support more advanced interaction modalities also considering new networked interactive media applications [7].

4 Conclusions

Games@Large is implementing an innovative architecture, transparent to legacy game code, that allows distribution of a cross-platform gaming and entertainment on a variety of low-cost networked devices. Virtually extending the capabilities of such devices the Games@Large system is opening important opportunities for new services and experiences in a variety of fields and in particular for the entertainment in the home and other popular environments.

Acknowledgement

This work has been carried out in the IST Games@Large project (http://www.gamesatlarge.eu) [1], which is an Integrated Project under contract no IST038453 and is partially funded by the European Commission.

References

[1] Tzruya, Y., Shani, A., Bellotti, F., Jurgelionis, A.: Games@Large – a new platform for ubiquitous gaming. In: Proc. BBEurope, Geneva, Switzerland, December 11-14 (2006)

[2] Jurgelionis, A., et al.: Platform for Distributed 3D Gaming. International Journal of Computer Games Technology, Article ID 231863, 2009

[3] Eisert, P., Fechteler, P.: Remote rendering of computer games. In: SIGMAP 2007, Barcelona, Spain, July 28-31 (2007)

[4] Bouras, Poulopoulos, Sengounis, Tsogkas: Networking Aspects for Gaming Systems. In: ICIW 2008, Athens, Greece, June 8-13 (2008)

[5] Laulajainen, J.-P.: Implementing QoS Support in a Wireless Home Network. In: WCNC 2008, Las Vegas, USA, 31 March-3 April (2008)

[6] Eisert, P., Fechteler, P.: Low Delay Streaming of Computer Graphics. In: Proc. Intern. Conf. on Image Processing (ICIP), October 2008 (2008)

[7] Jurgelionis, A., et al.: Distributed video game streaming system for pervasive gaming. In: STreaming Day 2009, Genoa, September 21 (2009)

[8] Laikari, A., Fechteler, P., Eisert, P., Jurgelionis, A., Bellotti, F.: Games@Large Distributed Gaming System. In: 2009 NEM Summit, Saint-Malo, France, September 28-30 (2009)

TrustVWS 2009

Session 1

Virtual Persons + Virtual Goods = Real Problems

Kai Erenli

Fachhochschule des bfi Wien
Wohlmuthstr. 22, 1020 Vienna, Austria
kai.erenli@fh-vie.ac.at

Abstract. Virtual Worlds have become serious business models and thus gained the attention of law professionals. The legal problems arising out of Virtual Worlds have started a discussion which will be summarized in this article. Moreover arguments will be delivered which can be used to protect users of those Virtual Realities.

Keywords: Virtual Worlds, Legal Problems, E-Commerce, Virtual Person.

1 Introduction and Definition

Virtual Worlds, Augmented Reality and Web 2.0 have had a huge impact on modern society. It is estimated that about 550 million avatars "live" in Virtual Worlds[1]. With about 40 million real people controlling these avatars one can imagine that the interactions need legal attention to ensure that every participant has legal security while acting as a virtual person. In its 2008 Report, the Internal Revenue Service (IRS) states: *"Economic activities in virtual worlds may present an emerging area of tax noncompliance, in part because the IRS has not provided guidance about whether and how taxpayers should report such activities."* The IRS identified over US$1 billion in revenues directly made in Virtual Worlds in 2005. Virtual Goods, also known as "items", will be worth nearly US$2.5 billion by 2013[2]. This paper focuses on the most important legal issues witnessed in Virtual Worlds at the moment. First of all, it will address issues regarding contract law, and second copyright issues arising from disputes on the question "Who owns the virtual sword?". Therefore a brief glance at legal disputes regarding Virtual Worlds will be provided. The conclusion will summarize the findings and lead to a proposal to lawmakers around the world.

Virtual Worlds are gaming. Period. Most Virtual Worlds are "Massively Multiplayer Online Role-Playing Games" (MMORPG) by definition and have been very popular for generations. The history of Virtual Worlds goes back to the first "Multi User Dungeons" where players could fight dragons together and gathered their first items by looting the dragon's corpse. Today we can still find the terminology in the "DKP system" – "DKP" meaning "Dragon Kill Points" – used by players to

---

[1] http://www.kzero.co.uk/blog/?page_id=2563
[2] http://www.emarketer.com/Article.aspx?R=1007226

P. Daras and O. Mayora (Eds.): UCMedia 2009, LNICST 40, pp. 265–270, 2010.

divide the loot when successfully winning an encounter in any Virtual World. The most popular virtual gaming world is the MMORPG "World of Warcraft" with more than 11.5 million paying subscribers[3]. The popularity of the game has caused many an addiction, which has already alerted governments[4] and the media[5]. But the evolution of virtual worlds into social virtual worlds began with Linden Lab's Second Life[6] entering the field and has proceeded to Virtual Worlds for working professionals[7]. The research done by consultants such as KZERO[8] shows that socializing virtual worlds are on the rise and infiltrate society just as Facebook and Youtube are doing right now. To separate Virtual Worlds from these social networking websites I will try to define measurement categories for Virtual Worlds:

1. coded
2. ability to interact with others (through voice or chat)
3. 2-D or 3-D visualized reality
4. persistency
5. run in real-time
6. client- or browser-based
All these factors have to be seen as cumulative.

Virtual Worlds themselves have to be separated into two major categories: **virtual gaming worlds (VGW)** and **social virtual worlds (SVW)**, which can again be subdivided within the respective categories. The distinction is important to apply legal opinions, because e.g. the "murder" of another player may be the main objective of a virtual gaming world, while doing so in a social virtual world can be considered "mobbing". The average age of Virtual World users is 14[9]. Taking into account what has been said in the introduction, these under-age citizens are primarily responsible for the more than US$ 1 billion in revenue from Virtual Worlds in the U.S. alone. This amount mostly comes from the creation or trading of virtual items, which is linked to real-life money. Sometimes it is linked legally, like in Entropia[10], where the Virtual Currency "PE Dollar" is exchanged to the US Dollar with a rate of 10:1, or illegally, like the gold trading in World of Warcraft. The last example is a legal issue commonly known as "gold farming" and provides work for about 400,000 people mostly living in China[11]. Having shown that many people are playing, interacting, trading or working in Virtual Worlds, unaccounted for if in VGWs or SGWs, we have to determine which kind of legal rules should be applied to their behavior and how their actions are protected by law.

---

[3] http://www.wow.com/2008/12/23/world-of-warcraft-hits-11-5-million-subscribers
[4] China opened a net addiction center in 2005, see:
    http://www.theregister.co.uk/2005/07/05/china_net_addicts
[5] http://edition.cnn.com/2008/BUSINESS/01/29/digital.addiction/index.html
[6] http://secondlife.com
[7] See for example: http://www.teleplace.com or https://lg3d-wonderland.dev.java.net
[8] http://www.kzero.co.uk/blog
[9] http://www.virtualworldsnews.com/2009/07/virtual-world-popularity-spikes.html
[10] http://www.entropiauniverse.com
[11] http://www.sed.manchester.ac.uk/idpm/research/publications/wp/di/di_wp32.htm

2 Virtual Persons

Virtual Worlds are non-real. Period. As *Koster*[12] states, *"Someday there won't be any admins. Someday it's gonna be your bank records and your grocery shopping and your credit report and yes, your virtual homepage with data that exists nowhere else."* [13] When at the beginning of Virtual Worlds VGWs were in the majority and killing dragons or solving adventures collaboratively online were the main purposes with no or just minimal ingame trades which could be linked to real-life money, the players did not need any legal protection. They were just enjoying some leisure-time. But when with the beginning of SVWs players evolved into users, the leisure aspect was about to disappear and serious money could be made. As stated, since the huge amounts made in Virtual Worlds are recognized by tax authorities, it must be clear that the entities participating in this business are in need of legal certainty. It is obvious that there is a big difference between a trade on the internet and a trade in a Virtual World. On the internet the host provider is not as powerful as the provider of a Virtual World. The reason is that the host provider in the internet provides the framework in which the content provider can create his Webspace. This Webspace must obey the rules and regulations set by the technical framework of the internet itself, which gives the power of creation to the user. In Virtual Worlds this is different. The provider has the power to determine which ways of creation a user can choose from. While in most SVWs this freedom is very advanced – e.g. in Second Life the user can create content freely with the *"Linden Scripting Language"* – in VGWs the possibility tends to zero. The "freedom" of creation is regulated in the *"End User License Agreements"* (EULA), also called *"Terms of Use"* (ToU) or "Terms of Service" (ToS). The contractual determination of rights can often be classified as autocratic and therefore the user is strongly dependent on the provider[14]. Therefore the liability of a user towards a contractual relationship to another user must be judged in consideration of the General Terms and Conditions established by the provider. Right now protection for the real person controlling the virtual person is very weak.

3 Virtual Goods

Virtual Worlds are business. Period. Trades related to Virtual Worlds have found their way to the classic internet. Auctioneers of virtual goods like FatFoogo[15] have established platforms which are often threatened by lawsuits from the respective providers. eBay has closed all auctions[16] regarding Virtual Worlds (let alone Second

---

[12] *Koster*, "Declaring the Rights of Players" in State of Play (2006) 55.

[13] See also: *Erenli/Sammer*, „Der Gnom zahlt nicht – Muss die Rechtsordnung der Zukunft die virtuelle Person anerkennen?" in *Schweighofer*, Komplexitätsgrenzen der Rechtinformatik (2008).

[14] Just think what would happen, if Blizzard canceled all subscriptions and abandoned World of Warcraft all of a sudden. The outrage of the community would most likely end in real-life riots.

[15] http://www.fatfoogoo.com

[16] http://news.cnet.com/eBay-bans-auctions-of-virtual-goods/2100-1043_3-6154372.html or http://games.slashdot.org/article.pl?sid=07/01/26/2026257

Life) and *Hani Durzy*, spokesman for eBay[17], has stated that *"the seller must be the owner of the underlying intellectual property, or authorized to distribute it by the intellectual property owner"*. Therefore the question: "Who owns the virtual sword?[18]" has been answered taking into account copyright law. The owner of the Virtual World is the provider who determines which kind of copyright is given to a player or user through the General Terms and Conditions (EULA, ToU, ToS, etc). Since copyright law follows the principle of territoriality we can then discuss which copyright law applies to items created by a user of a Virtual World. This question is most often answered by the General Terms and Conditions as well, which determine the applicable law. In the absence of such an agreement the copyright law of the permanent establishment of the provider should apply. This solution is used with regard to the "Virtual Society". Since Virtual Worlds can usually be accessed by anyone around the world it would lead to legal uncertainty if the applicable law of the respective user had to apply. Instead of understanding just one copyright law, each user would have to understand multiple ones. Despite the fact that copyright law gives the necessary legal protection to the provider, it leaves the player or user powerless. This problem is a general one. While the World Wide Web is decentralized, a Virtual World is controlled by its provider. Therefore the statement *"I create the games that you will play, I make the rules you have to obey"* is a very true one. Lawmakers have to consider how to react in such a way that both parties – provider and player or user – are protected equally.

4 Real Problems

Virtual Worlds are dangerous. Period. Most societies fear Virtual Worlds as a place which is responsible for the degeneration of the world's youth. Some fear that Virtual Worlds are a place where laws do not apply. Regardless of individual personal opinions on Virtual Worlds the fact remains that disputes in Virtual Worlds have ended up in real-life courts often enough. Therefore some of the most popular disputes related to intellectual property rights are described hereafter.

4.1 Eros vs Leatherwood[19]

The first big dispute in Second Life was about a copyright infringement. The plaintiff Eros, LLC, owned by *Kevin Alderman*, a maker and seller of virtual adult-themed objects within the Second Life platform, sued defendants and alleged that they had made and sold unauthorized copies of plaintiff's virtual products within Second Life using the plaintiff's trademark. At first it was uncertain who was in control of the avatar *"Volkov Cattaneo"*, who was "stealing" the source code of *Alderman's* products and selling the cloned products to other avatars for a price far less than requested by Eros. The case was won by Eros in 2008 after the court had found the right defendant. While in the US a complaint can be made against an unknown defendant – addressed as *"John Doe"* – in Europe such a complaint would be dismissed by the court from

---

[17] http://pages.ebay.com/choosingformats/digitalitems/faqs/#3
[18] Also see: *Benkler*, There Is No Spoon, in *Belkin/Novecek*, The State of Play 180.
[19] http://news.justia.com/cases/featured/florida/flmdce/8:2007cv01158/202603

the start. *Kevin Alderman* also was the plaintiff in the next big case regarding Second Life. He launched a class-action lawsuit against Linden Lab itself in September 2009, alleging that (among other things) it profits from negligence and delay in dealing with trademark and copyright infringement issues, and that it does so knowingly. The case has just begun and it will be very interesting to witness its outcome.

4.2 Marvel vs City of Heroes

Marvel Comics is the publisher of popular comics such as Spider-Man, The Hulk or X-Men[20]. Marvel filed for trademark infringement after taking a look at the Virtual World "City of Heroes". In this Virtual World users have the possibility to create their avatars to appear as look-a-likes of the Marvel comic heroes. The case was settled in 2005 with a confidential agreement. Marvel recognized the users' power and therefore did not push for a court decision.

4.3 Blizzard vs MDY Industries[21]

Blizzard Entertainment, the creator of World of Warcraft, filed a complaint against MDY Industries, LLC, creator of the WoWGlider program. WoWGlider was a third-party program created by the owner of MDY, *Donnelly*, to circumvent the need for a player to be present during World of Warcraft sessions. He achieved this by setting elaborate scripts to automatically perform quests and hunts. *Donnelly* charged US$25 for the key to unlock the program. Blizzard successfully proved that *Donnelly* breached the agreement made in the EULA.

These lawsuits demonstrate that problems arising out of Virtual Worlds can be solved by applying already existing laws. So do we need to focus on Virtual Worlds from a legal point of view?

5 Conclusion

Business models and user behavior are changing. The legal rules which have to be applied to these changing realities already exist. The task at hand therefore is not to ask for new rules but to change the existing ones to equally protect providers, players and users. Therefore Virtual Worlds should be categorized in VGWs and SVWs. Moreover a supervisory body should be implemented to help harmonize the different intellectual property laws and the different solutions law systems provide. The power of the provider once was limited to providing the technical infrastructure and possibilities for users to interact. By now these interactions have advanced to a life that can be lived parallel to the real one. Therefore the power of the provider has increased in a way that can harm society in both worlds. Hence a discussion on the protection of the Virtual Person is strongly needed. For Virtual Worlds are just like real life. Period.

---

[20] http://www.marvel.com
[21] http://www.wowglider.com/Legal/Feb%5F16%5F2007

References

1. KZERO - Resident experts in virtual worlds,
   http://www.kzero.co.uk/blog/?page_id=2563
2. eMarketer, http://www.emarketer.com/Article.aspx?R=1007226
3. World of Warcraft,
   http://www.wow.com/2008/12/23/
   world-of-warcraft-hits-11-5-million-subscribers
4. China's net addiction center,
   http://www.theregister.co.uk/2005/07/05/china_net_addicts
5. CNN on China's net addiction center,
   http://edition.cnn.com/2008/BUSINESS/01/29/
   digital.addiction/index.html
6. Second Life, http://secondlife.com
7. Teleplace (formerly known as Qwaq), http://www.teleplace.com
8. SUN's Wonderland, https://lg3d-wonderland.dev.java.net
9. KZERO - Resident experts in virtual worlds, http://www.kzero.co.uk/blog
10. Virtual World News, http://www.virtualworldsnews.com/2009/07/
    virtual-world-popularity-spikes.html
11. Entropia Universe, http://www.entropiauniverse.com
12. Institute for Development Policy and Management,
    http://www.sed.manchester.ac.uk/idpm/research/publications/
    wp/di/di_wp32.htm
13. Koster: Declaring the Rights of Players. Belkin/Novecek State of Play 55 (2006)
14. Erenli/Sammer, Der Gnom zahlt nicht – Muss die Rechtsordnung der Zukunft die virtuelle
    Person anerkennen? in Schweighofer, Komplexitätsgrenzen der Rechtinformatik (2008)
15. Fatfoogo, http://www.fatfoogoo.com
16. Cnet,
    http://news.cnet.com/eBay-bans-auctions-of-virtual-goods/
    2100-1043_3-6154372.html
17. Slashdot,
    http://games.slashdot.org/article.pl?sid=07/01/26/2026257
18. Ebay's policy on virtual goods,
    http://pages.ebay.com/choosingformats/digitalitems/faqs/#3
19. Benkler.: There Is No Spoon, in Belkin/Novecek. The State of Play 180 (2006)
20. Justitia.com on Eros vs Leatherwood,
    http://news.justia.com/cases/featured/florida/flmdce/
    8:2007cv01158/202603
21. Eros vs Linden Lab, http://media.taterunino.net/
    eros-vs-lri-Complaint_-_FINAL.pdf
22. Marvel's Homepage, http://www.marvel.com
23. Wowglider's Homepage,
    http://www.wowglider.com/Legal/Feb%5F16%5F2007

TrustVWS 2009

Session 2

A Comparison of Three Virtual World Platforms for the Purposes of Learning Support in VirtualLife

Kristina Lapin

Vilnius University, Faculty of Mathematics and Informatics,
Naugarduko 24, LT-03225 Vilnius, Lithuania
kristina.lapin@mif.vu.lt

Abstract. The paper addresses three 3D immersive collaborative virtual environments for their utility to learning support. We analyze the needs of a lecturer that intends to supplement face-to-face teaching with computer mediated collaboration. The analysis aims at informing the design of a new virtual environment that is being developed in the FP7 ICT VirtualLife project. We examine the modern learners' and tutor's needs as well as existing 3D virtual world platforms, which are free to download. We explore the usability features to include in the VirtualLife bundle intended for educational use.

Keywords: 3D virtual world platform, blended education, collaborative virtual environment, usability.

1 Introduction

VirtualLife is an ongoing 36 month project awarded by the European Commission to 7 small enterprises and 2 universities[1]. The project is aimed at developing of 3D immersive collaborative virtual environment with a number of innovative features: secure and trusted communication, virtual legal system, dispute resolution mechanism, user reputation management system and a peer-to-peer network communication architecture. The platform is intended as a serious virtual world for business and education.

Collaborative virtual environment is software that allows users to share the virtual environment and collaborate. Such an environment has to be consistent and scalable [1]. Consistency refers to a characteristic of a system to maintain the sole state of environment. Any change of avatar's state is visible for all users. Scalability is a characteristic that ensures effective consistency control even when many users enter the environment. VirtualLife is based on a peer-to-peer communication architecture that spreads the load between different clients to face the scalability challenge.

The 3D immersive virtual collaboration environments are often called virtual worlds. In the following we refer to the VirtualLife platform as a virtual world. Virtual worlds encompass the following features [2]:

---

[1] FP7 ICT VirtualLife - Secured, Trusted and Legally Ruled Collaboration Environment in Virtual Life, 2008-2010, http://www.ict-virtuallife.eu/

P. Daras and O. Mayora (Eds.): UCMedia 2009, LNICST 40, pp. 273–278, 2010.

- The world allows many users (avatars) to participate simultaneously.
- Interaction takes place in real time.
- The virtual world allows users to develop, alter and submit customized content.
- When an individual logs out, the virtual world continues to exist.
- The virtual world encourages the formation of in-world social groups.

This paper addresses the usage of virtual worlds for educational purposes. We analyze how a single tutor could complement a mere face-to-face learning with distance education elements. Such an education mode is called a blended education. The evidence shows that distance education is a fast growing sector of higher education and its technological demand constantly grows [3].

The following sections are organized as follows. The second section deals with educational needs of tutors and students in a blended education mode. The third section explores freely available 3D virtual world platforms that have a potential to implement educational needs. The fourth section provides the design solutions for a restricted educational suite of VirtualLife. At the end some conclusions are drawn.

2 Educational Needs

Today's students know more about technology than any generation before them. For communication with peers they fully use various features of cell phones and the Internet. A lot of students have profiles in social networks and avatars in virtual worlds. Web-based education is their natural expectation. They prefer the searchable learning materials where they find the information they need at the moment. Web 2.0 based learning tools showed their adequacy for modern learners' needs [4]. Experiments show that in the blended education students achieve better results than in a pure face-to-face education [5].

The amount of conveyed skills is constantly increasing. The large amount of the material does not leave much time for a discussion. Therefore after face-to-face classes conversation can be shifted to a virtual world. A learner receives there an instant feedback from other learners and the tutor. The feedback helps students to achieve learning goals and encourages them to continue learning as young individuals do like to stay and interact. Interactive 3D didactical means demonstrate the learner interesting features. Then the learner is directed to other places for further content. Such an interaction in an immersive environment creates a long-term retention [6].

In order to complement the face-to-face education with interaction in a virtual world the educator needs an available platform which does not require deep technical knowledge. Such a platform should facilitate the creation of interactive content.

3 Analysis of Existing Virtual World Platforms

Further we tackle some available platforms from the individual tutor's perspective. The following aspects are important for our exploration:

- a platform is freely available,
- the terms of usage are not restrictive and allow the user to create the content,

- hardware and Internet access requirements do not exceed an average level,
- a system does not require third-party commercial products such as graphics engines,
- the developer has a full control over the developed virtual world.

A long list of virtual platforms can be used for educational purposes [7]. A majority of these tools allows free registration whereas the creation of content is charged; see e.g. Second Life and Active Worlds Educational Universe. The Crocket Project and Open Source Metaverse Project are under development.

When a virtual world is placed on a service provider server, the user has to obey the rules provided by the supplier. For example, the terms of usage might note that the supplier "has the right at any time for any reason or no reason to suspend or terminate your account". We treat such a platform as inappropriate for educational purposes. In the case the user does not have full control over the environment it is not worth putting effort into the creation of teaching materials.

The above mentioned requirements are met in the three virtual world platforms: Multiverse (3D), OpenSim (3D), and Metaplace (2.5D). We find that Second Life, Active Worlds, The Crocket Project, Open Simulator and Open Source Metaverse Project do not satisfy our requirements. Further we explore the following issues:

- installation efforts,
- allowed actions,
- content development,
- import of content from outside tools,
- creation of interactive learning objects.

Multiverse provides a platform to create 3D virtual worlds on both the user's computer and a Multiverse hosting server [8]. The server installation and configuration efforts are similar to HTTP server installation and maintenance – command-line operations and textual configuration files. Such a task can be inconvenient for the user. Wizards are provided for supplementary tools only. The world visitors can move and chat. Additional functionalities should be developed by world creators. The content generation requires external tools for each type of content. A 3D modeling tool is required to create an object model. A graphical editor is needed to develop textures. Object libraries and the world editor facilitate static content creation whereas interactive objects are programmed with Python. We summarize that Multiverse is easy to access and is flexible to create a virtual world. The content generation is rather difficult for a novice user.

OpenSim, also called OpenSimulator, is an open source project with BSD license. The user has to install a browser, for example, Second Life Viewer. A visitor can move, fly and communicate by textual chat and gestures. The visitor is allowed to create objects in the case the world owner permits.

The server installation efforts can be compared with an installation of a HTTP server. The server is configured during the first activation by the way of answering questions. Such an installation is sufficient for a minimal usage. An advanced usage requires editing the configuration files.

The virtual world is comprised of grids that consist of regions. A region is a certain size square of earth and sea area that has the owner. He concedes the rights for visitors. The in-world content is developed from geometrical primitives. Therefore the creation of simple objects is easy. But the creation of elaborated objects is cumbered. An external content can be loaded using convenient in-world interfaces. Interactive objects are programmed in SecondLife LSL, C# or any .NET language.

To summarize, OpenSim is an easy accessible platform. The creation of static user content is simple whereas interactive objects have to be programmed. The installation and maintenance requires server administration skills.

Metaplace [10] is a platform based on Flash technology. Therefore it is rather slow. The world is 2.5D – a three-dimensional illusion is attained showing 2D images from the perspective. It looks like an animated world. Therefore kids enjoy it. A virtual world can be placed on a HTTP server as well as on a Metaplace hosting server. The user can move and chat.

Users and creators do not need anything to install. Each user can create content inside the world. An internal object library is provided. A multimedia object can be created with an external tool and imported into the world. Interactive objects are programmed in Metascript, a modification of LUA script language.

In conclusion, this platform is the easiest to use and generate content while interactiveness requires extensive programming skills.

Table 1. A comparison of three platforms

Feature	Multiverse	OpenSim	Metaplace
User actions	Move and chat	Move, fly, chat	Move, sit, chat
Installation	Alike HTTP server	Automatic	Alike HTTP server
Content generation	External tools	Inside the world	Inside the world
Interactiveness	Python scripts	LS or .NET scripts	Metascript

Table 1 presents a comparison of the explored platforms. Only one platform diminishes installation efforts by offering automatic installation. The full control over the world requires a significant effort to install and configure the server. The simplest content generation is in Metaplace, although animated graphics is too infantile for young adult learners. Implementing interactive features in all the three platforms requires significant programming skills. The documentation of scripting languages is provided in the form of manuals and function catalogues. A novice user could hardly use them in order to learn scripting.

The analysis shows that it is difficult to adopt the virtual world for the purpose of learning. To make learning attractive, the world should provide some learning content. Without interactive content a virtual world does not spread its full potential. However, the creation of such content requires significant programming skills. The generation of the searchable learning materials from any object of virtual worlds is desirable feature but not present in existing systems.

4   VirtualLife Design Decisions

VirtualLife allows creating virtual worlds and having a full control over it. Considering the problems with the explored platforms, a VirtualLife bundle for educational usage should encompass automatic installers for world creators (tutors) and users (students). Each user creates the content using graphical user interface. The developed assets are stored in the user's computer.

Our study shows that effective creation of interactive learning means still poses a challenge. Facilitating the creation of interactive objects, a rich library of default interactive objects, such as avatar motions, opening windows, rotating wheels, etc. is available. VirtualLife platform contains many ready-made virtual tools, like the web-board which visualizes web-pages in the 3D environment. The framework permits the creation of new objects, using an internal editor or importing them.

VirtualLife platform allows a high level of interaction both between avatars and between avatars and objects. The video demo[2] shows the simulation of a geometry lesson where the tutor presents complex geometrical solids. The teacher is using an interactive web-board, a virtual pointer and a set of interactive polyhedrons that enables her to introduce complex concepts (see Fig. 1). Static information is presented on in-world blackboards in the form of a webpage.

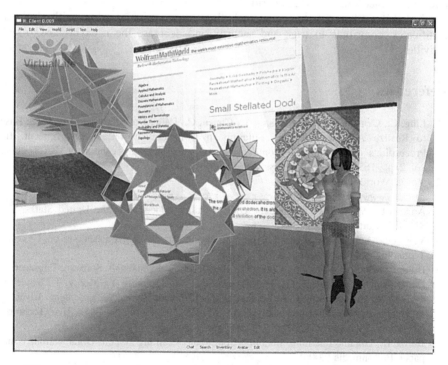

Fig. 1. Interactive geometrical solids and static learning material in a virtual world

A powerful and intuitive scripting language allows for the creation of complex interactive tools. The user is able to import web-pages and consult them in-world. A web-page generator function automatically creates searchable web pages for entities in the virtual world containing useful and descriptive information.

---

[2] FP7 ICT VirtualLife http://www.ict-virtuallife.eu/

5 Conclusions

We maintain that current technologies are difficult to adopt for education purposes by a single tutor without institutional support. Our analysis shows that creation of interactive 3D learning objects in existing open source environments is still time-consuming and requires high programming competence. The platforms have been analyzed with regard to installation and maintenance efforts. These activities require a significant level of technical knowledge. We argue that contemporary virtual world platforms require the tutor to concentrate on the tools instead of didactics. Considering the findings we argue that the usage of virtual worlds should not require deep technical knowledge.

VirtualLife has a potential to supplement face-to-face learning with on-line interactions. It provides simplified installation, content generation and a rich interactive object library. A web page generator facilitates the creation of searchable learning materials that create the searchable learning materials with links to interactive objects in the virtual world.

VirtualLife advanced features, including strong authorization with external certificates, contribute to trust and security and decrease the need of face-to-face activities.

References

1. Ling, C., Gen-Cai, C., Chen-Guang, Y., Chuen, C.: Using Collaborative Knowledge Base to Realize Adaptive Message Filtering in Collaborative Virtual Environment. In: Proceedings of International Conference on Communication Technology, ICCT 2003, vol. 2, pp. 1655–1661. IEEE Press, Los Alamitos (2003)
2. Virtual World Review,
   http://www.virtualworldsreview.com/info/whatis.shtml
3. Allen, I.E., Seaman, J.: Staying the course: Online education in the United States, The Sloan Consortium (2008)
4. Vassileva, J.: Toward Social Learning Environments. IEEE Transactions on Learning Technologies 1(4), 199–214 (2008)
5. Bule, J.: Adaptive E-learning Courses at Riga Technical University Software Engineering Department. In: Grundspenkis, J., Kirikova, M., Manolopoulos, Y., Morzy, T., Novickis, L., Vassen, G. (eds.) Local proceedings of the 13th East-European Conference Advances in Databases and Information Systems, ADBIS 2009, pp. 238–245. Riga Technical University, Riga, Latvia (2009)
6. Burns, R.: Organic Learning: Enabling Continuous Learning in Your Organization. White paper. Protonmedia (2006)
7. de Freitas, S.: Serious VirtualWorlds: A scoping study. Technical report, The Higher Education Funding Council for England, HEFCE (2008),
   http://www.jisc.ac.uk/media/documents/publications/
   seriousvirtualworldsv1.pdf
8. The Multiverse Network, http://www.multiverse.net/
9. OpenSim, http://opensimulator.org/wiki/Main_Page
10. Metaplace, http://www.metaplace.com

Transforming Legal Rules into Online Virtual World Rules: A Case Study in the VirtualLife Platform*

Vytautas Čyras

Vilnius University, Faculty of Mathematics and Informatics,
Naugarduko 24, 03225 Vilnius, Lithuania
vytautas.cyras@mif.vu.lt

Abstract. The paper addresses the implementation of legal rules in online virtual world software. The development is performed within a peer-to-peer virtual world platform in the frame of the FP7 VirtualLife project. The goal of the project is to create a serious, secure and legally ruled collaboration environment. The novelty of the platform is an in-world legal framework, which is real world compliant. The approach "From rules in law to rules in artifact" is followed. The development accords with the conception "Code is law" advocated by Lawrence Lessig. The approach implies the transformation of legal rules (that are formulated in a natural language) into machine-readable format. Such a transformation can be viewed as a kind of translation. Automating the translation requires human expert abilities. This is needed in both the interpretation of legal rules and legal knowledge representation.

Keywords: Legal rules, computer implementation, translation, legally ruled collaboration, Ought to Be inworld reality.

1 Introduction

The paper is devoted to the operational implementation of legal rules in software. The issues arose while developing an online virtual world platform within the FP7 project "Secure, Trusted and Legally Ruled Collaboration Environment in Virtual Life" (VirtualLife). The purpose is to create a serious virtual world – not a game.

The legal rules of a VirtualLife virtual world initially are formulated in a natural language, for example, 'Keep off the grass'. This rule can be paraphrased 'The subject – an avatar – is forbidden the action – walking on the grass'. The rule demonstrates that the "Ought to Be reality" concept (see, e.g., [7]) can be extended from the real world to an online virtual world.

While developing software, a further translation of the rules into machine-readable format is required. Such a translation seeks capabilities of human experts. A team includes an expert in law, virtual world developer and programmer.

We argue that the translation of legal rules requires natural intelligence. A translator faces the following problems (including but not limited to):

---

* Supported by the EU FP7 ICT VirtualLife project, 2008-2010, http://www.ict-virtuallife.eu

P. Daras and O. Mayora (Eds.): UCMedia 2009, LNICST 40, pp. 279–284, 2010.

1. **Abstractness of rules**. Legal rules are formulated in abstract terms.
2. **Open texture**; see e.g. Hart's example of "Vehicles are forbidden in the park" [1].
3. **Legal interpretation methods**. The meaning of a legal rule cannot be extracted from the sole text. Apart from the grammatical interpretation, other methods can be invoked such as systemic and teleological interpretation.
4. **Legal teleology**. The purpose of a legal rule can be achieved by various ways.
5. **Heuristics** – the ability to translate high level concepts and invent new ones.
6. **Consciousness of the society**. Law enforcement is a complex social phenomenon.

The transformation we explore accords with the conception "Code is law" [6]. Our analysis of the transformation tackles the problems above. We proceed with an introduction to VirtualLife and its legal framework.

Fig. 1. Interaction with an object, a complex solid, in a lesson within a virtual world [4]

2 About VirtualLife

The goal of VirtualLife software is a new form of civil organization, realized by the creation of secure and ruled places within a virtual world, where important transactions can occur (where transactions are those that normally occur in real life) [2]. At present VirtualLife is targeted at learning support scenarios, such as a university virtual campus. A professor avatar gives a lesson whereas student avatars listen (Fig. 1). An avatar can interact with other avatars and inworld objects.

3   Focus on VirtualLife Legal Framework

VirtualLife's legal framework consists of three tiers (Fig. 2):

1. A 'Supreme Constitution';
2. A 'Virtual Nation Constitution'. E.g., Constitution VN1, …, Constitution VNn;
3. A set of different sample contracts.

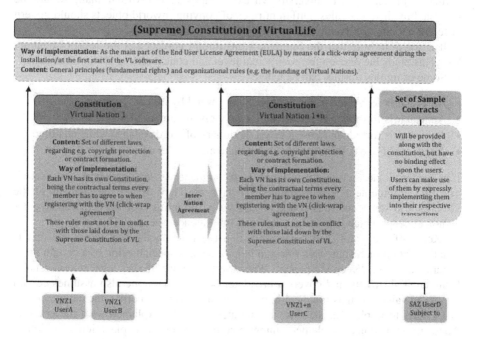

Fig. 2. The three tiers of VirtualLife legal framework – a Supreme Constitution, a Virtual Nation Constitution and a set of different sample contracts[1]

Initially the legal framework was elaborated in project deliverables[2]; see also [8]. Further the elaboration was in the form of the technical specification of Virtual Nation laws. The editor of rules is comprised – it serves as a tool to compose laws.

A virtual world is quite different from a standard video game, where there is a story, a final purpose, and the system only allows for a limited set of actions. In a virtual world there is not a determined purpose and there is not a game over. People move their avatar and establish their second life, driven by different purposes. Thus

---

[1] Spindler, G., Prill, A., Schwanz, J., Wille, S. D7.1 Preliminary Report of the Legal Fundamentals Concerning Virtual Worlds. VirtualLife deliverable, 2008.
[2] Spindler, G., Prill, A., Anton, K., Schwanz, J. D7.2 "Constitution" of a Virtual Nation. Model of a Contract about a Legal System of a Virtual Community. VirtualLife deliverable, 2009.

the rules of play should be replaced by a sophisticated legal framework, which is considered to be essential in order to guarantee the existence of a secure and safe virtual world. In VirtualLife, the legal system takes into account both real life values and real world laws [2].

3.1 From Rules in Law to Rules in Artifact

A Virtual Nation Constitution contains special provisions as regards, for example, the protection of objects used in that Virtual Nation under copyright law or the authentication procedure required to become a member of that nation.[3] Distinct Virtual Nations, e.g., a university virtual campus and a virtual mall, should be governed by different rules. Different rules of copying inworld objects should govern different nations, for example, CopyRight and CopyLeft nations.

It is worth to note that the Supreme Constitution is placed at the level of contract law. This binds the user on the contractual level and contributes to law enforcement.

Some examples of rules can be listed:[4]

- An avatar is forbidden to touch objects not owned by him or a certain group.
- An avatar is forbidden to create more than a given number of objects.
- An avatar is forbidden to use a given dictionary of words (slang) while chatting.
- An avatar of age is forbidden to chat with avatars under age.

If an avatar violates a rule, for instance, steps on the grass, his reputation is decreased. The rule enforcement is implemented by triggers. They trigger the changes of the virtual world states and thus invoke avatar script programs. The triggers implement a demon concept which is known in artificial intelligence; see, e.g., [3].

'Keep off the grass' can be viewed as a toy rule. In the real world the text is usually written on a sign. This rule can be treated as a specialisation of a certain more general rule, e.g., 'Keep the peace', 'Preserve the nature'. Every legal rule is usually formulated abstractly and covers a broad set of cases. The rule also assumes certain preserved values such as life, health, property, the established order, the nature, etc.

In order to implement the rule, a computer has to be explained what the grass is. Apart from this objective element, intentional factors need an explanation, too. Hence, the legal concepts, such as fairness, malice and negligence, have to be explained. For example, in the case an avatar is pushed by another avatar, the fault analysis can invoke distinguished causation theories[5]. Here we can ask: How far can be moved forward in representing all this?

We note the difference of interpretation between a legal rule and technical system rule. The legal rule allows a freedom of interpretation, the technical rule – does not. A certain degree of freedom in the legal rule is even intentionally introduced when designing the legal system. The technical system cannot afford such a freedom.

The open source problem [1] is characteristic of the legal system. The reason to formulate a legal norm abstractly is inherent in the complexity of the real world. A

---

[3] Cf. footnote 1, p. 11.
[4] Cordier, G., Zuliani, F. D7.3 Virtual Nation Laws – Technical Specifications. VirtualLife deliverable, 2009.
[5] Hart, H.L.A., Honoré, T. Causation in the Law, 2nd edition, Oxford 1985.

variety of potential cases cannot be foreseen. A variety of factors is not known in advance. The real world changes permanently; see, e.g., Web 2.0 applications.

A technical system, on the contrary, is required to foresee its future behavior. The system is designed to decide 'yes' or 'no'. Therefore the technical system is tested, verified and validated. Within the established social system, deciding a legal case for yes/no can be hard. The judge can also take a middle decision – yes/no/other.

3.2 The Editor of Rules

The rule concept follows the approach of Vázquez-Salceda et al. [9]. Similarly, laws are expressed in VirtualLife in the form of Norms.[6] A Norm is composed by: (1) NORM_CONDITION, (2) VIOLATION_CONDITION, (3) DETECTION_ MECHANISM, (4) SANCTION, and (5) REPAIR.

A NORM_CONDITION is expressed by:

- TYPE – Obliged, Permitted, or Forbidden
- SUBJECT – Avatar, Zone, or Nation
- ACTION – ENTER, LEAVE, CREATE, MODIFY, MOVE, CREATE, TRADE, SELL, BUY, CHAT, etc.
- COMPLEMENT – AREA, AVATAR, OBJECT, etc.
- IF {LOGICAL EXPRESSION USING SUBJECTS Properties}
  An example of Norm, referring to Norm composition:
1. Condition: FORBIDDEN   Student_Avatar   ENTER   Library IF Student_Avatar.age < 18
2. Violation   condition: NOT   over_age(Student_Avatar)   AND admit(Student_Avatar, Library)
3. Detection mechanism: call over_age(Student_Avatar) when Student_Avatar enters Library
4. Sanction: decrease_reputation(Student_Avatar); notify avatar
5. Repair: log and roll back if applicable

3.3 Values Protected by VirtualLife Law and the Law of Avatars

VirtualLife laws – like laws in general – identify purposes and protected values. These are the values of a Virtual Nation (VN). The values shall be enforced by code – a set of technologically implemented rules and laws.

The purpose of a virtual nation is described at the beginning of the Virtual Nation's constitution. Examples of values, which are immanent in a real-world constitution of a state, are democracy, life, legal certainty, etc. In legal theory, norm is a basic element of the law. Values can be worded explicitly, but mainly they are implicit. For example, the Code of Conduct within the Supreme Constitution identifies equality (non-discrimination), avatars integrity, honor, reputation, privacy, free movement, freedom of thought, sanctity of property, etc. Such explicit representation contributes to detect violations of the Virtual Nation laws by the users.

We argue that the behavior of artificial agents (including avatars) shall be governed by law, too. We call this kind of ruling "virtual law". Thus the Ought to Be inworld

---

[6] Cf. footnote 4, p. 14.

reality is identified. An example of a rule is that an avatar is forbidden to harm (kill, hit, etc.) another avatar. Thus we approach a code of avatars [5].

4 Conclusions

We focus on two activities which are implied by the implementation of a legal framework. The first is the observance of the law by software users. This is a classical function of law as identified in legal theory. The second activity is the transformation of legal rules into machine code.

The latter is attributed to informatics. It contributes to the enforcement of avatars law. The editor of rules serves as a tool for human-driven translation. A human translator has to distinguish between the methods of law and informatics. Just to mention a few differences, they are the abstractness of legal norms, the open texture problem in law, methods of legal interpretation, legal teleology, and heuristics to invent low level concepts.

Acknowledgments. The work is stipulated by the whole VirtualLife consortium.

References

1. Bench-Capon, T.: The Missing Link Revisited: the Role of Teleology in Representing Legal Argument. Artificial Intelligence and Law 10(1-3), 79–94 (2002)
2. Bogdanov, D., Crispino, M.V., Čyras, V., Glass, K., Lapin, K., Panebarco, M., Todesco, G.M., Zuliani, F.: VirtualLife Virtual World Platform: Peer-to-Peer, Security and Rule of Law. In: eBook Proceedings of 2009 NEM Summit Towards Future Media Internet, Saint-Malo, France, September 28-30, pp. 124–129. Eurescom GmbH (2009)
3. Brachman, R., Levesque, H.: Knowledge Representation and Reasoning. The Morgan Kaufmann Series in Artificial Intelligence, San Francisco (2004)
4. Čyras, V., Lapin, K.: Learning Support and Legally Ruled Collaboration in the VirtualLife Virtual World Platform. In: Grundspenkis, J., Kirikova, M., Manolopoulos, Y., Novickis, L. (eds.) Advances in Databases and Information Systems, Associated Workshops and Doctoral Consortium of the 13th East-European Conference, ADBIS 2009, Riga, Latvia, September 7-10 (2009) (in press)
5. Koster, R.A.: Declaration of the Rights of Avatars (2000), http://www.raphkoster.com/gaming/playerrights.shtml
6. Lessig, L.: Code version 2.0. Basic Books, New York (2006)
7. Pattaro, E.: The Law and the Right: A Reappraisal of the Reality that Ought to Be. A Treatise of Legal Philosophy and General Jurisprudence, vol. 1. Springer, Heidelberg (2007)
8. Spindler, G., Anton, K., Wehage, J.: Overview of the Legal Issues in Virtual Worlds. In: Proceedings of the First International Conference on User-Centric Media, UCMedia 2009, Venice, December 9-11 (2009)
9. Vázquez-Salceda, J., Aldewereld, H., Grossi, D., Dignum, F.: From Human Regulations to Regulated Software Agents Behavior. Connecting the Abstract Declarative Norms with the Concrete Operational Implementation. Artificial Intelligence and Law 16(1), 73–87 (2008)

Reverie and Socialization for the 'Electro-Nomadic Cyborg'

Duncan P.R. Patterson

University of Waterloo School of Architecture,
Cambridge ON, Canada
dprpatte@engmail.uwaterloo.ca

Abstract. This paper addresses the question of domesticity in the context of the increasingly nomadic condition of the world population. The history of the idea of domesticity is traced and the particular case of the hearth is addressed in detail. The paper speculates as to how user-centric information and communication technology might simulate the social and cultural value held by the traditional hearth.

Keywords: Nomadism, domesticity, ICT, user-centric media, hearth, actor-network theory, Faustian bargain.

1 'Nomadic Domesticity'

The notion of 'nomadic domesticity', is of course a contradiction. The word nomad implies, from its Greek root, *nome*, territory, coming from a Proto-Indo-European word for 'to allot', the nomad being someone who wanders about the territory putting their livestock to pasture. The word domesticity, coming from the Greek word, *domos*, meaning 'house', also implies territory, but this time the enclosed territory of a localized social group, the household, often bound together by family ties. It would seem that what is at stake then is the question of enclosure – what are the options between openness and closedness? Both openness and closedness have obvious appeals and attendant problems. Closedness implies inclusiveness as well as exclusivity. The closed territory of the domos for instance is the realm of the household, a structurally supportive social group. This seems to be very much a central issue here - while it is perfectly possible for the nomad to have a local social group within which they might situate themselves, a band, let's say, of jet-setting businessmen roaming the desert of the airport, it is unlikely: it is less the 'wandering' quality of the nomad that is troublesome so much as the 'loneliness'; less the 'place to hang your hat' that is missed in the domos, as the other hats that are also hanging there. Openness, as in the territory of the nomad, and closedness, as in the territory of the domos, are spatial ideas with important social and cultural implications.

An important symbol of the social value of the domos is the hearth, the place of fire within the house, generating light and heat. In the early middle ages in Europe, the hearth was simply an open fire in the middle of the room. Above this was often a lantern-type structure that directed the smoke through a hole in the roof. Sometime

P. Daras and O. Mayora (Eds.): UCMedia 2009, LNICST 40, pp. 287–292, 2010.

around the twelfth century, however, the masonry chimney came into more common acceptance, allowing the hearth to be nestled up next to the wall and the smoke to be more effectively evacuated. When this happened, the fireplace became an important place of social gathering where the group would congregate, focused upon the fire, taking advantage of its heat and light without having to worry about the smoke. This poetic image of congregation epitomizes the social idea of domesticity and so in an era when domesticity seems to have become increasingly uncertain, it is one we should pay close attention to.

2 History of Domesticity

As Witold Rybczynski has shown, however, the idea of 'domesticity' actually originates as recently as the seventeenth century, initially in the Netherlands and then spreading across Western Europe [1]. It originates in a time when, due to prosperity, the number of occupants in a house was shrinking and work was beginning to be separated from the living environment. Previous to this, in the Middle Ages, it was common for as many as twenty-five people to live in a relatively small house in the city and there would have been little separation between work-space and living-space. In the seventeenth century, though, amongst the bourgeois it became the norm for work-space to be aggregated together, separate from the living space. This reinforced a gendering of occupation and of space as the women stayed home and worked on the house and men went off to work. These ideas spread across Europe and laid the ground-work for what we now think of as the house, along with its stubborn gender associations. A clear schema of the world was contained in the idea of the house. The house was the location of family, hygiene and nourishment, of attending to the basic conditions of life. In the morning the nuclear structure of the family would fragment, the men going off to work, and at the end of the day the social group would reassemble. Thus the house was not just a material enclosure, but a symbol of the stable family unit: the bedrock, in most political figurations since Aristotle, of the whole social construct.

By the beginning of the twentieth century, however, the idea of the house had begun to undergo considerable change. Mechanization brought on significant changes as the care of the house became easier to perform. Servants were on their way out and appliances on their way in. Women began to enter the broader workforce in greater numbers. The fireplace began to be replaced by complex distributed means of tempering the internal environment. Fireplaces are inefficient both in their fuel consumption and in their ability to distribute heat, and they're also dangerous. Seriously beginning around mid-century in Western Europe and North America, the fireplace became decorative, kept alive by its social and cultural role. Meanwhile, however, this social and cultural significance was also being supplanted by technologies such as the radio and the television, technologies that are now in turn being replaced.

3 Wirelessness, Nomadism and Alienation

Today, technology has completely transformed our lives, who we are, how we relate to one another, how we understand our place in the world. Technology, as it becomes miniaturized, wirelessly networked, and immediately responsive to our will, especially within the emergent user-centric paradigm, tends towards empowerment of the individual. Read closely, technology can be seen clearly to extend our agency in the world and liberate us in one way or another from the confines of the contingency of our existence. Often, however, it seems that we fail to notice what we lose through our technology. As Neil Postman, celebrated and controversial critic of technology, put it, "the question 'what will a new technology do' is no more important than the question 'what will a new technology undo?'"[2] Every technology, he observed, comes with a Faustian bargain. Technological change is ecological by nature and no change can occur without causing broader ripples in both the material and cultural conditions of our lives.

Our powerful means of transport combined with our miniaturized, wireless technology, empowered to respond immediately to our will and enact change at very great distances, and the new possibilities of engaging with virtual worlds simultaneous to the real world or instead of the real world, increase our potential for mobility vastly. We no longer have need to be as 'situated' as we used to in order to be effective. As Leonard Kleinrock, pioneer of the Internet, put it, much of our new infrastructure constitutes "the system support needed to provide a rich set of computing and communication capabilities and services to nomads as they move from place to place in a way that is transparent, integrated, convenient, and adaptive" [3]. With the aid of, or perhaps because of our new technology, we are truly becoming what the architect William Mitchell has described as 'electro-nomadic cyborgs'. Says Mitchell, "in an electronically nomadicized world I have become a two-legged terminal, an ambulatory IP address, maybe even a wireless router in an ad hoc mobile network. I am inscribed not within a single Vitruvian circle, but within radiating electro-magnetic wavefronts"[3]. And this nomadism is not simply a theoretical fancy, but very real. The human population is on the move! According to the United States Current Popular Survey of 2008, 12% of respondents had moved within the previous year. The average American currently travels 22,300 km/year, and it is predicted that by 2050, the average American will be travelling 2.6 times this: 58,000 km/year [4]. 32% of Americans live outside of the state where they were born, while 9% of all European workers are migrants [5]. And technology is playing a strong role in this increasing nomadism. Cellular phones have reached 85% penetration of the adult market in the US [6], and according to Verizon Wireless, the number of text messages sent between 7:30 AM and 10:00 AM jumped by 50% in the past year alone [7]. People are rapidly adopting emerging technologies that facilitate this nomadism.

But yet, this emerging nomadism has its Faustian bargain. As we become unrooted, living more in 'open territory', the relationships we have with one another are destabilized. People are become increasingly alienated from one another and from the world about them. As the Pew Internet and American Life project has shown, from 1985 to 2004 the average number of intimate friends reported by survey respondents dropped from three to two. People who use social networking sites like Facebook it turns out are 30% less likely to know their neighbours [6]. Clearly information and communication technology, which emphasizes the mind over the body and shifts our concentration to virtual information, in supporting our ocular- and logo-centric orientation, seems to increase our alienation.

Our things are by no means neutral. Simple technology like eye-glasses, that have been around since the 1700s, frame and alter our visual confrontation with the world.

ICTs are even more intrinsically involved in our being. As Mitchell has put it, they "constitute and structure my channels of perception and agency – my means of knowing and acting upon the world" [3]. But as Foucault observed, we would be amiss to think of such material conditions causing cultural phenomena [8]. Cultural and material change, such as the dissolution of the hearth and social alienation are interconnected. 'Actor-network' theory can be instructive for understanding these ecological phenomena, positing the key role played by things as well as people in social networks. Things too are actors, affecting social and cultural change [9]. As the MIT psychologist Sherry Turkle has observed, "objects help us make our minds, reaching out to us to form active partnerships" [10]. It is of vital importance then to keep technology in mind when thinking about social and cultural change, and likewise to keep social and cultural implications constantly in mind when thinking about technology.

4 The Technological Hearth

If we think seriously about the fireplace, we see that it has been an important cultural site. Fire both resembles life itself and threatens death. It both nourishes and endangers. Through its uncontrolled action in forests and cities it both causes immense destruction and makes room for new vitality. The fireplace is a tool for controlling the unpredictability of fire. In their union the fire and the fireplace become a symbol of love and of home [11]. As a place of gathering, the hearth has been a site of shared narrative and negotiated truth. People would congregate and tell stories, cosmologically important stories about the shape of the world, the hierarchical importance of things, the position of the individual as they relate to varying scales of social group and the world outside. The hearthstone after all had a gravity to it that alluded to the earth, while the smoke coiled up the chimney towards the sky, placing us securely between these cosmological realms. Incorporating all of this symbolic value, the hearth was an important site of reverie [12] as well as a focal point (the Latin word for hearth, remember, was *focus*) for 'socialization', the sort of conditioning that occurs intersubjectively amongst engaged groups of individuals. The ancient Greeks had an interesting take on the hearth which they figured as young maiden goddess named Hestia. Hestia was virginal and innocent; she represented stability and serenity. According to Vernant, in the Greek pantheon Hestia was paired with Hermes, the itinerant god who was associated with the threshold [13]. The hearth and the threshold - doors and windows - may be seen as the key symbolic ingredients of the house then, the domos stretched between Hermes and Hestia, two strong forces, one pulling outwards while the other held the centre.

But now the hearth has transformed completely, and so has our social behaviour. We gather less than we used to as localized social groups. Our social networks are increasingly abstract, based on commonality of 'interest' and other perceived and declared similarities. Our primary communication is increasingly mediated by our technology, be it cell phones, email, instant messaging, etc. The television replaced the fireplace as the new hearth, situated in a prominent location in suburban living rooms, a focal point for local gathering. And it did provide for some negotiation of truth-claims and the sharing of important ideas about self and world. Unfortunately, this sharing of narratives occurred decidedly in a one-way direction: narratives and ideas were received, the television watcher an inert consumer. For a while it seemed like the television was being replaced by the home computer, especially in the days, not so long ago, of large towers and CRT monitors, but today these too have been supplanted by smaller means of computing such as notebooks, netbooks, and even smaller personal electronic technology such as smartphones. If the television had effectively replaced the fireplace as the hearth, then these technologies may be read as also being heir to its significance. This new miniaturized electronic technology allows for active socialization. Maybe then our large communication network is actually becoming like a world-sized fire accessed through 6 billion tiny hearths.

In terms of providing either a place of reverie or a focal point of solidarity for the local social ecology, however, these new hearths fail. As we've observed, our new technology can be very alienating, separating us from our immediate environment and people for the sake of connecting us to larger networks. But as Felix Guattari once reminded us, "it would be absurd to want to return to the past in order to reconstruct former ways of living"[14]. We cannot expect either ourselves or the world to be as they were. It is simply nostalgic to mourn the quiet reverie in front of the fire available to Rene Descartes in the seventeenth century. We have moved on, and as Foucault pointed out, we cannot simply blame these changes on technology. The hearth had provided us with a focus for socialization, for grounding us in the present time and place, for reverie. If we wish to retain some of these things, we must figure out how to incorporate them into our new technology. In the description of this workshop it was asked: "what is required to feel at home? Can this be digitized and delivered-on-demand to a mobile phone?" Here is the crux of the issue: can we write an app for existential belonging?

One consideration, given what has been said thus far, has to do with the Greek notion of Hestia and Hermes. Hestia focuses, brings in, while Hermes distracts, pulls out. Penelope concentrated on her loom while Odysseus wandered the surface of the earth. In the traditional Western house, the number and size of windows were limited and thus the relation between interior and exterior was clearly calibrated. With the intrusion of digital technologies into the house, the proliferation of screens amounted to a sudden proliferation of windows and with it a proliferation of this hermetic principle. When the television replaced the fireplace, according to this line of reasoning, this was paramount to the uprooting of the hestian almost completely by the hermetic. This is one major obstacle to the conversion of the smartphone into a hearth – it is in fact a window, a window through which we can see all kinds of useful and interesting things, through which we can be seen and through which we can communicate. Still, however, it is a window, not a hearth.

Perhaps then it could be a window to a hearth. The hearth at 'home', whatever form it may take in the future, could be visible through the small technological window that we carry in our pockets. Whenever you were lonely, perhaps in a hotel in far-flung location or waiting at an airport, and you were tired of your nomadism, you could set up your phone's built-in projector targeted upon a nearby surface and bring your far-off hearth into focus. Alternately, perhaps our smartphones could become a focus for localized socialization, a modern version of the innkeeper's hearth around which weary travelers could gather and swap stories. Maybe several nomads together on a commuter train, rather than reverting to private worlds accessed through private technology, could assemble their miniature hearths into something new, something communal.

We can see the dilemma at hand. Domesticity is a social and cultural phenomenon not easily quantified or digitized. This is an issue that architects deal with perpetually as we try to work with the social and the cultural in our very physical mode of operation. I hope that this case study of the hearth, from an architect's perspective, has helped to elucidate some of the intricacies of this dilemma and will aid in framing the pursuant discussion.

References

1. Rybczynski, W.: Home: A short history of an idea. Penguin Books, New York (1986)
2. Postman, N.: 5 Things We Need to Know About Technology. Denver, Colorado (1998)
3. Mitchell, W.J.: Me++: The cyborg self and the networked city. MIT Press, Cambridge (2003)
4. Schafer, A., Victor, D.G.: The Future Mobility of the World Population. Transportation Research 3(34), 171–205 (2000)
5. Eurofound, Krieger, H.: European Population Mobility. Technical Report, Eurofound, Dublin (2008)
6. Pew Internet and American Life Project, http://www.pewinternet.org/
7. Stone, B.: Breakfast Can Wait: The Day's First Stop Is Online. New York Times, New York, August 9 (2009)
8. Foucault, M.: Space, Power, Knowledge. In: During, S. (ed.) The Cultural Studies Reader, pp. 134–141. Routledge, New York (1993)
9. Verbeek, P.: What things do: Philosophical reflections on technology, agency, and design. Pennsylvania State University Press, University Park (2005)
10. Turkle, S.: Evocative Objects. MIT Press, Cambridge (2007)
11. Cirlot, J.E.: A Dictionary of Symbols. Routledge, London (1971)
12. Bachelard, G.: The Psychoanalysis of Fire. Beacon Press, USA (1968)
13. Vernant, J.P., Vernant, Myth, J.P.: Thought among the Greeks. Routledge, London (1983)
14. Guattari, F.: The Three Ecologies. Transaction Publishers, Somerset (2005)

Augmented Public Transit: Integrating Approaches to Interface Design for a Digitally Augmented City

J. Alejandro Lopez Hernandez

John H. Daniels Faculty of Architecture Landscape & Design,
University of Toronto, 230 College St.Toronto, ON
M5T 1R2, Canada
alejandro.lopezhernandez@utoronto.ca

Abstract. The contemporary body is a connected body that can instantly access home, work, leisure, and information via rapid mobility and communication networks, a "tethered" body as Sherry Turkle would put it. The ever-intrusive reach of technology into our social and physical environments and our bodies has given birth to the idea of the cybernetic organism (cyborg) or post-human (Hayles, 1999). In the discipline of Architecture the cyborg has provoked wild speculation and disembodied fantasies, yet little attention has been given to the micro realities of digital interfaces and to the discipline of user centric design, or to the empirical study of vast interactive information systems already in place across the urban environment. This paper will argue that integrated approaches between architecture, interaction design and other disciplines are key to meaningful interventions within an augmented urban environment. This argument will be illustrated by a recent interface design project dealing with usability and public transit, conducted at the Knowledge Media Design Institute at the University of Toronto.

Keywords: Interface design, Architecture, portable information, public transit information systems.

1 Introduction

Mobile technology has proved to be a powerful means for extending the subject's domestic sphere into the public domain and vice versa. Given the complex and contested relationship between public and private realms, Architects and designers often face the complex question of how to intervene in a way that is both meaningful and responsive to user's needs. One approach would be to engage readily available technologies such as the PC and the mobile phone and to establish new connections with existing urban information systems. Rather than focusing on novel concepts that require both new technology and new ways of gathering and accessing information, urban designers, architects and interface designers could collaborate on opening up existing datasets of information latent within urban systems to the individual user interface. This very approach was used in a recent project: Toronto Transit Commission (TTC) Trip Planner and User Profile. Our team designed and tested an online trip planner and user profile designed to improve the information gathering experience of users, which would thus

P. Daras and O. Mayora (Eds.): UCMedia 2009, LNICST 40, pp. 293–297, 2010.

decrease users' uncertainty of schedules, routes and wait times for user of public transit in Toronto. Central to this project was the idea of public Transit as an urban interactive information system connecting a central website/database of transit information to public interfaces such as bus stops, and private interfaces such as personal computers and mobile devices. The project was developed using questionnaires, iteration of high and low fidelity prototypes and current usability testing methods.

2   TTC Trip Planner and User Profile

The Toronto Transit Comission (TTC) Trip Planner and User Profile is the product of an interface design challenge addressing the issue of sustainable transportation. We targeted a young professional demographic living in the central core of the city as the end-users of this interface. The only requirement was that they be non-students between the ages of 19-44. We reasoned that this demographic would be the most likely to be making long-term decisions on means of transportation (i.e. purchasing an automobile versus using public transit or a bicycle). We created an online survey to gather information regarding the use of public transit by the target end-users. An important finding of this survey was that the TTC website was the primary source of information in planning unfamiliar journeys. Secondly, the questionnaire demonstrated that users were discouraged by the lack of comprehensive information regarding schedules and wait times, resulting in an overall feeling of uncertainty when using the system.

The group brainstormed a series of possible scenarios responding to the issue of uncertainty. The final sketch illustrated a GPS based vehicle information system that is accessed via three interfaces: web, PDA, and interactive bus stops. This diagram demonstrated the potentially expansive scope of the project and recognized that the existing GPS technology of the TTC's vehicles can be accessed through various information interfaces. The focus was consequently narrowed to one of the three suggested interfaces: a web-based trip planner. A review of current literature on the design of Trip-Planners helped guide our approach to the design of the interface. "Design of a Map-Based Transit Itinerary Planner" (Cherry, Hickman & Garg, 2006) provided a broad review of useful points regarding the design of a trip planner. Of particular relevance was the discussion on displaying graphical information and the importance of systems mimicking user decision-making processes. "Understanding the relationship between physical and virtual representations of transit agencies" (Yi, Rasmussen, & Rodrigues, 2007) provided additional context on transit website design in North America. It argues that the user's cognitive costs are reduced by good interface and design; i.e. if the user is only required to input minimal information and the system does the heavy cognitive lifting. The article argues that the time and effort saved by the user when searching for transit information on a website serves as an incentive to use public transit more frequently, which relates directly to the goal of this project. These insights helped refine the design of our low fidelity prototypes alongside diagrams describing the flow of tasks to be carried out.

The third low-fidelity prototype, similar to a website wire-frame, was used for Wizard of Oz testing. This version was designed with the intent to test two tasks: planning a trip and creating a user profile. The trip planner mimics a typical decision making process, and is modeled after airlines websites and other existing trip planners. The public Home Page of the Trip Planner is composed of three stages: (1) enter origin and destination; (2)

Refine search by travel method, time, walking time; (3) view search results. The Search Results page provides options that distinguish our prototype from currently existing solutions, namely a "save to my user profile," "send to text msg" and "send to PDA" functions. It was an important design goal to allow for the creation of a personal profile and to allow for the portability of query results via portable interfaces such as cell-phones and hand-held devices. The personal Home Page of this prototype included a field showing saved trips and travel warnings. The use of a personal profile would also allow users to save their PDA or cell phone number and store current location information via an IP address locator. This latter feature sets our prototype apart from other trip planners, such as the ones in place in Montreal and San Francisco.

The first high fidelity prototype, a website, was used to conduct a usability test with three end-users deemed representative of our target demographic. The testing was facilitated by the four members of the team, one serving as the computer, and three others as observers noting various aspects of the user reactions. Various aspects of the trip planner were static including pull-down menus in the Refine Search Results page as well as some radio buttons. Other functions such as the "send to pda," "send to text," "e-mail trip," and "locate me" via IP functions were included as buttons but were not fully implemented at this stage. This made the computer role a crucial one, requiring verbal prompting and guidance in the flow of some tasks. Moreover, this limited the scope of the usability test to tasks related to the core Trip-Planner and personal Home Page. The first task was designed to assess if users would enter the origin and destination via the Home Page or chose to create a Personal Profile and conduct this task within it. This task was also designed to evaluate the usability of the registration process. The second task focused on the user profile, with the objective being to evaluate the layout and organization of the information, including location and wording of icons and buttons. The tests lasted 20 minutes. One observer noted the body language as well as on-screen mouse movements, another observer noted the comments of the participants, and the fourth observer recorded the session on video in order to facilitate the transcription of comments.

The observations made during the usability tests provided insights into potential problems regarding the flow of the tasks vis-à-vis the decision making process of the user, and were therefore used to make changes to the visual organization and hierarchy of the Website interface. The design team felt that the TTC Trip Planner project met the requirements laid out in the initial proposal and that its design through iterative cycles and attention to user requirements and input produced a valuable and unique tool that could positively impact the use of sustainable transportation. Having addressed the Trip Planner web interface and its associated personal Profile, it was noted that the next possible steps in meeting the vision of an public transit information System could be the design of an integrated smart-phone application as well as integrated interactive bus stop information system. The methods of usability testing allowed us to evaluate how a fully functioning website could be implemented without investing heavily in full-functionality at the early design stage. Moreover, this approach, weighed more heavily towards Wizard of Oz techniques, proved manageable for a multidisciplinary team of Information Studies students and an Architecture student, all lacking the technical know-how to implement a fully working prototype. This method proved both economic in terms of time expended and technical resources. Its results can be seen as a tested and workable framework onto

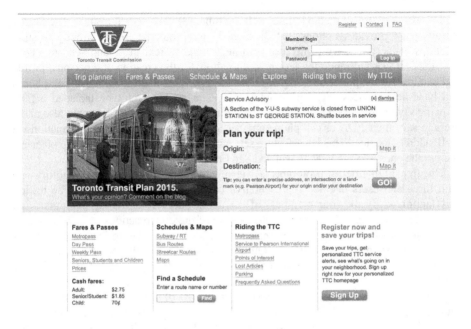

Fig. 1. Trip Planner Interface Home Page

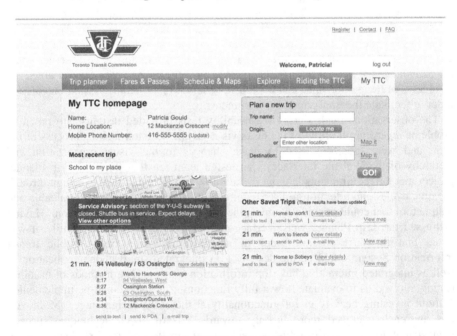

Fig. 2. TTC User Profile Page

which fully functioning features, particularly those regarding to data portability, can be subsequently incorporated.

3 Conclusion

This project is an example of how a multidisciplinary group can collaborate using the methods of user centric design to provide a more integrated and user-friendly public transit information system. The presence of different disciplines within the design team proved fruitful particularly in the brainstorming and research phases of the project. It is at this stage that potentially innovative approaches were identified, in this case that of an umbrella public transit information system that is accessible through a range of information interfaces public and private, stationary and portable. Information is not an elusive ether of 0's and 1's that is independent of its context. Rather, information is embedded in a temporal, spatial and cultural environment of distributed cognition, which we access and contribute to in a myriad of ways. This type of work poses a unique multidisciplinary approach to Interface Design research that considers information as a spatial-temporal phenomenon engaged in a feedback loop with humans wielding technological prostheses, one that can also be embedded and accessed in urban spaces.

References

1. Turkle, S.: Sensorium: embodied experience, technology, and contemporary art / edited by Caroline A. Jones. MIT Press, Cambridge (2006)
2. Hayles, N.K.: How we became posthuman: virtual bodies in cybernetics, literature, and informatics xiv, 350 p. (1999)
3. Ratti, C., Berry, D.: Sense of the City: Wireless and the Emergence of Real-Time Urban Systems. In: Châtelet, V. (ed.) Interactive Cities. Orléans, Editions HYX (2007)
4. Spatial Information Design Lab,
   http://www.spatialinformationdesignlab.org/
5. Yi, C., Rasmussen, B., Rodrigues, D.: Understanding the relationship between physical and virtual representations of transit agencies. Transportation Planning and Technology 30(2), 225–247 (2007)
6. Bork, C., Francis, B.: Developing Effective Questionnaires. Physical Therapy 65(6), 907–911 (1985)
7. IBM Design. (n.d.). Design Principles Checklist,
   https://portal.utoronto.ca/webapps/portal/
   frameset.jsp?tab_id=_2_1&url=%2Fwebapps%2Fblackboard%
   2Fexecute%2Flauncher%3Ftype%3DCourse%26id%3D_460946_1%26url%3D
8. STM Tous Azimuts,
   http://www2.stm.info/azimuts/carte.wcs?eff=OD&lng=a
9. Municipal Transportation Agency Home,
   http://www.sfmta.com/cms/home/sfmta.php
10. Holzinger, A.: Usability Engineering Methods for Software Developers. Communications of the ACH 48(1), 71–74 (2005)
11. Cherry, C., Hickman, M., Garg, A.: Design of a Map-Based Transit Itinerary Planner. Journal of Public Transportation 9(2), 45–68 (2006)
12. Bélanger, Lopez Hernandez, Mori, McCann.: Toronto Transit Commission Trip Planner and User Profile. KMDI, University of Toronto (2008)

Memoirs of Togetherness from Audio Logs

Danil Korchagin

Idiap Research Institute,
P.O. Box 592, CH-1920 Martigny, Switzerland
Danil.Korchagin@idiap.ch

Abstract. In this paper, we propose a new concept how tempo-social information about moments of togetherness within a social group of people can be retrieved in the palm of the hand from social context. The social context is digitised by audio logging of the same user centric device such as mobile phone. Being asynchronously driven it allows automatically logging social events with involved parties and thus helps to feel at home anywhere anytime and to nurture user to group relationships. The core of the algorithm is based on perceptual time-frequency analysis via confidence estimate of dynamic cepstral pattern matching between audio logs of people within a social group. The results show robust retrieval and surpass the performance of cross correlation while keeping lower system requirements.

Keywords: Time-frequency analysis, pattern matching, confidence estimation.

1 Introduction

The TA2 project (Together Anywhere, Together Anytime) is concerned with investigation of how multimedia devices can be introduced into a family environment to break down technology, distance and time barriers. How can we feel at home in a world where millions of people are in continual movement all around the world? How can we help to nurture social relationships? This is something that current technology does not address well: modern media and communications serve individuals best, with phones, computers and electronic devices tending to be user centric and providing individual experiences. In this sense, we are interested in breaking down the barrier between user centric and group centric media, in creation of mobile domesticity which can automatically generate memoirs of social interactions and fill the gap between user centric media devices and social networks.

Many of our enduring experiences, holidays, festivals, celebrations, concerts and moments of fun are kept as social memoirs. Additional media about these memoirs can be easily retrieved via services-on-demand from social networks. How can we automatically filter out and map only relevant information for personal memoirs of togetherness?

Nowadays more and more users start to use audio logging available in many palm devices. Can we profit from audio logs to augment a distributed domestic space with memoirs of social interactions? Most of the people do not intend to disclose private information and the purpose of each of audio log is primarily personal.

P. Daras and O. Mayora (Eds.): UCMedia 2009, LNICST 40, pp. 298–301, 2010.

The present investigation concerns the possibility of multiple audio log (recorded by user centric devices such as mobile phones and camcoders) synchronisation for automatic generation of memoirs of togetherness for personal archives. User centric devices do not normally provide such functionality. Further, if people do not share the same acoustic field or the devices are used inside big buildings, the GPS information cannot be used to reliably log social interactions. This leaves us with the audio signal [1] from which to infer a social context [2].

2 Audio Log Processing

Audio logs from user centric devices can be up to 24 hours per day. It is normal in such situations to reduce the initial large quantity of raw audio data, retaining only useful information about social context. In our work we use Mel Frequency Cepstral Coefficients (MFCC) [3] with a 10 ms frame rate. MFCC is a perceptually motivated spectrum representation that is widely used in acoustic and speech processing.

Audio logs are resampled to mono 16 kHz and then processed in pairs within the group of socially connected people on per day basis. External audio log in each pair is divided into subsamples of 30 seconds length each. This has the effect of removing a clock skew problem between different devices (within possible 0.03% range). Presumably the long subsamples could become misaligned, in which case additional techniques such as dynamic time warping [4] should be taken into account. Though some information can be retrieved via high-level modelling [5], we consider only low-level operating modes, one the well-known cross correlation and the other pattern matching based on ASR-related features [6].

Cross correlation is a measure of similarity of two waveforms as a function of a time-lag applied to one of them. It can be used to search a long duration signal for a shorter. Corresponding confidence is taken as a proportion between maximum of cross correlation product and its standard deviation. To get real-time computational efficiency we apply the convolution theorem and the fast Fourier transform, also known as fast cross correlation.

In case of pattern matching based on ASR-related features audio is pre-emphasised to flatten the spectral envelope and 13 Mel Frequency Cepstral Coefficients are retrieved in steps of 10 ms. The mel-scale corresponds roughly to the response of the human ear. The truncation to the lower 13 dimensions retains the spectral envelope and discards excitation frequency. Next, cepstral mean normalisation is performed for removing convolutional channel effects. Finally, if the norm of a vector of the 13 mean normalised cepstral coefficients (energy along with 12 cepstra representing the general spectral envelope) is higher than 1, then the vector is normalised in Euclidean space. This gives us the reduced variance of the search distance space. Corresponding confidence is taken as a proportion between best and worst relative distances from expectation in Euclidean space between corresponding audio logs. While having real-time computational efficiency, this approach requires much less RAM (15 MB versus 3 GB for fast cross correlation per 1 hour log).

3 Experimental Results and Visualisation

All results presented in this paper were achieved on a real life dataset of 10 social events (up to 1.5 hour each) with total 236 recorded subsegments (superposition of these events gives us 2360 possible combinations on the level of subsegments analysis), group of 8 socially connected people (who were using audio logging within the events) with personal audio-enabled palm devices (mobile phones and camcorders from 7 different manufactures). In figure 1 we illustrate how the length of the test subsegments influences the performance (the number of correctly clustered subsegments divided by the total number of test subsegments).

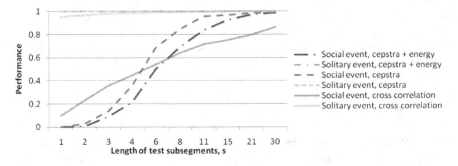

Fig. 1. Performance versus subsegment length of matching social context for the events with socially connected people and the events with no socially connected people involved

We were used the fixed confidence threshold equals to 50% of subjective confidence, which is higher than the worst confidence for solitary events. This has the effect of minimising false detection of social events, though the application of dynamic threshold selection can further increase the total performance. The performance of shorter subsegments is lower due the real world variability of the data (noise, reverberation, non-stationarity, etc).

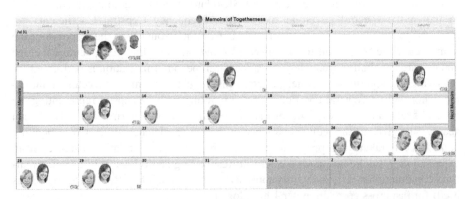

Fig. 2. Example of possible visualisation. Different socially connected people are automatically mapped on per day basis into personal memoirs of togetherness.

Figure 2 illustrates one of possible applications, targeting remote families (or any group of socially together people). When audio log is synchronised with an application the additional information about external media resources availability can be automatically retrieved via services-on-demand from social networks and mapped into the same personal memoirs, simplifying the navigation in tempo-social domain.

4 Conclusion

We have shown how the gap between user centric media devices and services-on-demand from social networks can be filled by automatic generation of tempo-social memoirs of togetherness. We found that the reliable memoirs can be generated from relatively short subsegments represented by small number of normalised cepstra. We have estimated that results surpass the performance of fast cross correlation, while requiring less resources. The achieved results give us the green light to further evaluate the presented concept from the privacy and the anxiety issues concerning being recorded everywhere and all the time, to concentrate on better understanding of the relations between psychoacoustic perception and social signal processing, to search for optimal ways of unobtrusive integration with existing applications.

Acknowledgments. The research leading to these results has received funding from the European Community's Seventh Framework Programme ICT Integrating Project "Together Anywhere, Together Anytime" (TA2, FP7/2007-2013) under grant agreement no. ICT-2007-214793. I should like to extend my gratitude to Philip N. Garner, Herve Bourlard and John Dines for their valuable help and support at various stages of this work.

References

1. Wyatt, D., Choudhury, T., Kautz, H.: Capturing Spontaneous Conversation and Social Dynamics: A Privacy-Sensitive Data Collection Effort. In: Proceedings of International Conference on Acoustics, Speech, and Signal Processing (ICASSP), Honolulu, USA (2007)
2. Farrahi, K., Gatica-Perez, D.: What Did You Do Today: Discovering Daily Routines from Large-Scale Mobile Data. In: Proceeding of the 16th ACM International Conference on Multimedia, Vancouver, Canada (2008)
3. Mermelstein, P.: Distance Measures for Speech Recognition, Psychological and Instrumental. In: Chen, C.H. (ed.) Pattern Recognition and Artificial Intelligence, pp. 374–388. Academic, New York (1976)
4. Ning, H., Roger, B.D., George, T.: Polyphonic Audio Matching and Alignment for Music Retrieval. In: 2003 IEEE Workshop on Applications of Signal Processing to Audio and Acoustics, New York, USA, pp. 185–188 (2003)
5. Choudhury, T., Pentland, A.: Sensing and Modeling Human Networks using the Sociometer. In: Proceedings of the 7th IEEE International Symposium on Wearable Computers, Washington, DC, USA (2003)
6. Korchagin, D.: Out-of-Scene AV Data Detection. In: Proceedings of the IADIS International Conference on Applied Computing, Rome, Italy (2009)

MinUCS 2009

Session 1

Automating Financial Surveillance

Maria Milosavljevic[1], Jean-Yves Delort[1,2], Ben Hachey[1,2],
Bavani Arunasalam[1], Will Radford[1,3], and James R. Curran[1,3]

[1] Capital Markets CRC Limited, 55 Harrington Street, Sydney, NSW 2000, Australia
[2] Centre for Language Technology, Macquarie University, NSW 2109, Australia
[3] School of Information Technologies, University of Sydney, NSW 2006, Australia
{maria,jydelort,bhachey,bavani,wradford,james}@cmcrc.com

Abstract. Financial surveillance technology alerts analysts to suspicious trading events. Our aim is to identify explainable false positives (e.g., caused by price-sensitive information in company news) and explainable true positives (e.g., caused by ramping in forums) by aligning these alerts with publicly available information. Our system aligns 99% of alerts, which will speed the analysts' task by helping them to eliminate false positives and gather evidence for true positives more rapidly.

Keywords: Financial Surveillance, Document Categorisation, Machine Learning, Sentiment Analysis.

1 Introduction

Systems for detecting trading fraud are currently used by exchanges to help manage market integrity and by trading houses to help manage compliance. These systems raise alerts based on trading history and heuristic patterns. For example, a rapid change in price with respect to historical trends might indicate market manipulation (e.g., ramping through forum posts or spam emails encouraging trades). On the other hand, these unexpected changes may be caused by legitimate price-sensitive information (e.g., earnings announcements to the exchange, macro-economic news). Exchanges and trading houses incur substantial expense employing analysts to determine whether alerts indicate unsanctioned trading that should be flagged for investigation or prosecution.

The vast majority of surveillance alerts are "explainable" via publicly-available information such as a company announcement, a news article or a post on a forum. That is, there is a high likelihood that particular information is responsible for causing the price change which led to the alert. For alerts that are not explainable, an analyst must decide whether the matter requires further investigation and is cause for prosecution.

We explore the extent to which information in the marketplace can be used to explain behavior which is identified by current alerting software. We find that approximately 29% of short-term price alerts are potentially explained by company announcements. A further 13% of alerts are aligned with company-specific news or forum postings. By analysing the relationships between companies, both in terms of sector influences and other forms of relationships, we can successfully align information to alerts in 99% of cases.

P. Daras and O. Mayora (Eds.): UCMedia 2009, LNICST 40, pp. 305–311, 2010.
© Institute for Computer Sciences, Social-Informatics and Telecommunications Engineering 2010

2   Background and Motivation

In an efficient market, informed investors must act on information quickly to be rewarded for their attentiveness. It has long been established that information drives investment decisions [6,8] and that informed individuals are compensated [7]. There is a recent growth of interest in measuring the impact of information on the financial markets, both retrospectively [2,13] and for prediction [9,14]. Language technologies such as sentiment detection ([4,3]) have become a popular area of research in this domain. In such cases, time is critical because stock prices effectively convey information from informed investors to the uninformed, that is, when informed investors observe information which they believe will drive the price up, they bid its price up [7]. Uninformed investors may observe this price change and act accordingly or may completely miss the opportunity to trade.

Surveillance analysts attempt to identify people behaving inappropriately with information in the marketplace. On the one hand, insiders trade on information which is not yet public which in turn affects the stock price prior to the public announcement [1]. On the other hand, investors manipulate the market by circulating unfounded information such as rumors [10]. Forums are a common venue for publishing inappropriate content and [5] has demonstrated the impact of such content on the market. Surveillance software (such as SMARTS[1]) identifies suspicious patterns in trading data and reports alerts to analysts.

We aim to automate some of the tasks which a surveillance analyst performs. A successful solution to this problem would involve supporting the analyst by:

- explaining false positive alerts, e.g. movement due to company announcements or macro-economic news, to eliminate the time spent by analysts on these;
- explaining true positive alerts, e.g. ramping in forums or spam emails, to expedite the collection of relevant information for further investigation;
- identifying market manipulation in text that cannot be detected from anomalous trading behaviour, e.g. unsuccessful or subtle ramping in forums.

We focus here on addressing the first two problems in the Australian market.

3   Data

A substantial component of our activities has been federating and processing the many sources of trade and text data, and meta-data, available in the finance domain into an experimental framework. This turned out to be surprisingly difficult because of the need to combine text and trading data at fine granularity and over such large scales. The remainder of this section describes the main data sources used in the experiments reported in this paper.

Alerts. Alerts represent unusual trading activity for a given financial instrument compared to an historical benchmark. We use Australian Securities Exchange (ASX) trading data from SIRCA's Taqtic service[2], which includes aggregated price

---

[1] http://www.smartsgroup.com/
[2] http://www.sirca.org.au/

and volume information for best bids (the price a purchaser is willing to pay) and best asks (the price a seller is willing to accept) at any given time. The alerts are generated using the SMARTS tool suite. In particular, short term price movements are generated if a price change over 15 minutes exceeds certain thresholds. This price change value is compared to 1) a minimum threshold (3%), 2) a scaled standard deviation threshold (4σ) based on historical data from the preceding 30 calendar days, and 3) a reissue threshold that governs when an alert is re-shown to the analysts. If these thresholds are exceeded, then an alert is generated indicating unusual price movement. The issue time associated with an alert is the same as the trade that triggered the alert.

Company Announcements. The first source of textual information we use is ASX company announcements. As a condition of listing on the ASX, companies are required to comply with various listing rules aimed at protecting market integrity. Among these is the continuous disclosure rule,[3] which states: "Once a company is or becomes aware of any information concerning it that a reasonable person would expect to have a material effect on the price or value of the company's securities, the entity must immediately tell ASX that information." Therefore, any unusual price-movement based on information from within a company should be preceded by an announcement. ASX announcements are obtained through SIRCA and have meta-data including broadcast time, associated ticker(s), and the announcement category, e.g. a change in directors notice. The ASX also labels announcements as price sensitive. However, we believe this labelling is oriented towards high recall because the ASX would not want to mark an announcement incorrectly as not being price sensitive. In Section 5, we report results on reproducing this labelling.

Reuters Newswire. The second source of textual information we use is news from the Reuters NewsScope Archive (RNA),[4] also obtained through SIRCA. Each RNA story is coded with extensive meta-data [12] including Reuters instrument codes (RICs), which are used to identify stocks, indices and tradeable instruments mentioned in a document. For instruments traded on the ASX, RICs are created by adding ".AX" to the end of the ASX ticker code (e.g., BHP.AX for BHP Billiton traded on the ASX). Each RNA story also has meta-data that indicates its relevance to Reuters topics (e.g., interest rates, corporate results), products (e.g., commodities) and entities (e.g., US equities diary). RNA stories comprise multiple broadcast events. For a typical story, this may consist of a news alert containing a concise statement of the key information followed by a story headline and body text, followed by further updates as the story unfolds [11].

Hot Copper Forum. The third source of information we use is content from Hot Copper, a discussion forum for the Australian stock market that currently

---

[3] http://www.asx.com.au/ListingRules/chapters/Chapter03.pdf
[4] http://thomsonreuters.com/products_services/financial/
financial_products/event_driven_trading/newsscope_archive

Table 1. Size of the alert, announcement, news and forum datasets by year

Year	Alerts	Announcements	RNA Events	Forum Posts
2003	5 365	65 233	92 419	
2004	7 043	80 570	86 955	246 338
2005	8 773	90 484	84 537	
2006	12 110	102 235		
2007		117 469		

has over 80,000 active members and more than 4,000 posts per day. We scraped the Hot Copper web site to obtain meta-data for each post including the time it was submitted, the ticker it is about, the poster and the thread it belongs to.

Table 1 shows the document counts for each type of data. The growth in market activity is evident from the substantial increases in alerts and official announcements between 2003 and 2007. This growth will need to be matched by greater resourcing of surveillance operations or smarter technology.

4   Aligning Information to Alerts

A document is aligned to an alert if there is a possibility that it is responsible for causing the price change which led to the alert. In other words, alignment characterises a potential causality relationship between a document and an alert. If causality is established then the document is said to contain "market sensitive" information or to be "price sensitive" for the ticker associated with the alert.

As noted previously, an efficient market adjusts to new information quickly, meaning that the price of a stock changes rapidly. The time period between the information being released and a resulting price movement is termed the document's "decay period". We have calculated that a one-hour decay period covers the behaviour of most stocks, so we use this as a cutoff for aligning documents to alerts. Our alignment strategies include the following:

Ticker alignment. A document and an alert are aligned if the document meta-data contains the alert ticker. This is the baseline alignment method.

Sector alignment. Many of the documents in our corpora do not have specific tickers listed in their meta-data. For example, 59% of RNA events which include the topic "Australia" are not associated with any ticker. We use statistical analysis to identify significant pairwise χ^2 correlations ($p < 0.01$) between RNA topic codes and sectors. Then, RNA documents which do not have tickers are labelled with multiple sectors according to the resulting rules. 5-fold cross-validation on the 2004 RNA data showed this technique achieves 90% precision, 94% recall, and 91% F-score. This resulted in 198 rules. The top four are:

> Telecommunications Services $\mapsto$ Telecommunication Services
> Pharmaceuticals, Health, Personal Care $\mapsto$ Health Care
> Non-Ferrous Metals $\mapsto$ Materials
> Banking $\mapsto$ Financials

Table 2. The coverage indicates the number of alerts that are aligned to at least one document of the given type following the given alignment strategy. The document types are company announcements (A), Reuters news (R) and Hotcopper forum posts (F).

Alignment scheme	Document type	Coverage (%)	Cumulative coverage (%)
Ticker	A	29	29
Ticker	R	2	29
Ticker	F	28	42
Sector	R	77	85
Firm	A+R+F	96	99

A document and an alert are aligned if the document sector matches the sector of the alert ticker.

Firm Relationships aligning alerts to documents which refer to related firms. Two firms can be related in many ways (e.g., partners, competitors, producer/ consumer, having board members in common). Consequently, a price sensitive announcement for a firm may impact the price of related firms. To date we have focused on identifying common sector (industry group) membership. A document and an alert are aligned if the document ticker and the alert ticker have the same sector.

The results for our alignment strategies are shown in Table 2. We can link 42% of alerts to ticker-specific documents. Adding in sector influences results in alignment of 85% of alerts to documents. Finally, by combining all three approaches, we can identify at least one document in the preceding hour which may be responsible for causing the market changes which led to an alert in 99% of cases. It is also worth mentioning that while Reuters news stories with tickers cannot be aligned to many alerts, Reuters news without tickers can be aligned to 50% of alerts using the sector-level information scheme.[5]

5   Price Sensitivity

We conducted an experiment on reproducing the price sensitivity labels for ASX announcements issued in 2004. We used Weka's Naïve Bayes classifier with unigram and bigrams from the title and body of the announcements as features. Infogain was used to select the top 2000 features that best discriminate between the price sensitive and non price sensitive announcements. A separate classifier was trained and tested for each of the ASX ANNOUNCEMENT TYPES. Results from 5-fold cross-validation are shown in Table 3. The overall F-measure achieved was 0.901, with good recall on the minority YES class.

---

[5] Evaluation of alignment accuracy depends on annotation of true positive and false positive alerts as well as annotation of alert-document alignments, which is a matter for future work.

Table 3. Results for price sensitivity classification on 2004 ASX announcements

Sensitive	Precision	Recall	F-Measure
YES	0.787	0.868	0.826
NO	0.950	0.914	0.931

6 Conclusion

This paper has presented some preliminary results towards our goal of automated financial surveillance. Our analysis demonstrates that automation will be critical for timely investigation as information sources and trade volumes in capital markets continue to grow rapidly.

We have identified the primary sources of textual information that can potentially explain, with up to 99% coverage, the alerts presented to ASX surveillance analysts. We have also shown that price sensitivity labels on ASX announcements can be reliably reproduced automatically. These are key stages in demonstrating that (semi-)automated financial surveillance is accurate and efficient.

References

1. Aitken, M., Czernkowski, R.: Information Leakage Prior to Takeover Announcements: The Effect of Media Reports. Accounting and Business Research 23(89), 3–20 (1992)
2. Antweiler, W., Frank, M.Z.: Is all that talk just noise? The information content of internet stock message boards. Journal of Finance 59(3), 1259–1294 (2004)
3. Chua, C.C., Milosavljevic, M., Curran, J.R.: A Sentiment Detection Engine for Internet Stock Message Boards. In: Proceedings of the Australasian Language Technology Workshop, ALTW (2009)
4. Das, S.R., Chen, M.Y.: Yahoo! for Amazon: Sentiment extraction from small talk on the web. Management Science 53(9), 1375–1388 (2007)
5. Delort, J.-Y., Arunasalam, B., Milosavljevic, M., Leung, H.: The Impact of Manipulation in Internet Stock Message Boards (submitted, 2009)
6. Fama, E.: Efficient Capital Markets: A Review of Theory and Empirical Work. Journal of Finance 25(2), 383–417 (1970)
7. Grossman, S.J., Stiglitz, J.E.: On the Impossibility of Informationally Efficient Markets. The American Economic Review 70(3), 393–408 (1980)
8. Mitchell, M.L., Mulherin, J.H.: The Impact of Public Information on the Stock Market. Journal of Finance 49(3), 923–950 (1994)
9. Mittermayer, M.: Forecasting Intraday Stock Price Trends with Text Mining Techniques. In: Proceedings of the 37th Annual Hawaii International Conference on System Sciences, HICSS 2004, p. 30064.2 (2004)
10. Pound, J., Zeckhauser, R.: Clearly Heard on the Street: The Effect of Takeover Rumors on Stock Prices. Journal of Business 63(3), 291–308 (1990)
11. Radford, W., Hachey, B., Curran, J.R., Milosavljevic, M.: Tracking Information Flow in Financial Text. In: Proceedings of the Australasian Language Technology Workshop, ALTW (2009)
12. Reuters NewsScope Archive v2.0: User Guide (2008)

13. Robertson, C., Geva, S., Wolff, R.: What types of events provide the strongest evidence that the stock market is affected by company specific news? In: Proceedings of the fifth Australasian conference on Data mining and Analytics, pp. 145–153 (2006)

14. Schumaker, R.P., Chen, H.: Textual analysis of stock market prediction using breaking financial news: The AZFin text system. ACM Transactions on Information Systems (TOIS) 27(2) 1–19 (2009)

MinUCS 2009

Session 2

Cross-Lingual Analysis of Concerns and Reports on Crimes in Blogs

Hiroyuki Nakasaki[1], Yusuke Abe[1], Takehito Utsuro[1],
Yasuhide Kawada[2], Tomohiro Fukuhara[3], Noriko Kando[4],
Masaharu Yoshioka[4], Hiroshi Nakagawa[3], and Yoji Kiyota[3]

[1] University of Tsukuba, Tsukuba, 305-8573, Japan
[2] Navix Co., Ltd., Tokyo, 141-0031, Japan
[3] University of Tokyo, Kashiwa 277-8568 / Tokyo, 113-0033, Japan
[4] National Institute of Informatics, Tokyo, 101-8430, Japan
[5] Hokkaido University, Sapporo, 060-0814, Japan
{nakasaki,utsuro}@nlp.iit.tsukuba.ac.jp

Abstract. Among other domains and topics on which some issues are frequently argued in the blogosphere, the domain of crime is one of the most seriously discussed by various kinds of bloggers. Such information on crimes in blogs is especially valuable for outsiders from abroad who are not familiar with cultures and crimes in foreign countries. This paper proposes a framework of cross-lingually analyzing people's concerns, reports, and experiences on crimes in their own blogs. In the retrieval of blog feeds/posts, we take two approaches, focusing on various types of bloggers such as experts in the crime domain and victims of criminal acts.

Keywords: cross-lingual blog analysis, blog feed retrieval, crime reports, Wikipedia.

1 Introduction

Weblogs or blogs are considered to be one of personal journals, market or product commentaries. Among other domains and topics on which some issues are frequently argued in the blogosphere, the domain of crime is one of the most seriously discussed by various kinds of bloggers. One type of such bloggers are those who have expert knowledge in the crime domain, and keep referring to news posts on criminal acts in their own blogs. Another type of bloggers who have expert knowledge also often post tips for how to prevent certain criminal acts. Furthermore, it is surprising that victims of certain criminal acts post blog articles on their own experiences. Blog posts by such various kinds of bloggers are actually very informative for both who are seeking for information on how to prevent certain criminal acts and who have been already victimized and are seeking for information on how to solve their own cases. Such information is especially valuable for outsiders from abroad who are not familiar with cultures and crimes in foreign countries. Based on this observation, this paper proposes a framework of cross-lingually analyzing people's concerns, reports, and experiences on crimes in their own blogs.

P. Daras and O. Mayora (Eds.): UCMedia 2009, LNICST 40, pp. 315–320, 2010.

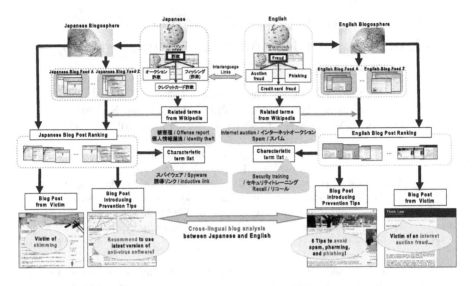

Fig. 1. Overall Framework of Cross-Lingual Blog Analysis

The overview of the proposed framework is shown in Figure 1. First, we start from terms which represent relatively narrow range of concepts of criminal acts. In this paper, we focus on *"fraud"* and *"Internet crime"*. Then, we refer to *Wikipedia* (English and Japanese versions[1]) as a multilingual terminological knowledge base, and search for Wikipedia entries describing criminal acts in the range of *"fraud"* or *"Internet crime"*. Then, from the collected Wikipedia entries, multilingual queries for retrieving blog feeds/posts (in this case *"auction fraud"*, *"credit card fraud"*, and *"phishing"*) are created, where interlanguage links are used for linking English and Japanese translated entries.

Next, in the retrieval of blog feeds/posts, we take two approaches. The first approach [1] focuses on collecting blog feeds rather than on directly collecting blog posts. Its major component is designed as the blog feed retrieval procedure[2] recently studied in the blog distillation task of TREC 2007 blog track [3]. In this first approach, we regard blog feeds as a larger information unit in the blogosphere. We intend to retrieve blog feeds which roughly follow the criterion studied in the blog distillation task, which can be summarized as *Find me a blog with a principle, recurring interest in X*. In the results of empirical analysis of this paper, we found that this first approach is mostly quite appropriate for discovering bloggers (i.e., blog feeds) who are deeply interested in watching news reports on cases of those criminal acts and warning others by referring to those

---

[1] http://{en,ja}.wikipedia.org/. The underlying motivation of employing Wikipedia is in linking a knowledge base of well known facts and relatively neutral opinions with rather raw, user generated media like blogs.

[2] Its detailed evaluation including improvement over the baseline as the original rankings returned by "Yahoo!" API and "Yahoo! Japan" API is published in [2].

news reports in their own blog posts. This first approach is also quite effective in discovering bloggers who introduce tips for how to prevent such criminal acts.

In the second approach, we simply use existing Web search engine APIs and follow the original ranking of blog posts given a query keyword. Generally speaking, we can not estimate the ranking strategy employed by Web search engine APIs. In principle, however, it is expected that this second approach generally returns blog posts which have more inlinks than other posts. Actually, throughout the empirical analysis of this paper, we found that this second approach returns blog posts by victims of those criminal acts more often than the first approach. Here, we observe that victims of criminal acts such as fraud and Internet crime sometimes post one or two articles to their own blogs just after they were victimized. Though, in most cases, those victims do not keep posting articles related to those crimes, and hence, their blog feeds are not ranked high in the result of blog feeds ranking by the first approach.

Finally, to retrieved English and Japanese blog feeds/posts, we apply our framework of mining cross-lingual/cross-cultural differences in characteristic terms within top ranked blog posts [1] (to be presented in section 3). This framework is originally designed for discovering cross-lingual/cross-cultural differences in concerns and opinions that are closely related to a given topic. With this framework, it becomes much easier for users to discover regional differences in criminal acts which originate from cultural differences. For example, recently, *"bank transfer scam"*, and especially *""it's me" fraud"*[3] is very frequent in Japan, while they are not very frequent in western countries.

2   Topics in the "Fraud / Internet Crime" Domain

In this paper, as topics in the domain of criminal acts, we focus on *"fraud"* and *"Internet crime"*. We first refer to Wikipedia and collect entries listed at the categories named as *"fraud"* and *"Internet crime"* as well as those listed at categories subordinate to *"fraud"* and *"Internet crime"*. Then, we require entry titles to have the number of hits in the blogosphere over 10,000 (at least for one language)[4]. Finally, for the category *"fraud"*, we have about 10 to 15 entries both for English and Japanese. For the category *"Internet crime"*, we have about 5 entries both for English and Japanese. Figure 2 shows samples selected from remaining entries[5]. In the figure, the category *"Internet fraud"* is an immediate descendant of both *"fraud"* and *"Internet crime"*, where three entries listed at

---

[3] The fraudster makes a phone call to the victim and pretends to be his/her child or grandchild, and then requests the victim to transfer funds to the fraudster's account.

[4] We use the search engine "Yahoo!" API (http://www.yahoo.com/) for English, and the Japanese search engine "Yahoo! Japan" API (http://www.yahoo.co.jp/) for Japanese. Blog hosts are limited to major ones, i.e., 12 for English and 11 for Japanese.

[5] For some of those selected samples, only English or Japanese term is listed as a Wikipedia entry. In such a case, translation is found in an English-Japanese translation lexicon Eijiro (http://www.eijiro.jp/, Ver.79, with 1.6M translation pairs).

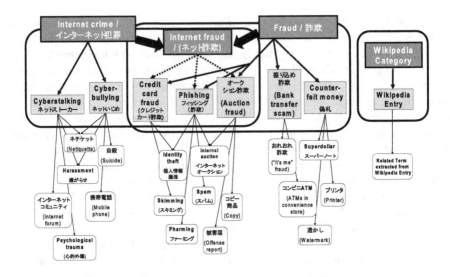

(Note: "A term t_x" ("a term t_y") in the nodes above indicates that t_y is not an entry in Wikipedia, nor extracted from any of Wikipedia entries, but translated from t_x by Eijiro.)

Fig. 2. Wikipedia Entries and Related Terms in the "Fraud / Internet Crime" Domain

Table 1. Statistics of "Fraud / Internet Crime"

ID	Wikipedia Entry	# of Cases (sent to the court in the Year of 2008)		# of Hits in the Blogosphere (checked at Sept. 2009)	
		U.S.A.	Japan	English	Japanese
1	Internet fraud	72,940	N/A	21,300	61,600
2	(Auction fraud)	18,600	1,140	1,760	44,700
3	(Credit card fraud)	6,600	N/A	43,900	8,590
4	(Phishing)	N/A		479,000	136,000
5	Bank transfer scam	N/A	4,400	30	349,000
6	Counterfeit money	N/A	395	16,800	40,500
7	Cyberstalking	N/A		20,300	32,100
8	Cyber-bullying	N/A		38,900	45,700

this category are selected as samples. The figure also shows samples of related terms automatically extracted from those selected Wikipedia entries. For those selected sample entries, Table 1 shows the number of cases actually sent to the court both in U.S.A and in Japan[6]. The table also shows the number of hits of those sample entry titles in the blogosphere.

---

[6] Statistics are taken from the Internet Crime Complaint Center (IC3), U.S.A. (http://www.ic3.gov/), and National Police Agency, Japan (http://www.npa.go.jp/english/index.htm)

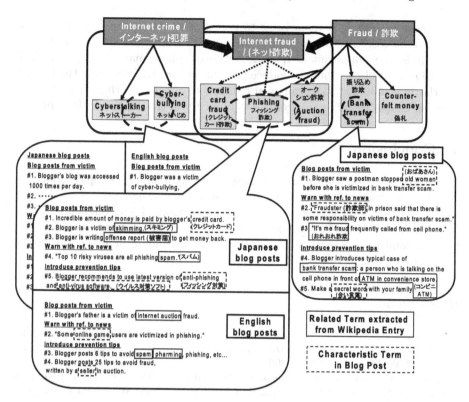

Fig. 3. Topics, Terms in Blogs, and Summaries of Blog Posts: Examples

3   Blog Analysis with Terms and Summaries of Blog Posts

For the sample Wikipedia entries shown in Figure 2, both English and Japanese blog feeds / posts are retrieved and ranked according to the first approach of blog feeds/posts ranking described in the previous section. Along with this result of blog feeds/posts ranking by the first approach, blog post ranking by the search engine "Yahoo!" API for English and the Japanese search engine "Yahoo! Japan" API for Japanese are used as the second approach of blog post ranking. Then, characteristics terms within top ranked blog posts are shown through an interface for mining cross-lingual/cross-cultural differences between English and Japanese [1]. The interface also has a facility of showing top ranked blog posts, which enables cross-lingual blog analysis quite efficiently. Figure 3 illustrates results of manual cross-lingual blog analysis, where summaries of blog posts are categorized into the following three types: (1) blog posts from a victim or one who personally knows a victim, (2) blog posts which warn others with reference to news posts on criminal acts, (3) blog posts which introduce how to prevent criminal acts. In the figure, samples of related terms automatically extracted from Wikipedia entries (those shown in Figure 2) are marked. Manually selected characteristic terms included in the blog posts are also marked.

In those results, it is remarkable to note that, both in the English and the Japanese blogosphere, many victims actually mention their cases in their own blogs. It is also important to note that we can collect many blog posts which refer to news posts or which introduce prevention tips. Differences in the English and the Japanese blog posts can be also detected, discovering that *"bank transfer scam"* is unique to Japan, and only was observed in the Japanese blogosphere[7].

4   Conclusion

This paper proposed a framework of cross-lingually analyzing people's concerns, reports, and experiences on crimes in their own blogs. There exist several works on studying analysis of concerns in multilingual news [4,5,6], but not in blogs. Future works for cross-lingual blog analysis on facts and opinions include incorporating multilingual sentiment analysis techniques and automatic extraction of reports or experiences of victims of crimes.

References

1. Nakasaki, H., Kawaba, M., Yamazaki, S., Utsuro, T., Fukuhara, T.: Visualizing Cross-Lingual/Cross-Cultural Differences in Concerns in Multilingual Blogs. In: Proc. ICWSM, pp. 270–273 (2009)
2. Kawaba, M., Nakasaki, H., Utsuro, T., Fukuhara, T.: Cross-Lingual Blog Analysis based on Multilingual Blog Distillation from Multilingual Wikipedia Entries. In: Proc.ICWSM, pp. 200–201 (2008)
3. Macdonald, C., Ounis, I., Soboroff, I.: Overview of the TREC-2007 Blog Track. In: Proc. TREC-2007 (Notebook), pp. 31–43 (2007)
4. Yangarber, R., Best, C., von Etter, P., Fuart, F., Horby, D., Steinberger, R.: Combining Information about Epidemic Threats from Multiple Sources. In: Proc. Workshop: Multi-source, Multilingual Information Extraction and Summarization, pp. 41–48 (2007)
5. Pouliquen, B., Steinberger, R., Belyaeva, J.: Multilingual Multi-document Continuously-updated Social Networks. In: Proc. Workshop: Multi-source, Multilingual Information Extraction and Summarization, pp. 25–32 (2007)
6. Yoshioka, M.: IR Interface for Contrasting Multiple News Sites. In: Li, H., Liu, T., Ma, W.-Y., Sakai, T., Wong, K.-F., Zhou, G. (eds.) AIRS 2008. LNCS, vol. 4993, pp. 508–513. Springer, Heidelberg (2008)

---

[7] We collect only 4 blog feeds and 13 blog posts from the English blogosphere, while we collect 132 blog feeds and 2617 blog posts from the Japanese blogosphere.

Automated Event Extraction
in the Domain of Border Security

Martin Atkinson[1], Jakub Piskorski[2], Hristo Tanev[1], Eric van der Goot[1],
Roman Yangarber[3], and Vanni Zavarella[1]

[1] Joint Research of the European Commission, 21027 Ispra (VA), Italy
{Firstname.Lastname,Firstname.Multi-word-lastname}@jrc.ec.europa.eu
[2] Frontex, Rondo ONZ 1, Warsaw, Poland
Firstname.Lastname@frontex.europa.eu
[3] University of Helsinki, P.O. Box 68, 00014 Helsinki, Finland
Firstname.Lastname@cs.helsinki.fi

Abstract. This paper gives an overview of an ongoing effort to construct tools for automating the process of extracting structured information about border-security related events from on-line news. The paper describes our overall approach to the problem, the system architecture and event information access and moderation.

Keywords: event extraction from on-line news, border security, text analysis, open source intelligence.

1   Introduction

Mining open sources for gathering intelligence for security purposes is becoming increasingly important. Recent advances in the field of automatic content extraction from natural-language text result in a growing application of text-mining technologies in the field of security, for extracting valuable structured knowledge from massive textual data sets on the Web. This paper gives an overview of an ongoing effort to construct tools for Frontex[1] for automating and facilitating the process of extracting structured information on border-security related events from on-line news articles. The topics in focus include illegal immigration incidents (e.g., illegal entry, illegal stay and exit, refusal of entry), cross-border criminal activities (e.g., trafficking, forced labour), terrorist attacks, other violent events (e.g., kidnappings), inland arrests in third countries, displacements, troop movements, man-made and natural disasters, outbreak of infectious disease, and other crisis-related events.

The need of strengthening capabilities for tracking the security situation in the source and target countries for illegal immigration into the EU has been identified and acknowledged by the European Commission (EC). Specifically, the Commission Communication COM (2008) 68 proposes the creation of an

---

[1] Frontex is the European Agency for the Management of Operational Cooperation at the External Borders of the Member States of the European Union.

P. Daras and O. Mayora (Eds.): UCMedia 2009, LNICST 40, pp. 321–326, 2010.

Integrated European Border Surveillance System (EUROSUR), where step 6 of Policy Option 1 suggests development and deployment of new tools for strategic information to be gathered by Frontex from various sources in order to recognize patterns and analyze trends, supporting the detection of migration routes and the prediction of risks for Common Pre-frontier Intelligence Picture (CPIP). The deployment of open-source intelligence tools for event extraction plays a significant role in this context. There are two end-users of such tools in Frontex, namely, the Risk Analysis Unit (RAU), whose task is to carry out border-security related risk analysis to drive the operational work of the agency, and Frontex Situation Centre (FSC) responsible for providing a constant and short-term picture of the situation, at the EU-external borders and beyond.

Taking into account RAU's and FSC's needs, there are some challenges to be tackled while developing tools for automated event extraction from on-line news. First, for the purpose of risk analysis it is crucial to extract as fine-grained event descriptions as possible, ideally including information on: event's type/subtype, date, precise location, perpetrators, victims, methods/instruments used (if applicable), number and characteristic of people affected, link to the source(s) from which the event was detected, and system's confidence in the automatically generated event description. Second, the capability of providing a live or near-live picture of border-security related events imposes that such tools allow for processing vast amount of news articles in real or almost real-time. Finally, it is crucial to be able to process news articles in many different languages, since a significant fraction of relevant events are only reported in non-English, local news, where Italian, Spanish, Greek, French, Turkish, Russian, Portuguese, and Arabic are the most important ones at the moment.

2   System Architecture

This Section gives an overview of the system architecture, which is depicted in Figure 1.

- First, news articles are gathered by a dedicated software platform for electronic media monitoring, the Europe Media Monitor (EMM)[2] developed at the JRC [1]. EMM currently retrieves 100,000 news articles per day from 2,000 news sources in 42 languages. Articles are classified according to about 700 categories and then scanned to identify known entities. Information about entity mentions is added as meta-data for each article.
- The news articles (harvested in a 4-hour time window) are grouped into clusters according to content similarity. Then each cluster is geo-located, and clusters describing events relevant for the border security domain are selected using keyword-based heuristics. These clusters constitute a small portion (between 1-2% on average) of the stream retrieved by EMM. The clustering process is performed every 10 minutes.
- Next, two event extraction systems are applied on the stream of news articles.

---

[2] http://press.jrc.it

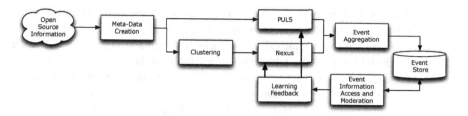

Fig. 1. Event extraction System Architecture

- The first extraction system, *NEXUS*, follows a cluster-centric approach [3,2]. Each cluster is processed by an event extraction engine, which performs shallow linguistic analysis and applies a simple cascade of finite-state extraction grammars[3] on each article in the cluster. The system processes only the top sentence and the title of each article for the following reasons: (a) news articles are written in the "inverted-pyramid" style, i.e., the most important parts of the story are placed in the beginning of the article and the least important facts are left toward the end; (b) processing the entire text might involve handling more complex language phenomena, which is hard and requires knowledge-intensive processing; (c) if some crucial information has not been captured from one article in the cluster, it might be extracted from other articles in the same cluster. Since the information about events is scattered over different articles, the last step consists of cross-article cluster-level information fusion in order to produce full-fledged event descriptions, i.e., information extracted locally from each single article in the same cluster is aggregated and validated. *NEXUS* detects and extracts only the main event for each cluster. It is capable of processing news in several languages, including, i.a., English, Italian, Spanish, French, Portuguese, and Russian. Due to a linguistically light-weight approach *NEXUS* can be adapted to processing texts in a new language in a relatively short time [7,8].
- The second system, *PULS*[4], uses a full-document approach, described in, e.g., [5,6]. This extraction component exploits a similar pattern-based technology to analyze the news articles. It *first* analyses one document at a time, to fill as many event templates as possible for each given document, and then attempts to unify the extracted events into "groups" in a subsequent phase, across different articles. The document-local analysis covers the entire text of each article, which aims to obtain not only the current information, but possibly links to background information—typically reported further down in the article—as well. PULS currently performs processing in English and French, with plans to extend to additional languages in the near future.

---

[3] The grammars are encoded and processed with *ExPRESS*, a highly efficient extraction pattern matching engine [4].
[4] http://puls.cs.helsinki.fi/medical/

NEXUS follows a cluster-centric approach, which makes it more suitable for extracting information from the entire cluster of topically-related articles. *PULS* performs a more thorough analysis of the full text of each news article, which allows us to handle events for which not much information has been reported.

The two approaches are deployed for event extraction in order to get richer coverage. Also, the two extraction components are (at present) tuned to detect sets of event types that are not entirely overlapping. In particular, *NEXUS* has been mainly deployed for detecting violent incidents and natural and man-made disasters, whereas *PULS* has been primarily customized for the epidemics domain. The combined system is designed to compare and experiment with different approaches to event extraction on the same data.

Although no evaluation of the combined system has been performed yet, information on the current performance levels of *NEXUS* and *PULS* can be found in [3,2,7,9].

3   Event Information Access and Moderation

Event moderation provides the bridge between the automated system and more in-depth situation analysis, which requires clean and reliable data. Moderated event structures also provide feedback to machine-learning components of the automated system. The moderation system enables selection and visualization

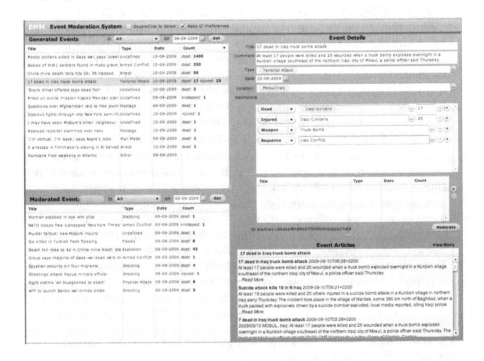

Fig. 2. Event moderation tool

of the generated events, verification of their slot values against the corresponding original document, comparison of new events with previously moderated events, in order to create event chains, and saving in a dedicated information store.

Figure 2 shows a preliminary front-end of the moderation system. The top-left pane shows automatically generated events; the bottom-left pane shows other moderated events from the same time period; the right-hand panes shows the detailed slot values of the selected event, and its link to other events and the original sources from which the events were generated.

In order to provide a near real-time geo-spatial visualisation of the 'current' events (which is important for the creation of the short-term situational picture) the system produces a stream of new event descriptions in Google's KML format every 10 minutes, which is then passed to a *Google Earth* visualisation.[5]

4 Outlook

We have outlined a system architecture based on leading-edge technologies in information extraction applied to the domain of Border Security Intelligence. The main challenge here is to minimize information overload without overlooking weak, but potentially important, signals.

Although automated news surveillance in the domain of border security is new (we are not aware of prior work in this domain), it bears some similarity to previously studied domain of epidemic surveillance, e.g., MedISys and PULS ([9]), HealthMap ([10]), and BioCaster ([11]). In fact, epidemic surveillance is in part (though not entirely) subsumed by the security domain, since the spread of epidemics impacts border security as well. The event schema used in epidemic surveillance is similar to that used in the border security domain, in that it tries to cover the victim descriptions, the cause of harm, etc. In the security domain, the schema is considerably more complex, and requires covering many similar and partly overlapping event types, as described in section 1. It therefore exhibits a higher level of complexity of text analysis.

Because the project is in an early phases, quantitative evaluations are not yet available. We believe that adopting a combination of different approaches to information extraction and aggregating this information via moderation into an information store, will provide an important step toward meeting this challenge. Technical challenges include automated detection of duplicate events, location of event boundaries and linking of event sequences. Thanks to the close collaboration with two groups of end-users at Frontex we will be able to closely monitor the extent to which these challenges are met, as well as the features and usability of the applications as they develop.

---

[5] A subset of the event descriptions automatically generated by the system is publicly accessible by starting *Google Earth* application with KML: http://press.jrc.it/geo?type=event&format=kml&language=en. For other languages change the value of the language attribute.

References

1. Atkinson, M., Van der Goot, E.: Near Real Time Information Mining in Multilingual News. In: Proceedings of the 18th World Wide Web Conference, Madrid, Spain (2009)
2. Piskorski, J., Tanev, H., Atkinson, M., Van der Goot, E.: Cluster-Centric Approach to News Event Extraction. In: Proceedings of the International Conference on Multimedia & Network Information Systems, Wroclaw, Poland. IOS Press, Amsterdam (2009)
3. Tanev, H., Piskorski, J., Atkinson, M.: Real-Time News Event Extraction for Global Crisis Monitoring. In: Kapetanios, E., Sugumaran, V., Spiliopoulou, M. (eds.) NLDB 2008. LNCS, vol. 5039, pp. 207–218. Springer, Heidelberg (2008)
4. Piskorski, J.: ExPRESS Extraction Pattern Recognition Engine and Specification Suite. In: Proceedings of the 6th International Workshop Finite-State Methods and Natural language Processing 2007 (FSMNLP 2007), Potsdam, Germany (2007)
5. Grishman, R., Huttunen, S., Yangarber, R.: Information Extraction for Enhanced Access to Disease Outbreak Reports. Journal of Biomedical Informatics 35(4) (2003)
6. Yangarber, R., Jokipii, L., Rauramo, A., Huttunen, S.: Extracting Information about Outbreaks of Infectious Epidemics. In: Proceedings of the HLT-EMNLP 2005, Vancouver, Canada (2005)
7. Zavarella, V., Tanev, H., Piskorski, J.: Event Extraction for Italian using a Cascade of Finite-State Grammars. In: Proceedings of the 7th International Workshop on Finite-State Machines and Natural Language Processing, Ispra, Italy (2008)
8. Tanev, H., Zavarella, V., Linge, J., Kabadjov, M., Piskorski, J., Atkinson, M., Steinberger, R.: Exploiting Machine Learning Techniques to Build an Event Extraction System for Portuguese and Spanish. Under submission to the Journal Linguamática: Revista para o Processamento Automático das Línguas Ibéricas
9. Steinberger, R., Fuart, F., Van der Goot, E., Best, C., Von Etter, P., Yangarber, R.: Text Mining from the Web for Medical Intelligence. In: Mining Massive Data Sets for Security. IOS Press, Amsterdam (2008)
10. Freifeld, C., Mandl, K., Reis, B., Brownstein, J.: HealthMap: Global Infectious Disease Monitoring through Automated Classification and Visualization of Internet Media Reports. Journal of American Medical Informatics Association 15(1) (2008)
11. Doan, S., Hung-Ngo, Q., Kawazoe, A., Collier, N.: Global Health Monitor—A Web-based System for Detecting and Mapping Infectious Diseases. In: Proc. International Joint Conf. on NLP, IJCNLP (2008)

MinUCS 2009

Session 3

Security Level Classification of Confidential Documents Written in Turkish

Erdem Alparslan and Hayretdin Bahsi

National Research'Institute of Electronics and Cryptology-TUBITAK, Turkey
{ealparslan,bahsi}@uekae.tubitak.gov.tr

Abstract. This article introduces a security level classification methodology of confidential documents written in Turkish language. Internal documents of TUBITAK UEKAE, holding various security levels (unclassified-restricted-secret) were classified within a methodology using Support Vector Machines (SVM's) [1] and naïve bayes classifiers [3][9]. To represent term-document relations a recommended metric "TF-IDF" [2] was chosen to construct a weight matrix. Turkic languages provide a very difficult natural language processing problem in comparison with English: "Stemming". A Turkish stemming tool "zemberek" was used to find out the features without suffix. At the end of the article some experimental results and success metrics are projected.

Keywords: document classification, security, Turkish, support vector machine, naïve bayes, TF-IDF, stemming, data loss prevention.

1 Introduction

In recent years, protecting secure information became a challenge for military and governmental organizations. As a result, well defined security level contents and rules are more preferable than in the past. Each piece of information has its own security level. Correct detection of this security level may lead to apply correct protection rules on information.

Document classification aims to assign predefined class labels to a new document that is not classified [2]. An associated classification framework provides training documents with existing class labels. Therefore, supervised machine learning algorithms are fitting as a solution to classification problems. Well-known machine learning tasks such as Bayesian methods, decision trees, neural networks and support vector machines (SVM) are widely used for classification [3].

Classification accuracy of textual data is highly related to preprocessing tasks of training and test data. [4] These tasks become more difficult in processing unstructured textual data than in structured data. Unstructured nature of data needs to be formatted in a relational and analytical form. In this study TF-IDF (term frequency-inverse document frequency) is preferred to represent text based contents of documents. TF-IDF representation holds each word stem as an attribute for classification; and each document represents a separated classification event.

P. Daras and O. Mayora (Eds.): UCMedia 2009, LNICST 40, pp. 329–334, 2010.

Another important task of formatting textual data is stemming. In this study of Turkish documents, stemming tasks are more difficult than in other studies based on English or in other Latin-based languages. Turkic languages involve diverse exceptional derivation rules. Therefore stemming of Turkish terms provides some unstable rules varying from structure to structure.

Each of the distinct terms mentioned in our text document set is a dimension of this TF-IDF representation. Hence this representation leads to a very high dimensional space, more than 10000s dimensions. It is mainly noted that feature selection tasks are critical to make the use of conventional learning methods possible, to improve generalization accuracy, and to avoid "overfitting" [2].

Recent studies on document classification are performed on text datasets, especially on news stories in English [2][6]. We previously performed a classification of Turkish news stories and obtained a classification accuracy of 90%. [5]

In this study, internal documents of TUBITAK UEKAE (National Research Institute of Electronics and Cryptology) are classified into three classification levels: "secret, restricted and unclassified" by using support vector machines and naïve bayes algorithms.

2   Experimental Settings

In this study, 222 internal documents of TUBITAK UEKAE (National Research Institute of Electronics and Cryptology) are used to develop a framework which has an ability to classify documents according to their security levels by using support vector machines and naïve bayes algorithms.

First, all of 222 internal documents are classified into correct security levels (secret, restricted, unclassified) according to the general policies of TUBITAK UEKAE with the help of an expert. (The numbers of secret, restricted and unclassified documents are 30, 165 and 27 respectively.) Then these classified documents are converted into UTF-8 encoded txt based file format. Training and test documents have totally about 2.5 millions of words except stopping words. All the documents are grouped and arranged in a relational database structure.

These 2.5 million words are unstemmed Turkish words. A comprehensive stemmer library zemberek [8] is used to find out roots of unstemmed words. Zemberek gives us all the possible stemming structures for a term. Our stemming system selects the structure that has the biggest probability of semantic and morphologic patterns of the Turkish language. Table-1 shows some stemming examples of zemberek.

Table 1. Stemming Examples of Zemberek [8]

word	root	suffixes
getirilebilmesi	getir	il + ebil + me + si
yayımladığı	yayım	la + dığ(k) + ı
göstergelerin	gösterge	ler + in
endeksteki	endeks	te + ki
işlevin	işlev	in

Performing the stemming process we obtained approximately 9000 distinct terms. These 9000 terms may cause a high dimensionality and a time consuming classification process. As we mentioned above SVM's are able to handle high dimensional TF-IDF matrix with the same accuracy; however the time complexity of the classification problem is also an important aspect of our study. Therefore a χ^2 statistics with a threshold (100) was performed to select important features of classification. By the feature selection, the size of selected features is reduced from ~9000 to ~2000. This is known as the corpus of classification process.

The final task performed for the preprocessing phase was constructing a TF-IDF value matrix of all the features for all documents. The application was calculating TF-IDF values for each feature-document pair in the corpus.

Preprocessing software of this study was developed in JAVA, an open-source powerful software development language, and data were stored on PostgreSQL, an open-source professional database.

3 Results and Discussions

SVM-multiclass [7], Joachim's new multiclass support vector machine implementation, was executed on 3 different train and test sets with standard linear kernel and learning parameters. A leave-one-out cross validation was performed to confirm classification accuracy and parameter selection. All the documents have been grouped randomly into 3 train/test set pairs as in table 2.

Table 2. Test sets applied to SVM and NB algorithms

	doc set 1	doc set 2	doc set 3
number of train docs	145	163	136
number of test docs	77	59	86
total	222	222	222

Parameter c of SVM, the trade-off between training error and margin [10] is defined as a high value like 1000. As shown in figure 1, classification accuracy varies respect to the parameter c. The best fitting c value for all these 3 document sets is 1000.

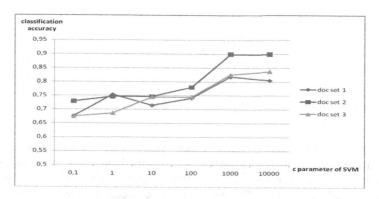

Fig. 1. Classification accuracy respect to the parameter c

Support vector machines and naïve bayes algorithms were performed on all 3 document sets which were randomly selected from 222 documents. Accuracy rates of both naïve bayes and support vector algorithms for 3 document sets are similar as shown in figures 2 and 3. Hence explaining the results of only one document set (ex doc set 2) may supply us the overall inferences for all 3 train/test pairs.

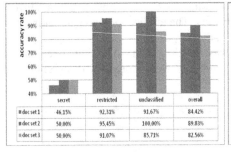

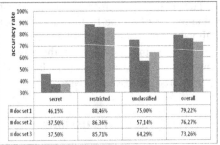

	secret	restricted	unclassified	overall
doc set 1	46.15%	92.31%	91.67%	84.42%
doc set 2	50.00%	95.45%	100.00%	89.83%
doc set 3	50.00%	91.07%	85.71%	82.56%

	secret	restricted	unclassified	overall
doc set 1	46.15%	88.46%	75.00%	79.22%
doc set 2	37.50%	86.36%	57.14%	76.27%
doc set 3	37.50%	85.71%	64.29%	73.26%

Fig. 2. SVM results for different doc. sets   **Fig. 3.** NB results for different doc. sets

Performing classification algorithms on **document set 2** the tables 3, 4 and 5 summarize overall accuracy rates for SVM, naïve bayes and sub-classification based classification by using SVM respectively. Rows in tables 3, 4 and 5 represent the actual/real class labels of documents and columns refer to the given class labels by classification algorithms on document set 2. For example in table 3 we infer that document set 2 provides 8 secret documents. (4+3+1) And SVM classifier assigned 4 of them as secret, 3 as restricted and 1 as unclassified. Hence the accuracy rate for secret documents in document set 2 is 50%. (4/8)

For 59 test documents, pure security level classification results were in a satisfactory level, but may be improved as shown in table 3.

Table 3. Classification with SVM of doc set 2   **Table 4.** Classification with NB of doc set 2

Predicted / Actual	Secret	Restricted	Unclassified	ACCR
Secret	4	3	1	50%
Restricted	0	42	2	95,5%
Unclassified	0	0	7	100%
Overall Accuracy Rate:	89,83% (53/59)			

Predicted / Actual	Secret	Restricted	Unclassified	ACCR
Secret	3	4	1	37,5%
Restricted	4	38	2	86,3%
Unclassified	0	3	4	57,1%
Overall Accuracy Rate:	76,27% (45/59)			

Regarding support vector classification results, we noticed that *restricted* documents are very well classified with 2 misclassified out of 44 documents. On the other hand SVM classifier is not very effective to detect *secret* labeled documents.

Another well known classifier widely used in document classification is naïve bayes classifier. In this study we used Weka as a naïve bayes classifier tool. The overall results obtained in naïve bayes classification are less preferable than in support vector classification as shown in table 4. Accuracy rates of naïve bayes classifier for all classes are lower than of support vector classification table.

In this study, we are classifying randomly selected test documents into 3 classes. It is a very interesting result that the misclassified document numbers of support vector classification are mostly matching with the misclassified document numbers of naïve bayes classification for all 3 document sets. For example in set 2, there are totally 6 documents which are misclassified in support vector classification. (DocIds: 1-11-12-13-14-25) 5 of these 6 documents are also misclassified in naïve bayes classification. (DocIds: 1-11-12-13-14) That means, classification errors in both two classification algorithms are nearly the same.

This study also aims to state a relation between class labels and sub-class labels implicating parent label. Sub-classification areas are other distinctive properties of documents like document type, area or format. Detecting class labels of internal documents of an organization depends on some interaction rules of sub-classes. For example, document area (military, private company, government) and document type (study report, travel report, meeting report, procedure) are sub-classes of security level classification (secret, restricted, unclassified) of TUBITAK UEKAE.

Some of rule based subclass – class interactions can be defined as:

If area: **military** and type: **spec. document** then level: **secret**
If area: **military** and type: **procedure** then level: **restricted**
If area: **government** and type: **travel** then level: **restricted**
If area: **general** and type: **tech. guide** then level: **unclassified**

All of 59 test documents are classified with support vector machines according to their area (military, private company, government, general etc.) and their type (study report, travel report, meeting report, procedure etc.) respectively. The results of two SVM sub-classifications are merged for each test document according to the subclass-class interaction rules mentioned above. Finally we obtained a success matrix as follows:

Table 5. SVM Classification results performing sub-classification logic

Predicted Actual	Secret	Restricted	Unclassified	ACCR
Secret	3	3	2	37,5%
Restricted	0	42	2	95,5%
Unclassified	0	1	6	85,7%
Overall Accuracy Rate:	76,27% (45/59)			

We expected that sub-classification based classification may increase security classification accuracy. Because we believe that learning the type and the area of a document is easier and much effective than learning its security level. But in this study, for this document set accuracy level of sub-classified solution is lower than the conventional solutions.

4 Conclusion and Future Work

In this study we have classified internal Turkish documents of TUBITAK UEKAE (a military-governmental organization) using support vector machines and naïve bayes classifiers. Obviously, support vector machines are more preferable than naïve bayes classifiers for text classification. It is also noticeable that subclass-class interaction based classification is no more successful than the conventional classification methods. Finally we highlight that the classification framework suggested in this study can be used to detect and prevent loss of confidential data of organizations via web platform as a data loss prevention solution.

The document set retrieved from the internal documents of TUBITAK UEKAE obviously provides "restricted" documents many more than "secret" and "unclassified" documents because of the military-governmental nature of the organization. In the future we aim to weight the instances of underrepresented "secret" and "unclassified" instances by using re-weighting techniques.

Another issue is about the structured representations of textual data. In this study, SVM's are trained with a TF-IDF form of data. Bag of words representation of textual data will also be used to train SVM algorithms in the future.

Finally we aim to extend our classification framework by using semi-supervised classification methods with unlabeled documents and machine learning techniques by means of knowledge engineering to obtain more accurate results from sub-classification issues.

References

1. Cortes, C., Vapnik, V.: Support-vector Networks. Machine Learning 20, 273–297 (1995)
2. Joachims, T.: Text Categorization with Support Vector Machines: Learning with Many Relevant Features. In: Nédellec, C., Rouveirol, C. (eds.) ECML 1998. LNCS, vol. 1398. Springer, Heidelberg (1998)
3. Feldman, R., Sanger, J.: Text Mining Handbook. Cambridge University Press, Cambridge (2007)
4. Han, J.W., Kamber, M.: Data Mining Concept and Techniques, 2nd edn. (2007)
5. Alparslan, E., Bahsi, B., Karahoca, A.: Classification of Turkish News Documents Using Support Vector Machines. INISTA (2009)
6. Cooley, R.: Classification of News Stories Using Support Vector Machines. In: IJCAI Workshop on Text Mining (1999)
7. http://svmlight.joachims.org/
8. http://code.google.com/p/zemberek/
9. Eyheramendy, S., Lewis, D., Madigan, D.: On the Naive Bayes Model for Text Categorization (2003)
10. Ageev, M., Dobrov, V.: Support Vector Machine Parameter Optimization for Text Categorization. In: International Conference on Information Systems Technology and its Applications (2003)

Signalling Events in Text Streams

Jelle J.P.C. Schühmacher and Cornelis H.A. Koster

Institute for Computing and Information Sciences (ICIS)
Radboud University Nijmegen, P.O. Box 9010, 6500 GL Nijmegen, The Netherlands
{j.schuhmacher,kees}@cs.ru.nl

Abstract. With the rise of Web 2.0 vast amounts of textual data are generated every day. Micro-blogging streams and forum posts are ideally suited for signalling suspicious events. We are investigating the use of classification techniques for recognition of suspicious passages.

We present *CBSSearch*, an experimental environment for text-stream analysis. Our aim is to develop an end-to-end solution for creating models of events and their application within forensic analysis of text-streams.

Keywords: Signalling Events, Interactive learning, Text Mining, Text Streams.

1 Introduction

We are investigating the use of classification techniques for signalling potentially suspicious events in passages coming from a stream of text. Micro-blogs and forum posts are ideally suited for signalling possibly suspicious events. It is difficult to get a grip on this kind of content, because texts are typically short. Furthermore, they need to be processed fast, since signalling an event after the fact is not very useful, and may even prove catastrophic.

Depending on the severity of the event to be signalled, a balance needs to be found between the economic feasibility of manually processing false positives and missing out on suspicious events. Therefore, it is important that a forensic text-stream mining application provides support for deciding on such a balance in an informed and verifiable way.

Due to the dynamic nature of user generated content, new types of suspicious events may frequently emerge. Therefore, it is not enough to be able to signal known events in text streams in an automated way. A truly useful forensic software support system will also provide methods to quickly and interactively build new models of emerging events.

2 CBSSearch

We are currently in the process of developing *CBSSearch* (Classifier Based Stream Search), an integrated environment for text-stream analysis. Our aim is to create an end-to-end solution for automated text-stream analysis.

P. Daras and O. Mayora (Eds.): UCMedia 2009, LNICST 40, pp. 335–339, 2010.
© Institute for Computer Sciences, Social-Informatics and Telecommunications Engineering 2010

Existing systems differ from CBSSearch in that they tend to use keyword-picking or hand crafted rules to find events in text-streams, opposed to using classification techniques to build a model of a particular event. An event model in CBSSearch is a collection of terms with statistical weights. Later this model will be extended to use dependency triples, recent experiments show this could prove to be helpful [1].

With CBSSearch it is possible to interactively train a event model of a particular event of interest. The resulting model is used to continuously search in streams of text for passages that match the event model.

2.1   System Overview

CBSSearch comprises four major components, a stream indexer, a stream analysis component, an interactive learning component, and a support system for sifting through suspicious events.

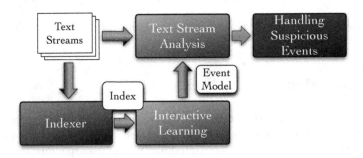

Fig. 1. Overview of CBSSearch

2.2   Stream Indexer

The *index* is used for interactively bootstrapping new event models from past examples. The Stream Indexer uses the Apache Lucene framework to efficiently create a database of examples. Apache Lucene is high-performance, scalable, full-featured, open-source, and written in Java. More information can be found on the official Lucene website[1]. In our testing Apache Lucene proves to be fast enough for our purposes.

2.3   Interactive Learning

The purpose of the *Interactive Learning* component is to train new event models with a minimum of effort. The component provides means for creating an event model by semi-supervised bootstrapping [3] from examples captured by the indexer.

---

[1] http://lucene.apache.org/java/

Training Bootstrap. A training bootstrap starts with a seed, a set of positive and negative examples provided by a human operator. An initial event model is trained using the *Linguistic Classification System*[2]. The initial model is then iteratively refined by letting the system choose the best next training document which is presented to the user for judgement.

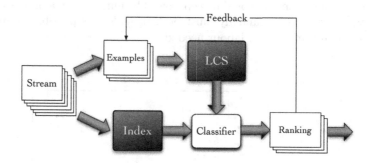

Fig. 2. The bootstrap process

The LCS is a basic component for all applications involving document classification, developed in the course of the ESPRIT projects DORO and PEKING, with a proven track record in the field of patent classification [2].

Active Learning. The number of manual judgements to be made is kept low by using active learning [4]. In CBSSearch active learning means the system makes a classification using the event model under construction, from this classification a the set of passages is selected for which the classifier is least certain they fit in this event model. These passages are presented to the operator for judgement using the same method as the stream analysis component. Passages that score above a certain threshold are assumed to be judged correctly by the classifier and are automatically added as training examples.

2.4   Stream Analysis

The resulting event models, in the form of linear classifiers, are used used to perform *on-the-fly classification* of stream passages. When a segment of text in the stream contains a combination of terms present in an event model for which the weights sum up to more than a threshold defined by the operator, the segment is signalled as suspect. The operator has control over precision and recall by adjusting the threshold.

The system reports suspicious passages to a human operator, who will then be able to see why a passage has been signalled as suspect by means of a document popup. The popup displays the segment with terms that occur in the event model highlighted.

---

[2] http://www.phasar.cs.ru.nl/LCS/

3   Preliminary Results

The CBSSearch system is still under development. We have a made preliminary analysis of the performance of stream classification using an artificial example derived from pre-classified documents. In order to investigate the best strategies for stream classification we measured whether event model complexity and the number of training examples could be kept low. The number of training examples which have to be judged manually must be kept as low as possible in order to reduce the amount of manual labour needed.

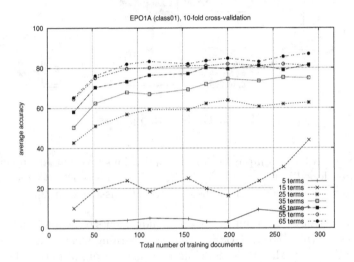

Fig. 3. Learning curves with increasing number of terms

This graph shows the trade-off between model complexity and the number of train documents. Fir this example a reasonable number of training examples (less than 100) suffices to achieve an acceptable accuracy (F1-value) as a starting point. Further training can then be performed using feed-back.

We are currently implementing CBSSearch and are collaborating with police authorities to find a suitable application for which training material can be easily obtained.

References

1. Koster, C.H.A., Beney, J.: Phrase-based document categorisation revisited. In: CIKM 2009: Proceedings of the 18th ACM Conference on Information and Knowledge Management (to appear, 2009)
2. Koster, C.H.A., Seutter, M., Beney, J.: Multi-classification of patent applications with winnow. In: Broy, M., Zamulin, A.V. (eds.) PSI 2003. LNCS, vol. 2890, pp. 546–555. Springer, Heidelberg (2004)

3. Koster, C.H.A., Jones, P., Vogel, M., Gietema, N.: The bootstrapping problem. In: 2nd Workshop on Operational Text Classification Systems (2002)
4. Lewis, D., Gale, W.: A sequential algorithm for training text classifiers. In: SIGIR 1994: Proceedings of the 17th annual international ACM SIGIR conference on Research and development in information retrieval, pp. 3–12. Springer, Heidelberg (1994)

MinUCS 2009

Session 4

A Proposal for a Multilingual Epidemic Surveillance System

Gaël Lejeune[1], Mohamed Hatmi[1], Antoine Doucet[1],
Silja Huttunen[2], and Nadine Lucas[1]

[1] GREYC, University of Caen 14032 Caen Cedex
Firstname.Lastname@info.unicaen.fr
[2] PULS Project, University of Helsinki, P.O. Box 68, 00014 Helsinki, Finland
Firstname.Lastname@cs.helsinki.fi

Abstract. In epidemic surveillance, monitoring numerous languages is an important issue. In this paper we present a system designed to work on French, Spanish and English. The originality of our system is that we use only a few resources to perform our information extraction tasks. Instead of using ontologies, we use structure patterns of newspapers articles. The results on these three languages are encouraging at the preliminary stage and we will present a few examples of interesting experiments in other languages.

1 Introduction

Recent epidemic events have demonstrated that health authorities have a great need for quick and precise information about the spreading of diseases. It is also known that the amount of on-line news in this domain is quite large and increasing. Therefore selecting really relevant documents for specialists and health authorities is becoming more and more important. Furthermore, as we are in a global world, it means that being able to monitor numerous languages with confidence should be a great improvement. As it is difficult to improve, even slightly, performance on monolingual systems, much effort is devoted to multilingual versions [2]. But the problem faced is that building an entire system for each new language is really time- and resource-consuming. For instance, we can guess that for major languages like English, Chinese, Spanish, Russian and Arabic one can have a valuable system. But they "only" represent 40% of world's population. For some parts of the world it will therefore take time for an epidemic to be detected by such systems.

Machine translation may be a way to fulfill these needs [10] although it has also its own limits. This paper will show an extension of an Information Extraction system for French to a Spanish and an English version, all three based on the same structural patterns. Its results will be compared to the state of the art to see how these results can be a road to an extension to other, and even rare, languages.

P. Daras and O. Mayora (Eds.): UCMedia 2009, LNICST 40, pp. 343–348, 2010.

2   Related Work

In epidemic surveillance, recent works use mostly two approaches: automatic processing with or without human post-analysis. For instance in Health Map [1], after potentially relevant documents have been selected automatically, experts have to check the relevance of the extracted events. By contrast, Helsinki CS Department's PULS system [11] combines Information Retrieval with Information Extraction to perform a full automatic processing. These systems are mostly based on keywords and linguistic analysis.

In contrast to these, our proposal is to use structure patterns [6] for extracting and selecting relevant content in newswire. The system worked at first for French, and then experiments were made on small sets of English and Spanish documents to check if the position of relevant content can also be compared in these two languages. Here, the system is in a way considered as a non-native speaker: as it does not know very well how the language function in details, it will only try to check if the terms searched are in some clearly identified positions. Details on this approach will be given in the next section.

3   Our Approach

Our goal is to see how we could build an Information extraction system with very low resources and without machine translation. The idea was therefore to use a different grain in analysis, following the state of the art for press articles [3,4]. Therefore the system uses the structure of the text to tell us where the relevant information may be. It is mostly based on the "5W rule" which, in press articles, tells that the answers to the main questions "What, Where, Who, When and Why" are to be easily found by the reader to help her check if the article is interesting for her. In order to find relevant content, the system divides the document in two parts:

- HEADER: title and first two sentences
- BODY: rest of the text

If a string is found in both "HEADER" and "BODY", using repeated string algorithm for [5], it is stored as a potentially relevant content. If one of these strings corresponds to a disease name then the document is considered potentially relevant.

In fact we assume that, according to relevance principle in human communication [7,9], there is only one important event by article and as we want to control redundancy we consider that secondary events have been treated elsewhere as primary events. Therefore if numerous diseases are found in the "potentially relevant content", it means that the document is not an important alert (in which case only one disease is mentioned) and by the way less interesting for our purpose. Finally it is not necessary to have many different names for each disease because in newspaper articles only a few are really used. Finally to be able to monitor new languages easily, resources are limited: 200 diseases names,

400 toponyms and a few words for date matching in each of our languages. The slots we need to fill in the database were What (disease), Where (location), Who (cases, people affected by disease) and When (date).

To extract the location the following algorithm is applied: the relevant location correspond to a string in the "relevant content". If numerous locations matches, the system compares frequencies in the whole document: if one location is more than twice as frequent as others it is considered as the relevant one. Concerning cases, they are defined as the first numeric information found in the document and not related to money or date. Furthermore the extracted cases are considered more relevant if they appear twice in the document. Thus the system uses regular expressions to round up and compare them (see example 4).

4 Some Results

In the experiments of Spanish and English (see example 1 and example 2), it has been found that the algorithm used for French still works. Documents for the experiments were selected at random and manually tagged. The combination between position and frequency seems as reliable in these two languages as we had found it to be in French. Some problems may occur concerning cases, see for instance in example 2 where the system will detect "50" and consider it relevant because of repetition. It is impossible at this stage for the system to detect which kind of case is concerned by the number (human or animals for instance). In previous experiments, results (Table 1) were slightly better than in Table 2 and Table 3. More precisely the disease extracted is relevant in 90% of documents while location seem more difficult to detect correctly: around 80% are good.

> **HONG KONG** , Sept. 21 (Xinhua) – **Hong Kong** 's Department of Health said Monday that it had advised eight primary schools and three secondary schools to suspend classes for seven days starting Tuesday to stop the possible spread of **A/H1N1 flu** in the schools. The advice was made following the outbreaks of the special flu in the schools involving 824 students aged between six and 17, a spokesman for the department said.[...] **Hong Kong** reported 446 newly confirmed cases of **influenza A/H1N1** in the 24 hours up to 2:30 p.m. Monday, bringing the number of the city's cumulative cases to 22,500. Meanwhile, the city reported another fatal case of **Influenza A/H1N1** , bringing to 16 the total number of fatal cases of the special **flu** in the city.

Example 1. English *disease* **country** cases

> Un cocinero de 50 años es la cuarta víctima mortal de la **gripe A** en Cataluña Unos familiares lo encontraron muerto en su casa Un cocinero de 50 años residente en L'Estartit, en Torroella de Montgrí (Girona), es la última víctima de la **gripe A** en **España**, según ha confirmado una portavoz de Salud. [...] El número de fallecidos en **España** por la enfermedad es ya de 33, a la espera de que mañana el Ministerio de Sanidad actualice las cifras.

Example 2. Spanish *disease* **country** cases

Table 1. Results on French

Manually tagged	Extracted	Rejected	Results
Relevant documents	196	14	Recall 93%
Non relevant documents	28	962	Precision 87.5%

Table 2. Results on English

Manually tagged	Extracted	Rejected	F-measure 84%
Relevant documents	44	6	Recall 88%
Non relevant documents	11	39	Precision 80%

Table 3. Results on Spanish

Manually tagged	Extracted	Rejected	F-measure 85%
Relevant documents	61	6	Recall 91%
Non relevant documents	15	25	Precision 80%

Work is in progress on Russian, Finnish and Turkish from documents of similar dates. At this stage, algorithms seem to keep good reliability (Examples 3, 4 and 5). String repetitions permit to extract interesting informations even in languages with declension. Using position also allow the system to prune the last sentence of the example in Russian which concerns Spanish flu. Then comparing different collections might quicken the acquisition of new terms by quasi alignments techniques [8].

« *Свиной грипп*» шествует по **миру**: уже 4379 заболевших в 29 странах Опубликована: 10 мая 2009 19:53:11 По данным Всемирной организации здравоохранения количество заболевания гриппом А/H1N1 увеличилось до 4379 в 29 странах **мира**.
Еще в субботу ВОЗ сообщал, что количество заболевших 3440 человек. НА сегодняшний момент 45 человек уже умерло от « *свиного грипп*а» в Мексике, 2 – в США, 1 – в Канаде, 1 – в Коста-Рике: итого – 49 человек. [...]
Ранее ученые неоднократно заявляли, что нынешняя эпидемия *гриппа* вряд ли повторит "испанку", которая в 1918-1920 годах унесла более 20 миллионов жизней, поскольку теперь медики и эпидемиологи намного больше знают о возбудителе *гриппа* и механизмах распространения болезни, сообщает РИА Новости.

Example 3. Russian *disease* country cases

The disease identified here is "swine flu" (repeated string "Свин грипп") worldwide (repeated string "мир"). The system extracts "4379" as the number of cases with confidence because it appears twice.

> **Kolera** tappanut jo yli 3000 **Zimbabwe**ssa
> julkaistu 28.01. klo 12:37, päivitetty 28.01. klo 14:10
> Eteläafrikkalaisessa **Zimbabwe**ssa **kolera**an kuolleiden määrä on noussut jo yli
> kolmen tuhannen. Maailman terveysjärjestön WHO:n mukaan kuolleita on 3 028.
> Lisäksi yli 57 000 ihmistä on sairastunut. Tiistain jälkeen on rekisteröity 57 uutta
> kuolemantapausta ja yli 1 500 uutta tartuntaa.
> Elokuussa puhjennut epidemia on Afrikan pahin 14 vuoteen.
> **Zimbabwe**n presidentti Robert Mugabe on väittänyt, ettei **Zimbabwe**ssa enää
> ole **kolera**a. [...]
> **Kolera** tarttuu bakteerin saastuttamasta ruoasta tai juomavedestä. Tauti aiheut-
> taa ripulia, oksentelua ja niistä johtuvan nestehukan. Pahimmassa tapauksessa
> kolera johtaa kuolemaan.

Example 4. Finnish *disease* **country** cases

This document tells us that there were 3028 cases of cholera identified in Zimbabwe, 3000 cases were mentioned in the title but it was identified as a round up of "3028". Repetition analysis permits the system to identify "Zimbabwe"(repeated string "Zimbabve") as the location of the main event of the document.

> **Zimbabve**'de **kolera** salgınında ölenlerin sayısı, 3868'e çıktı
> 25 Şubat 2009 Çarşamba 04:30
> CENEVRE -AA- **Zimbabve**'de **kolera** salgınında ölenlerin sayısı 3868'e
> yükseldi. Dünya Sağlık Örgütü yetkilileri, ağustostan bu yana görülen vak'a
> sayısının 83 bini aştığını bildirdi. Yetkililer, **kolera**dan ölüm oranının yüzde
> 4,7 olduğuna dikkat çekti. Dünya ortalaması ise yüzde 1'in altında. Hastalığın
> yayılmasında kirli sular baş rolü oynuyor.

Example 5. Turkish *disease* **country** cases

This document shows that the system does not extract Geneva (string "CENEVRE") as a relevant location because there is no repetition for the concerned string.

5 Conclusion

The results for English and Spanish seem to show comparable reliability to those for French. It means that this high-grain method can be useful. One can also see that first experiments on such different languages like Russian, Finnish and Turkish are interesting because they are from different families. The agglutinative aspect of Finnish would be for instance a great problem for a word-based approach. However, work is still needed to improve the case extraction in order to make a truthful comparison to the PULS English system. Our first goal will be to evaluate our system in a multi-event perspective. We shall then ponder on the trade-off between the subsequent loss and the possibility of monitoring new languages quickly and efficiently.

References

1. Freifeld, Mandl, Reis, Brownstein: HealthMap: Global infectious disease monitoring through automated classification and visualization of internet media reports. Med. Inform. Assoc. 15, 150–157 (2008)
2. Hull, Grefenstette: Querying across languages: a dictionary-based approach to multilingual information retrieval. In: Proceedings of the 19th annual international ACM SIGIR conference on Research and development in information retrieval (1996)
3. Itule, Anderson: News writing and reporting for today's media. Mcgraw-Hill College, New York (1991)
4. Kando: Text Structure Analysis as a Tool to Make Retrieved Documents Usable. In: 4th International Workshop on Information Retrieval with Asian Languages, Taipei, November 11-12, pp. 126–132 (1999)
5. Kärkkäinen, Sanders: Simple linear work suffix array construction. In: Baeten, J.C.M., Lenstra, J.K., Parrow, J., Woeginger, G.J. (eds.) ICALP 2003. LNCS, vol. 2719, pp. 943–955. Springer, Heidelberg (2003)
6. Lucas: The enunciative structure of news dispatches: A contrastive rhetorical approach. In: Ilie (ed.) Language, culture, rhetoric: Cultural and rhetorical perspectives on communication, ASLA, Stockholm, pp. 154–164 (2004)
7. Reboul, Moeschler: La pragmatique aujourd'hui. Une nouvelle science de la communication. Paris: Le Seuil (1998)
8. Riloff, Schafer, Yarowski: Inducing information extraction systems for new languages via cross language projection. In: 19th international conference on Computational linguistics, Taipei, Taiwan, vol. 1, pp. 1–7. Association for Computational Linguistics (2002)
9. Sperber, Wilson: Relevance: Communication and cognition. Blackwell Press (1998)
10. Linge, Steinberger, Weber, Yangarber, van der Goot, Al Khudhairy, Stilianakis.: Internet surveillance systems for early alerting of health threats in Eurosurveillance 14(13) (2009)
11. Steinberger, Fuart, Van der Goot, Best, von Etter, Yangarber.: Text mining from the Web for medical intelligence in Mining massive data sest for security. IOS Press, Amsterdam (2008)

Monitoring Social Attitudes
Using Rectitude Gains

Aleksander Wawer

Institute of Computer Science, Polish Academy of Science,
ul. J.K. Ordona 21, 01-237 Warszawa, Poland
axw@ipipan.waw.pl

Abstract. The article describes a prototype system aimed at monitoring attitudes toward any social group. The approach involves web mining and content analysis based on Rectitude Gain category from the Laswell dictionary of political values, extended into shallow predicate rules. The system requires no lexical sentiment resources and no training corpora. It has been designed, implemented and tested in Polish.

Keywords: content analysis, partial parsing, extremism detection.

1 Introduction: Extreme Movements and Security

Knowledge about attitudes that potentially give rise to radical extremist activity becomes all the more important. In the United States, for well-known reasons, the focus is on movements grounded in radical religious ideology. In Eastern and Central Europe, long-term threats to public safety seem to be more related to progressing nationalist radicalization[1]. This demands active monitoring of emerging nationalist ideas and their frequency in user generated content.

The goal of such monitoring is not only to identify movements with potential for radicalization and danger, which enables policy makers to undertake preventive steps, but also to measure the impact of their messages on internet audience. Arguably, the ideal moment when such formations should be tracked is their early age, when a group or movement has not yet reached a violent stage.

Identifying extremist ideas at their early stage is not an easy task. The main reason of this difficulty is that extreme and thus potentially dangerous ideas span over a wide spectrum of social movements and groups, emerging from various religious, political and social backgrounds [2]. Because it is not possible to identify

---

[1] Perhaps amplified by economic downturn, nationalist ideology is gaining momentum as it became evident in the last Europarlament Elections. Examples of extreme nationalist movements include Hungarian Guard, Slovakia's SNS, German NPD and many more.

[2] Domestic Extremism Lexicon (http://www.fas.org/irp/eprint/lexicon.pdf) compiled by the Strategic Analysis Group and the Extremism and Radicalization Branch, U.S. Department of Homeland Security, includes (among many others) anarchist, animal rights, antiabortion, antitechnology, environmental, Jewish and neo-Nazi movements.

P. Daras and O. Mayora (Eds.): UCMedia 2009, LNICST 40, pp. 349–354, 2010.

one ideology or belief system which defines extreme formations, early detection of extreme views should go beyond keyword recognition. Such diversity indicates also that corpora-trained systems may be ineffective at recognizing emerging, new extreme groups.

2 Previous Work

Examples of related works include research on text categorization using machine learning, applied to automated hate speech detection as in [4]. Authors experiment with bag of words and POS feature vectors to train SVM classifiers on racist text corpora. Another relevant research is [2], which mixes social network analysis with sentiment analysis (scores calculated using SentiWordNet-based algorithm) to detect online jihadist radicalisation on YouTube. Both mentioned works rely either on hate speech corpora or existing sentiment lexicons.

3 Content Analysis and Rectitude

3.1 Laswell Rectitude Gains

In the Laswell dictionary [5] of political values[3], one category is of specific interest: RECTITUDE. Among the five RECTITUDE subcategories, two are substantive (ETHICS and RELIGIOUS) and three refer to transitions (GAINS, LOSSES, ENDS). Substantive categories address the problem of rectitude sources, either established as *the social order and its demands as a justifying ground* (ETHICS) or *transcendental, mystical or supernatural grounds* (RELIGIOUS)[4].

The subcategory around which the proposed system has been designed is RECTITUDE GAIN, a very abstract category which has to be carefully interpreted. Specifically, RECTITUDE GAIN must not be confused with perhaps too obvious situations like court sentences or apologies. Instead, RECTITUDE GAIN is better explained as states or situations when unethical, immoral or bad conduct **is revealed** (*blame, breach, double cross, lying etc.*) and states or situations referring to **actions aimed at rectitude growth** (*root out, condone, forgive, justify, reparation, restitution, vindicate*).

Detection of extremism, as proposed in this article, is not based on direct attacks or expressions of hate, as in hate speech, but instead turns toward somehow less offensive statements — if offensive at all, expressing (the need of) rectitude gain related to a social group, like nation, race, ethnical, sexual or religious minority. The need of rectitude growth is the driver or rationale of actions against

---

[3] Integrated with the General Inquirer Harvard IV dictionary [7]. Keeping the convention, references to rectitude category names in the General Inquirer are typed uppercase.

[4] It is remarkable that Laswell dictionary does not mention nation among substantive RECTITUDE subcategories. It seems that several 20th century extreme European social movements, including the cruelest ones, were grounded on nationalism rather than ethics or religion.

this social group, explanation of the need for acting as the group does not conform to rules of moral conduct, historical sense of justice, transcendental or spiritual rules, or any other rectitude grounding substance. Perhaps rectitude gains, as used in the sense of this paper, can be compared to a hypothetical zero step on the Allport scale [1].

3.2   Rectitude Gains in Polish

Translating senses in RECTITUDE GAIN category into Polish has been done manually with special attention to ensure maximum quality. Out of 30 entries in the English Laswell dictionary we selected 6 ones, presented with their Polish translation and sentiment in Table 1, which are likely to appear in the context of social group names.

Table 1. Selected Rectitude Gain entries and translation

General Inq.	POS	Meaning	Polish
BLAME#1	noun	Reproof, culpability	są winni (-)
BLAME#2	verb	To hold responsible for failure, to censure	obwiniać (-)
CLEAR#6	verb	To absolve from blame	rozgrzeszać (+)
CONDONE	verb	To forgive, to excuse	przebaczać (+)
EVEN#3	idiom verb	Get even; to have one's revenge	wyrównać
ROOT#4	idiom verb	To root out, to extirpate	wyplenić (-)

The RECTITUDE GAIN category contains entries marked as negative, as well as positive on the Osgood evaluative dimension. Taking this into consideration, we distinguished between good (+) and bad (-) rectitude entries in Table 1 and the remaining part of the paper [5].

4   Web Mining

Mining the social web and meanings it generates for monitoring extreme ideas requires as wide access to the influx of user generated content as possible. In Poland, where this research has been carried out, Google has dominant market position in the search engine market, and because of this fact alone it is more than likely that no other engine indexes as much web content as Google. Thus, we have decided to collect the data by regularly submitting queries to Google and analyzing obtained results. The queries are Polish translations of Laswell entries as in Table 1, wrapped in double quotes for exact phrase matching.

---

[5] "good/bad" was chosen over "positive/negative" to avoid confusion with Osgood's categories.

5 Spejd Rules

The corpus of 4747 web pages has been analyzed using Spejd[6], originally a tool for partial parsing and rule-based morphosyntactic disambiguation, used in the context of this work to find names of social groups — subjects of Laswell rectitude gains. 36 Spejd rules were manually constructed by extending translated terms into subject-predicate structure, to capture subject appearances. The recall has yet to be verified, but due to Google queries, they are likely to be exhaustive in covering the most common syntactical structures involved.

Full description of Spejd formalism is out of the scope of this paper. As an example, we describe 6 rules, the translation of BLAME#1 extended into subject-predicate structure. Rules take into account inverted syntax and possible insertions of modifiers. Extracted subjects are base forms (`base~"word"`) of plural noun forms (`pos~"subst" && number~"pl"`), sometimes restricted to specific grammatical case (for example, `case~"nom"` means forms with possible nominative interpretation), referred to by the last two `group` arguments in `Eval` rule part. Originally, this notation has been introduced to mark semantic and syntactic heads of a group, in our system we use it to point to appropriate token number. The first three rules (handling negation) are as follows:

```
Rule    "X, which are not to be blamed"
Match:  [pos~"subst" && case~"nom" && number~"pl"] ns? [pos~"interp"]?
        [base~"kto"]? []? [orth~"nie"] [base~"być"]
        [base~"winny" && negation!~"neg"];
Eval:   group(kt_nie_sa_winni_1, 1, 1);

Rule    "are not to be blamed X (inverted)"
Match:  [orth~"nie"] [base~"być"]? []? [base~"winny" && negation!~"neg"]
        [pos~"subst" && case~"nom" && number~"pl"];
Eval:   group(nie_sa_winni_1, 5, 5);

Rule    "X are not to be blamed (negated adjective form)"
Match:  [pos~"subst" && case~"nom" && number~"pl"] [base~"być"]?
        [base~"winny" && negation~"neg"];
Eval:   group(sa_niewinni_1, 1, 1);
```

Another three rules for BLAME#1 are analogous to the three presented above, but do not involve negation:

```
Rule   "X, which are to be blamed"
Match: [pos~"subst"  && case~"nom" && number~"pl"] ns? [base~"kto"]? []?
       [base~"być"] [pos~"adv"]? [base~"winny" && negation!~"neg"];
```

---

[6] http://nlp.ipipan.waw.pl/Spejd/ [6]

```
Eval:   group(kt_sa_winni_0, 1, 1);

Rule    "are to be blamed X (inverted)"
Match:  [base~"być"]? []? [base~"winny"]
        [pos~"subst" && case~"nom" && number~"pl"];
Eval:   group(sa_winni_0, 4, 4);

Rule    "X are to be blamed"
Match:  [pos~"subst" && case~"nom" && number~"pl"] []?
        [base~"być"]? [base~"winny"];
Eval:   group(sa_winni_0, 1, 1);
```

Because Spejd is a cascade of grammars, more specific rules like those with negation have to be run prior to more general ones. Frequency counting inverts rectitude polarity when extracted by negated rules (from bad to good in the case of presented BLAME#1 rules).

6 Results

As anticipated, extraction rules yielded different types of subjects with regard to which rectitude raises, not only names of nationalities, minorities and social groups. For reporting purposes, the list has been limited to manually selected names, appearing in the initial corpus. Yet, lists of subjects have to be reviewed periodically for social group names. While this approach may seem labour-intensive, it ensures that no social group name is omitted. Unfortunately, at the moment no exhaustive list of nations and social groups in Polish appears to be publicly available. Automated methods have yet to be implemented.

Good and bad rectitude frequencies toward social groups and nations extracted from the gathered data have been presented in Table 2 below. While the interpretation of these results is out of the scope of this paper, they seem to be rather confirming the general intuition on how Poles feel about particular

Table 2. Good and bad rectitude frequencies toward nations and social groups

	Good rect.	Bad rect.
Jews	0	10
Israeli	0	8
Germans	6	9
Russians	8	9
Poles	0	9
Iraqui	0	2
Afghans	0	4
homosexuals	0	4
masons	0	1

nations and groups. Low frequencies are the consequence of not using nation and group names in web queries. As it has been argued above, querying using explicit names is less likely to produce reliable results.

7   Conclusions and Future Work

We presented a prototype system aimed at detecting attitudes toward any social group and tested it on a sample of web pages. Initial results indicate that rectitude gains extended into shallow rules can be a promising way of mining sentiment toward nations and minorities. The proposed method has advantages over existing approaches. Firstly, it does not require sentiment lexical resources like General Inquirer's evaluative categories or SentiWordNet [3] and does not rely on training corpora. Secondly, it does not require apriori knowledge of social group names and is easy to implement in any language. In the final version, queries will be issued monthly and limited to results indexed within the last month to enable continuous monitoring. Additional work has to be conducted to further evaluate the rules with regard to complex sentence structures.

References

1. Allport, G.: The Nature of Prejudice. Addison-Wesley, Reading (1954)
2. Bermingham, A., Conway, M., McInerney, L., O'Hare, N., Smeaton, A.F.: Combining social network analysis and sentiment analysis to explore the potential for online radicalisation. In: International Conference on Advances in Social Network Analysis and Mining, pp. 231–236 (2009)
3. Esuli, A., Sebastiani, F.: Sentiwordnet: A publicly available lexical resource for opinion mining. In: Proceedings of LREC (2006)
4. Greevy, E., Smeaton, A.: Text categorisation of racist texts using a support vector machine. In: SIGIR 2004: Proceedings of the 27th Annual International ACM SIGIR conference on Research and Development in Information Retrieval, SIGIR 2004: Proceedings of the 27th Annual International ACM SIGIR conference on Research and Development in Information Retrieval, pp. 468–469 (2004)
5. Namenwirth, Z.J., Weber, R.P.: Dynamics of Culture. Allen & Unwin, Inc. (1987)
6. Przepiórkowski, A., Buczyński, A.: spade: Shallow parsing and disambiguation engine. In: Proceedings of the 3rd Language & Technology Conference, Poznań (2007)
7. Stone, P.J., Dunphy, D.C., Ogilvie, D.M., Smith, M.S.: The General Inquirer: A Computer Approach to Content Analysis. MIT Press, Cambridge (1966)

ExpDes 2009

Is the Social Television Experience Similar to the Social Online Video Experience? Extending Domain-Specific Sociability Heuristics to a New Domain

David Geerts

Centre for User Experience Research (CUO), IBBT / K.U.Leuven
Parkstraat 45 Bus 3605, 3000 Leuven, Belgium
david.geerts@soc.kuleuven.be

Abstract. Online social video applications, combining video watching with social media, are gaining in popularity. However, evaluation tools that focus on optimally supporting social interaction (i.e. sociability) when designing social video applications are still lacking. We have created a list of sociability heuristics for social interactive television applications, but it is unsure if these can be applied to similar applications but on a different platform, the web. In this position paper, we discuss the possibilities and limitations of applying sociability heuristics in a new – but related – domain, and we highlight opportunities for future work in this area.

Keywords: social TV, sociability, heuristics, social media, video, evaluation.

1 Introduction

In recent years, web applications with video content (e.g. YouTube, Current TV, Joost) as well as television sets including network connections offering IPTV services and online widgets (such as Yahoo! Connected TV or Opera for Connected TVs) are becoming increasingly popular. The social nature of watching video and television programs, as well as the success of social media on the web, drives developers of both types of applications to explore the integration of social media with streaming video. This leads to new kinds of services such as integrating twitter updates during live video streaming (e.g. at the inauguration of Barack Obama as president of the USA), Facebook apps that allow chatting while watching video content (e.g. Clipsync) or even applications that are purposefully built to chat around television programs (e.g. CBS Watch & Chat). These kinds of applications can be called social A/V media, allowing remote viewers to interact with each other via the television set or via PC. Features include remote talking or chatting while watching television, sending recommendations, viewing buddy lists with information on which television shows each buddy is watching or sharing video clips, but the range of possibilities is still broader than that.

P. Daras and O. Mayora (Eds.): UCMedia 2009, LNICST 40, pp. 357–360, 2010.

2 Sociability Heuristics for Interactive Television

As is good practice in user-centered design, evaluating these systems early and often is important to design for an optimal user experience. Several guidelines for evaluating the usability of online video and interactive television exist, and heuristic evaluation [1] is a well-known and often practiced technique to perform early in the development process in order to uncover problems as soon as possible and with a relative low-cost. However, for applications being used in a social context offering a shared user experience, such as the applications mentioned above, evaluating only usability is not enough. Even if these applications are evaluated to improve their usability, it doesn't mean that the social interactions they are supposed to enable are well supported. The term 'sociability' is used to indicate these interface aspects that support and enhance social interaction with and through new technologies and applications [2]. Several research areas already focused for quite some time on this phenomenon, such as Computer Mediated Communication (CMC) and Computer Supported Cooperative Work (CSCW). However, specifically evaluating how well these social interactions are supported has become only recently an area of research, e.g. in online communities or groupware. Similar to these last examples, social A/V media applications can benefit from sociability guidelines or heuristics. The specific nature of social television watching, such as enjoying a television program while communicating or using television content as conversation starter [3], warrant the use of domain-specific sociability heuristics for social television. This will help interface developers to design well-functioning applications that support social interaction in a non-work related environment.

We created such a list of sociability heuristics for social television, by testing five social interactive television systems in a lab environment with in total 149 users. The systems that were tested are AmigoTV [4], Windows Media Center, Social TV [5], Communication Systems on Interactive TV (CoSe) and Ambulant Annotator [6]. The results of these user tests were analysed using a grounded theory approach, and were complemented with reports from other lab and field studies of similar systems, in order to increase the ecological validity of the guidelines. This resulted in a list of twelve sociability heuristics that can guide the design as well as the evaluation of social television systems [7]. The list of sociability heuristics includes important aspects to take into account when designing social interactive television systems, such as offering different options for communicating, guaranteeing personal as well as group privacy, or adapting the programs and services to specific television genres. Although the heuristics are aimed at social interactive television in particular, they can also be used to make traditional interactive television services such as Electronic Programme Guides (EPG) more social. Therefore, these sociability heuristics can lead to interactive television programs and services that support the social uses of (interactive) television.

3 Extending Sociability Heuristics to Social Online Video

As social video watching on the web and combining social media with watching online video is gaining popularity, the question remains in how far these sociability heuristics, created specifically with social television in mind, can be applied to a new, but related, domain. The interactive TV applications that have served as a basis for creating the heuristics are all applications whose main function was to foster social interaction in some way. Some of these systems had a limited range of social features and others a very wide range of those features, but the core functionality remained the same. It is thus clear that the main applicability of the heuristics lays with social interactive television systems. Although there are many similarities with watching video on the web, as for many people this is starting to replace traditional television watching, the platform as well as the viewing experience and social context of Internet based systems are often substantially different from set-top box based systems, e.g. because of the difference in screen size, making it unsure in which degree certain heuristics are applicable or not. As Internet based systems were not included in the analysis, we cannot claim that the heuristics are in any case applicable to these systems as well. However, certain heuristics can be assumed to be as relevant for these systems as for set-top box based systems, for example the heuristic "adapt to appropriate television genres". Other heuristics, such as "support remote as well as co-located sociability" lose part of their meaning as watching television programs on a PC is more often a solitary experience than with a normal television set. It is therefore important to be able to assess which heuristics can be applied directly, which should be discarded or reformulated, or if there is a need for new heuristics which are not needed for social interactive television.

4   Conclusion and Further Work

Although we think that some heuristics have an impact on Internet based systems, we think this is an area where further research is needed. Two main approaches can be taken. One could try to validate the existing set of sociability heuristics using a field-study of Internet based systems, which are easily accessible. For validating the heuristics this way, a long-term field study should first be conducted with one or more online social video applications, including a wide range of features, to detect the sociability problems that actual users have with these applications in a real use situation. At the same time, the same systems should be evaluated by a number of evaluators – with one group using the sociability heuristics and the other group not. The problems found by the evaluators can then be compared with the problems found during the field tests. Although it is not necessary for the heuristics to uncover all problems found in the field – as this is not possible due to the nature of heuristic evaluation – if the evaluators using the sociability heuristics uncover more real problems than the evaluators using no heuristics, we can conclude that the heuristics work at uncovering the sociability issues in online social video applications [8]. However, this would probably lead to the elimination or reformulation of several of the heuristics, while at the same time it would be harder to create new heuristics. A better approach might be to start from scratch and conduct a competitive analysis of several online social video applications in order to create a new set of sociability guidelines, similar to how we created the original sociability heuristics for interactive television. These guidelines can consequently be compared to the sociability guidelines presented here, to see what the similarities are. One of the benefits of taking this approach is that not only existing heuristics can be confirmed or refuted, but also new sociability heuristics could come up during this analysis.

As we have discussed in the first sections of this position paper, when enhancing video watching with social features, it is important to make sure these social features are well implemented, supporting social interaction between different users as optimally as possible. A heuristic evaluation based on sociability heuristics is a fast and efficient way to detect and repair sociability problems early in the design process. We hope the sociability heuristics created for social interactive television are a starting point for creating heuristics specifically targeted at online social video applications.

References

1. Nielsen, J., Molich, R.: Heuristic evaluation of user interfaces. In: Proc. CHI 1990, pp. 249–256. ACM, New York (1990)
2. Preece, J.: Online Communities: Designing Usability, Supporting Sociability. John Wiley & Sons, Chichester (2000)
3. Lull, J.: The Social Uses of Television. Human Communication Research 6(3), 197–209 (Spring 1980)
4. Coppens, T., Trappeniers, L., Godon, M.: AmigoTV: towards a social TV experience. In: Masthoff, J., Griffiths, R., Pemberton, L. (eds.) Proc. EuroITV 2003, University of Brighton (2004)
5. Harboe, G., Massey, N., Metcalf, C., Wheatley, D., Romano, G.: The uses of social television. ACM Computers in Entertainment (CIE) 6(1) (2008)
6. Cesar, P., Bulterman, D.C.A., Geerts, D., Jansen, A.J., Knoche, H., Seager, W.: Enhancing Social Sharing of Videos: Fragment, Annotate, Enrich, and Share. In: Proc. of ACM MM 2008. ACM, New York (2008)
7. Geerts, D., De Grooff, D.: Supporting the Social Uses of Television: Sociability Heuristics for Social TV. In: Proceeding of the Twenty-Seventh Annual SIGCHI Conference on Human Factors in Computing Systems, CHI 2009, Boston, MA, USA, April 04 - 09. ACM, New York (2009)
8. Hartson, H., Andre, T., Williges, R.: Criteria for evaluating usability evaluation methods. International Journal of Human-Computer Interaction 13(4), 373–410 (2001)

Using Social Science in Design

Magnus Eriksson

The Interactive Institute,
C/O ITUniversity of Gothenburg, SE-41296, Gothenburg, Sweden
magnuse@tii.se

Abstract. This paper deals with ways in which designers of user-centered systems and products can make use of other disciplines, primarily social science, to inform design practice. It discusses various ways in which theories, models and frameworks from social science can be integrated into design research and the problems that can arise when translating concepts, theories and methods from one discipline to another. The paper argues for not only using social science before or after the design process to provide context but letting it intervene in all stages of the design process. The paper draws from experiences with the TA2 project for social group communication.

Keywords: Design methods, social science, social technologies.

1 Purpose

Designers of technologies for social communication are intervening in social processes and practices that the design is in no control over. These practices and processes must be understood and using theory from other disciplines is a way to accomplish that. For design in general this is nothing new and especially the field of HCI and interaction design has always borrowed heavily from other disciplines, for example psychology and cognitive science [1].

Often this has had to do with creating usability and a part of the design process has involved user trials to make the interaction with the design more efficient [2]. When it comes to technologies for social communication that resides within an already existing social situation, things quickly become complicated [3]. Within interaction design, there have been attempts at borrowing from social science, for example from ethnomethodology [4]. However, methods from social science often require long periods of research to make sense and are based on foundations that can't simply be translated over to design research.

This paper will describe methods for using social science in design and how problems, solutions and concepts differ between the two disciplines. It argues for the use of social science as an attempt to manage the uncertainties involved in the design process. In doing this, it draws from experiences gained in TA2, which is a research project aiming to increase togetherness in groups with strong ties separated across different locations.

P. Daras and O. Mayora (Eds.): UCMedia 2009, LNICST 40, pp. 361–365, 2010.

The main question this paper addresses is:

How can knowledge from other disciplines be incorporated into design research in a way that does not only become superficial add-ons on top of an already formated design process?

2 Three Ways Social Science Can Impact Design

We can consider three ways of using social science in design which also corresponds to different dimensions of social science.

1. *Social science as confirming the foundation which design is using as a basis.*
A designer can use social science as a way to create a foundation for the assumptions made during the design process, in order to make them explicit as well as confirm or justify these assumptions. Social sciences will then function as an analytic basis for the design decisions, as a way of creating certainty that the process are on the right track. This is the way for example sociology has traditionally informed political decisions. Social science focusing on the analysis and description of empirical data can be used in this way.

2. *Social science functioning as a catalyst for the design process.*
Another case where social science can inform the design process is when it can inspire design to think in creative and radical ways. This resembles how social science have influenced radical thinkers and social movements to change power structures and inequalities. The kind of social science useful here is not about making statements about the fundamental fact of social life but function more as cultural probes trying to discover new ways of being. Social science here functions as a catalyst, in the sense that it is an element that gets a process going, but that is not involved in the transformation process itself [5]. Valuable here is not only empirical data confirming what goes on in the world, but also the sociological sensitivity that Fraser [6] writes about: Sociology's ability to not only look at the past and the present but to get a feel for the trajectories, hinting to a state beyond the present.

3. *Social science as a tool.*
This is a somewhat different approach to the previous paragraphs where social science was used in its unchanged form, as something external to the design process, although they did it in two opposite ways. This third approach would break away from this and instead integrate the social science way of thinking into the very design process. This would mean having to change, tweak and hack the social science concepts and theories in order to make them function in the design process. It would not be a way of telling design what the world is like or how it should be, but a method to create thicker descriptions, richer concepts and a new kind of subtle sensibility in the design practice.

3 Empirical Science in Design Methodology

Using social science in this way differs from the way Dourish describes how design often only borrows a method from another discipline without considering the proper sensibility and understanding of the core fundament behind the method itself. This makes the actual value of the results from the method unclear.

According to Hallnäs and Redström, this creates problems when the distinction between the analytical-empirical scientific practice and the constructive and creative design process are blurred [1]. It is somewhat of a paradox to use descriptive and analytic theories about the state of the world today, such as the social sciences, as a basis for building the designs of tomorrow - as this is a practice of synthesis rather than analysis.

The social sciences may very well study use of technological artifacts as an empirical study, and there exist many such studies. In design practice on the other hand, the future use and even the future user itself is not something that can be studied in the present but something that is firstly defined as intended use by the designer, and secondly defined in use by the real users – and this actual use might deviate from the original intended use. There will always be a gap between the empirical results of social science and the design choices made. Empirical sciences can provide variables to work with, but the design will then have to be an interpretation of that variable that has no correlation in the present. Central to the use of social science in design is then to turn analytical and empirical results into definitions and concepts, to inspire hypotheses and ideas that can be explored, or to interpret "analytical information in terms of design choices" [1].

This gap between empirical results and design choice can however be made explicit together with making explicit assumptions are based on empirical science and what is the designers own design choices and then begin to investigate what assumptions they are based on.

4   Problems and Solutions in Social Science and Design

Because of this gap, a foundation for the design practice cannot be built from empirical sciences that study the actual, that is what is currently given. However design can make use of the virtual in social science, which would allow us to see trajectories and give us concepts to help describe and define the world and the intended use the designer wants to create, and perhaps also to challenge it.

Social science is traditionally considered to be analytical and descriptive. This means that the scientist typically wants to affect the subject of the research as little as possible as opposed to design that want to intervene in the lives and habits of users.

However, recent developments in social science are questioning this approach and are approaching an understanding of knowledge production that is more similar to the interventionist methods of design research [7].

Social science change the object of study through performativity, by the activity of defining it with concepts and making claims about what the world is made up of. Law says that this means that social science is caught up in "ontological politics" {Law, 2004}. He proposes a version of social science that acknowledges that the scientist affects the world and embraces it by explicitly thinking about the worlds it wants to help realize, thus assuming a normative stance similar to that of designers The difference with design, which is also a discipline that wants to be relevant and engaged, is that social science only has discourse as its means of changing the world while design has a number of material and discursive practices.

Fraser makes the proposition for sociology to deal with virtual structures rather than actual structures [6]. The virtual/actual distinction comes from the philosopher Gilles Deleuze [8] where the actual is the given situation and the virtual is the possible realities that can be created by the forces that traverse the situation. The task of sociology would in this case be to discover the potentialities and possibilities of a situation rather than just describing its actual state. When using these kinds of concepts from social science in design practice it is not a matter of finding answers in social science to use as a basis for research, rather it is about entering into a problem space with several possible outcomes. Social science evolves problems rather than finding direct answers and in this space of problems the design practice can enter.

5 The Theoretical Loop

Design is often thought of as applied science, resting on theoretical foundations from other subjects [9]. In the perspective proposed in this paper however, all throughout the design process, new insights are gained that reformulate the way the gap between the design and the foundations are negotiated. Theory in design research is part of a loop. Not a progression from theory to practice. As Schön puts it, design process is a continuous dialogue between problem formulation and solution [10].

6 What Does Social Science Allow Designers to Do?

By using social science in the correct way, design can achieve the following:

- By using concepts from social science, design can make a bigger impact with less effort by bridging the gap between the actual, its trajectories and the design. The designer can better understand the space where the design operates and and able to fit the design there without forcing the technology into it.
- Social science allows the designer to be more precise and explicit about the hypothesis and assumption that goes into the design process. Instead of basing the assumptions on common sense or intuition and therefore keeping them as facts, basing them on empirical science enables the designer to postulate them as explicit and open hypothesis that can be evaluated and challenged.
- Designs will be able to contain a larger spectrum of use cases, by making use of the rich descriptions of the social world that social science provides. Social science has made the same mistakes designers will do when making assumptions about the social for a hundred years and at a much bigger scale. Social science can give designers things to pay attention to and ways of thinking.
- Avoid previous mistakes made when design research borrows methods from empirical science without considering the fundaments behind the theory.

References

1. Hallnäs, L., Redström, J.: Interaction design: foundations, experiments (2006)
2. Carroll, J.M.: Human-Computer Interaction: Psychology as a Science of Design. Annual Review of Psychology 48, 61–83 (1997)
3. Hutchins, E.: Imagining the cognitive life of things. workshop, The cognitive life of things: recasting the boundaries of the mind. McDonald Institute for Archaeological Research, Cambridge (2006)
4. Dourish, P.: Implications for design. In: Proceedings of the SIGCHI conference on Human Factors in computing systems, p. 550 (2006)
5. Delanda, M., Landa, M.D.: A Thousand Years of Nonlinear History, Zone Books (2000)
6. Fraser, M.: Experiencing Sociology. European Journal of Social Theory 12, 63 (2009)
7. Latour, B.: How to talk about the body? The normative dimension of science studies. Body and Society 10, 205–229 (2004)
8. Deleuze, G., Guattari, F.: A thousand plateaus. Continuum International Publishing Group (2004)
9. Ludvigsen, M.: Designing for social interaction: an experimental design research project. In: Proceedings of the 6th conference on Designing Interactive systems, pp. 348–349. ACM, University Park (2006)
10. Schön, D.A.: The reflective practitioner. Basic Books (1983)

User-Centric Evaluation Framework for Multimedia Recommender Systems

Bart Knijnenburg[1], Lydia Meesters[1], Paul Marrow[2], and Don Bouwhuis[1]

[1] Technische Universiteit Eindhoven
P.O. Box 513, 5600 MB Eindhoven, The Netherlands
{B.P.Knijnenburg,L.M.J.Meesters,D.G.Bouwhuis}@tue.nl
[2] BT Innovate
Orion 3 Room 7 PP 1, Ipswich, IP5 3RE, UK
Paul.Marrow@bt.com

Abstract. Providing useful recommendations is an important challenge for user-centric media systems. Whereas current recommender systems research mainly focuses on predictive accuracy, we contend that a truly user-centric approach to media recommendations requires the inclusion of user experience measurement. For a good experience, predictive accuracy is not enough. What users like and dislike about our systems is also determined by usage context and individual user characteristics. We therefore propose a generic framework for evaluating the user experience using both subjective and objective measures of user experience. We envision the framework, which will be tested and validated in the large-scale field trials of the FP7 MyMedia project, to be a fundamental step beyond accuracy of algorithms, towards usability of recommender systems.

1 Introduction

The common practice in recommender systems research mainly focuses on algorithm performance [9]. Measures like mean absolute error or root mean square error measure the accuracy of an algorithm in predicting user ratings [3]. The implicit assumption is that better predictions lead to better recommendations, and that better recommendations lead to higher user satisfaction.

There are several flaws in this approach to recommender system evaluation. First of all, algorithm accuracy is not necessarily correlated with user experience [6]. Furthermore, the user experience of an interactive product such as a recommender system does not depend exclusively on prediction accuracy but is also influenced by factors such as the ease of use, usefulness, engagement, transparency, satisfaction, novelty and enjoyment of the system itself and the items retrieved [4, 10]. This list of factors is far from complete, and current literature lacks a unified understanding of what experience factors need to be measured or how to measure them. Although it is commonly understood that user experience is best measured using a combination of objective (user behaviour) and subjective (questionnaire) measures [2], there have been few attempts at a structural approach towards user experience measurement for media recommenders [7]. A framework is therefore needed for evaluating the user experience of media recommender systems. Specifically, the framework should:

P. Daras and O. Mayora (Eds.): UCMedia 2009, LNICST 40, pp. 366–369, 2010.

- provide a generic set of measurable concepts related to experience that is applicable to all types of media recommender systems,
- link the subjective experience measurements to objective recommender system aspects on the one hand, and objective user behaviour on the other hand,
- take into account personal and situational characteristics that may also influence the user experience [4, 8, 9],
- be validated in both qualitative studies (user diaries, interviews and focus groups) and quantitative studies (questionnaires).

2 Evaluation Framework

In order to construct our user-centric evaluation framework, we take the proven concept of the technology acceptance model (TAM) as a starting point [1], but also include hedonic experiences such as appeal, fun, and user emotions [2]. The basic framework, shown in figure 1, covers objective recommender system aspects, subjective evaluations, subjective experiences and objective behaviours, and relates them in a structural way. Furthermore, as the particular situation and the particular user may matter for the evaluation, it embeds the framework in situational and personal characteristics. A brief description of each component is given below, for a detailed description see [7].

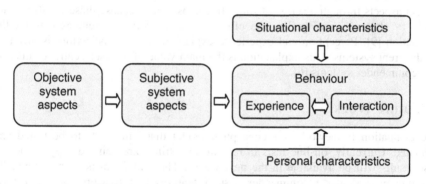

Fig. 1. User-centric evaluation framework for Multimedia recommender systems

The main assumption in our framework is that the *objective system aspects* (qualities of the recommender system) eventually influence the user experience [9]. These aspects include the visual and interaction design, the recommender algorithm, the presentation of recommended items, explanations and additional system features such as social networking and profile control. To measure the impact of a specific system aspect on the user, one would ideally manipulate the aspect across two or more conditions. For example, to test the impact of profile control, one would test a system that provides no profile control against a system that provides some profile control. By keeping all other system aspects the same, one can observe the effect of the manipulated aspect on the users' perception, experience and behaviour.

The *subjective system aspects* are users' evaluations of the objective system aspects. Although these evaluations may differ between users, they should typically not be influenced by personal characteristics and situations. Subjective system aspects include system usability, perceived item quality and visual attractiveness.

The *situational* and *personal characteristics* are beyond the influence of the system itself, but may significantly influence the user experience and interaction. In different situations users may use the system for different tasks, and in these tasks the choice goal and the domain knowledge may differ [4]. Personal characteristics that are typically stable across tasks include trust (in general, not in terms of the system), control and social factors [8, 9].

The *experience* also represents an evaluation of the system by the user. However, in contrast to the subjective system aspects, it is the users' behaviour and not the system that is the main focus of the experience measures. The experience typically depends on personal characteristics as the system aspects may influence different users in different ways. Also, the experience may change over time and across different choice situations. Experience is divided conceptually between hedonic qualities, usefulness, trust and outcome evaluations [2].

The *interaction* itself can also become an object of investigation of recommender systems research. Interaction is objectively measurable, and determines the 'final step' of the actual evaluation of the system: a system that is evaluated positively will be used more extensively. We explicitly link the objective interaction to objective system aspects through a series of subjective constructs, because these constructs are likely to attenuate and qualify the effect of the objective system aspects on the interaction [9]. Perception and experience explain why users' behaviour is different for different systems. This explanation is the main value of the user-centric evaluation of recommender systems [6].

3 Future Work

The evaluation framework is a conceptual model that still needs to be tested and validated. Currently, on-line tests in real-world settings are being deployed and the first results will be available in the next weeks. The initial aim is to systematically manipulate particular recommender system features and measure their effect on selected components in the framework. Specifically we are currently testing:

- the effect of personalized recommendations on the users' evaluation, experience and interaction with a media recommender system,
- the effect of recommendation variety on the users' evaluation, experience and interaction with a media recommender system,
- the effect of recommendation quality on the users' willingness to share more information with the system,
- the effect of personal trust factors on the willingness to share information with a media recommender system.

These effects are tested in on-line experiments across three different MyMedia recommender system prototypes in three different countries. The results will be used to refine the framework and to develop a set of evaluation metrics that can discriminate between the various recommender system features and user variables.

Consecutively, the refined framework will be used in a large scale, long-term field trial, which will measure how the relative importance of the different user experience components (e.g. pragmatic and hedonic components) develops over time. The trial will be conducted in three countries (Germany, Spain and UK) in 2010. Each trial is based on the same underlying MyMedia software framework but the company branding, user interface design, system functionalities and content catalogues result in quite different front-ends. The aim is to compare the four Multimedia recommender systems on a set of evaluation criteria that is partially shared and partially tailored to the specific system. This allows for comparison across recommender systems, as well as the exploration of unique system features (e.g. social networking and recommender algorithms) or user group characteristics (e.g. age and cultural background).

The generic nature of the proposed evaluation framework allows it to be applied to any recommender system. Acknowledging the importance of user experience related factors that transcend algorithmic performance, it aspires to take a next step in the road towards usable and useful recommenders. We therefore encourage designers and researchers to use our framework as an integrative guideline for the design and evaluation of new generations of recommender systems.

Acknowledgement

This research is funded by the European Commission FP7 project MyMedia (www.mymediaproject.org) under the grant agreement no. 215006.

References

1. Davis, F.D., Bagozzi, R.P., Warshaw, P.R.: User Acceptance of Computer Technology: A Comparison of two theoretical models. Management Science 35(8), 982–1003 (1989)
2. Hassenzahl, M.: User Experience (UX): Towards an experiential perspective on product quality. In: IHM 2008, Metz, France, September 2-5, pp. 11–15 (2008)
3. Herlocker, J.L., Konstan, J.A., Terveen, L.G., Riedl, J.T.: Evaluating collaborative filtering recommender systems. ACM Transactions on Information Systems 22(1), 5–53 (2004)
4. Knijnenburg, B.P., Willemsen, M.C.: Understanding the Effect of Adaptive Preference Elicitation Methods on User Satisfaction of a Recommender System. In: ACM Conference on Recommender Systems 2009, New York (2009)
5. McNee, S.M., Riedl, J., Konstan, J.A.: Making recommendations better: An analytic model for human-recommender interaction. In: CHI 2006, Montreal Canada (2006)
6. McNee, S.M., Riedl, J., Konstan, J.A.: Accurate is not Always Good: How Accuracy Metrics have hurt Recommender Systems. In: CHI 2006, Montreal, Canada (2006)
7. Meesters, L., Marrow, P., Knijnenburg, B., Bouwhuis, D., Glancy, M.: Deliverable 1.5 End-user recommendation evaluation metrics. FP7 MyMedia project, http://www.mymediaproject.org
8. Spiekermann, S.: Online Information Search with Electronic Agents: Drivers, Impediments, and Privacy Issues. PhD Thesis, Humboldt University, Berlin (2001)
9. Xiao, B., Benbasat, I.: E-Commerce Product Recommendation Agents: Use, Characteristics, and Impact. MIS Quarterly 31(1), 137–209 (2007)
10. Zins, A., Bauernfeind, U.: Explaining Online Purchase Planning Experiences with Recommender Websites. In: Information and Communication Technologies in Tourism 2005, Innsbruck, Austria, pp. 137–148 (2005)

Author Index